PERFORMANCE OF SOLAR ENERGY CONVERTERS: THERMAL COLLECTORS AND PHOTOVOLTAIC CELLS

ON ENERGY SYSTEMS AND TECHNOLOGY

A series devoted to the publication of courses and educational seminars given at the
Joint Research Centre, Ispra Establishment, as part of its education and training program.
Published for the Commission of the European Communities,
Directorate-General Information Market and Innovation.

Already published:

ENERGY STORAGE AND TRANSPORTATION

PROSPECTS FOR NEW TECHNOLOGIES

Edited by G. BEGHI

1981, x + 497 pp.   ISBN 90-277-1166-6

THERMAL ENERGY STORAGE

Edited by G. BEGHI

1982, vi + 505 pp.   ISBN 90-277-1428-2

Additional volumes in preparation.

The publisher will accept continuation orders for this series which may be cancelled at any time and
which provide for automatic billing and shipping of each title in the series upon publication. Please
write for details.

PERFORMANCE OF
SOLAR ENERGY CONVERTERS:
THERMAL COLLECTORS
AND PHOTOVOLTAIC CELLS

*Lectures of a Course held at the Joint Research Centre,
Ispra, Italy, November 11-18, 1981*

Edited by

G. BEGHI

Joint Research Centre, Ispra, Italy

D. REIDEL PUBLISHING COMPANY
Dordrecht : Holland / Boston : U.S.A.
London : England

Library of Congress Cataloging in Publication Data
Main entry under title:

Performance of solar energy converters. Thermal collectors and photo-
    voltaic cells

    (Ispra courses on energy systems and technology)
    Published for the Commission of the European Communities.
    Includes index.
        1. Solar collectors—Addresses, essays, lectures. 2. Photoelectric
cells—Addresses, essays, lectures. I. Beghi, G. II. Commission of the
European Communities. III. Series.
TJ810.P393   1983   621.47   82–24154
ISBN 978-94-011-9815-8   ISBN 978-94-011-9813-4 (eBook)
DOI 10.1007/978-94-011-9813-4

Commission of the European Communities   Joint Research Centre Ispra (Varese), Italy

Publication arrangements by
Commission of the European Communities
Directorate-General Information Market and Innovation, Luxembourg

EUR 8400 EN
Copyright © 1983 ECSC, EEC, EAEC, Brussels and Luxembourg
Softcover reprint of the hardcover 1st edition 1983
LEGAL NOTICE
Neither the Commission of the European Communities nor any person acting on behalf of the
Commission is responsible for the use which might be made of the following information.

Published by D. Reidel Publishing Company
P.O. Box 17, 3300 AA Dordrecht, Holland

Sold and distributed in the U.S.A. and Canada
by Kluwer Boston Inc.,
190 Old Derby Street, Hingham, MA 02043, U.S.A.

In all other countries, sold and distributed
by Kluwer Academic Publishers Group,
P.O. Box 322, 3300 AH Dordrecht, Holland

D. Reidel Publishing Company is a member of the Kluwer Group

TABLE OF CONTENTS

INTRODUCTION TO THE ISPRA COURSE

G. Beghi

Commission of the European Communities
Joint Research Centre - Ispra Establishment
I-21020 Ispra (Va), Italy

I am pleased to welcome you, on behalf of the Joint Research Centre of the Commission of the European Communities.

The problems of energy have always been the main objective of the research and development programme of our Centre, from the time of its birth. In recent years increasing attention has been given to the renewable energies, and in particular to solar energy. This is in harmony with the policy of the Commission to help decrease the energy dependency of the European Community. Oil is the main primary energy source upon which Europe depends. Thanks to trends encouraged by the European Community, oil represented 59% of primary energy consumption in the 9 countries of the EC in 1973, decreased to 52% by 1980, and to 50% in 1981. But even with this reduction (573 Mio.ton in 1979, 475 Mio.ton in 1979, 475 Mio.ton in 1980), the "oil bill" increased greatly, due to rises in price. 10 Billions ECU (European Unity of Account) in 1973, 50 Billions ECU in 1979, 77 Billions ECU in 1980.

The importance of finding efficient and economical alternatives is obvious and the efforts of the Commission in this field are being redoubled. The funds made available to the research programmes on solar energy, following decisions taken by the Council of Ministers of the European Community, are as follows:

G. Beghi (ed.), Performance of Solar Energy Converters: Thermal Collectors and Photovoltaic Cells, vii–viii.
Copyright © 1983 ECSC, EEC, EAEC, Brussels and Luxembourg

- for the Contractual Action (partial funding from the Commission within the framework of the Energy R&D Programme of the Directorate General Research Science Development, Brussels):

| 1975 - 79 | 17. 5 million ECU |
| 1979 - 83 | 46   "   " |

- for the Direct Action (executed in the Joint Research Centre):

| 1977 - 80 | 14. 5 million ECU |
| 1978 - 83 | 22.9   "   " |

In relation to the overall expenses for energy, and for energy research, these funds are somewhat limited, but do at least show an increase of interest. It is true that it is impossible to foresee, at least in the short and medium term, a significant contribution by solar energy to the solution of our energy problems in Europe. Nevertheless, efforts must be made to evaluate the potential and to prepare suitable technologies.

Many applications of solar energy are possible in various fields and all are under consideration. We might mention, for instance, some applications already of interest, such as solar heating of domestic hot water or the photovoltaic electricity for isolated users. It is useful to have a detailed understanding of the potential applications, and to be able to make a realistic assessment of their benefits. The economics of these potential uses depend upon several factors: the available solar radiation, the necessary investments, the total amount of energy which can be obtained, depending upon efficiency and the lifetime of the installations.

It is to these two latter parameters, their study, evaluation, standardised measurements, that the JRC is paying attention, and all aspects of these parameters, their performances and qualifications, form the subject of this Ispra Course.

We hope that this Course, with your participation and contributions, will be useful for each participant and also fulfil your needs. This is my wish to you for the coming days.

MEASUREMENT AND ANALYSIS OF SOLAR RADIATION DATA

Fritz Kasten

Deutscher Wetterdienst
Meteorologisches Observatorium Hamburg

1. EXTRATERRESTRIAL SOLAR RADIATION

The radiation of the sun is the primary natural
energy source of the planet Earth. Other natural energy
sources are the cosmic radiation, the natural terrestrial
radioactivity and the geothermal heat flux from the interior to the surface of the earth but these sources are
energetically negligible as compared to solar radiation.

From the sun, the planet Earth receives electromagnetic as well as corpuscular radiation. The latter, the
so-called solar wind, induces important physical effects in
the high atmosphere but is of minor importance for the
lower atmosphere and negligible for its energy budget.

When we speak of solar radiation we mean the electromagnetic radiation of the sun. The energy distribution of
electromagnetic radiation over different wavelengths λ is
called spectrum. The electromagnetic spectrum is divided
into several spectral ranges as shown on Fig.1.

The solar radiation incident on the upper border of
the terrestrial atmosphere is called extraterrestrial solar
radiation. The spectrum of the irradiance = power per unit
area, of extraterrestrial solar radiation is tabulated on
Table 1. The accuracy of these spectral data is better than
1.5 %. Below λ = 290 nm = 0.29 μm and above λ = 4 000 nm =
4 μm, less than 1 % each of the total solar power are
present. At λ = 730 nm = 0.73 μm which wavelength, inciden-
tally, is the upper limit of the visible spectral range,

- 1 -

G. Beghi (ed.), Performance of Solar Energy Converters: Thermal Collectors and Photovoltaic Cells, 1–64.
Copyright © 1983 ECSC, EEC, EAEC, Brussels and Luxembourg

λ	$\Delta \overline{I}_0$	$\overline{I}_{0\lambda}$	λ	$\Delta \overline{I}_0$	$\overline{I}_{0\lambda}$	λ	$\Delta \overline{I}_0$	$\overline{I}_{0\lambda}$
0- 250	2,51	-	358-360	1,72	0,860	468-470	4,02	2,010
250- 252	0,13	0,065	360-362	2,04	1,020	470-472	3,90	1,950
252- 254	0,12	0,060	362-364	2,16	1,080	472-474	4,04	2,020
254- 256	0,17	0,085	364-366	2,19	1,095	474-476	4,03	2,015
256- 258	0,27	0,135	366-368	2,46	1,230	476-478	4,03	2,015
258- 260	0,27	0,135	368-370	2,35	1,175	478-480	4,12	2,060
260- 262	0,22	0,110	370-372	2,31	1,155	480-482	4,10	2,050
262- 264	0,32	0,160	372-374	1,89	0,945	482-484	4,01	2,005
264- 266	0,57	0,285	374-376	2,07	1,035	484-486	3,83	1,915
266- 268	0,57	0,285	376-378	2,62	1,310	486-488	3,54	1,770
268- 270	0,55	0,275	378-380	2,51	1,255	488-490	3,88	1,940
270- 272	0,55	0,275	380-382	2,33	1,165	490-492	3,80	1,900
272- 274	0,44	0,220	382-384	1,48	0,740	492-494	3,84	1,920
274- 276	0,37	0,185	384-386	2,04	1,020	494-496	3,95	1,975
276- 278	0,52	0,260	386-388	1,93	0,965	496-498	3,98	1,990
278- 280	0,28	0,140	388-390	2,24	1,120	498-500	3,82	1,910
280- 282	0,37	0,185	390-392	2,56	1,280	500-502	3,63	1,815
282- 284	0,67	0,335	392-394	1,49	0,745	502-504	3,84	1,920
284- 286	0,45	0,225	394-396	2,53	1,265	504-506	3,93	1,965
286- 288	0,72	0,360	396-398	1,72	0,860	506-508	3,85	1,925
288- 290	0,90	0,450	398-400	3,21	1,605	508-510	3,88	1,940
290- 292	1,24	0,620	400-402	3,60	1,800	510-512	3,93	1,965
292- 294	1,13	0,565	402-404	3,36	1,680	512-514	3,76	1,880
294- 296	1,14	0,570	404-406	3,27	1,635	514-516	3,77	1,885
296- 298	1,07	0,535	406-408	3,25	1,625	516-518	3,40	1,700
298- 300	0,99	0,495	408-410	3,37	1,685	518-520	3,45	1,725
300- 302	1,12	0,560	410-412	3,45	1,725	520-522	3,76	1,880
302- 304	1,06	0,530	412-414	3,52	1,760	522-524	3,84	1,920
304- 306	1,09	0,545	414-416	3,45	1,725	524-526	3,88	1,940
306- 308	1,22	0,610	416-418	3,56	1,780	526-528	3,56	1,780
308- 310	1,18	0,590	418-420	3,32	1,660	528-530	3,84	1,920
310- 312	1,31	0,655	420-422	3,67	1,835	530-532	3,95	1,975
312- 314	1,36	0,680	422-424	3,29	1,645	532-534	3,76	1,880
314- 316	1,42	0,710	424-426	3,50	1,750	534-536	3,92	1,960
316- 318	1,44	0,720	426-428	3,25	1,625	536-538	3,81	1,905
318- 320	1,43	0,715	428-430	3,05	1,525	538-540	3,76	1,880
320- 322	1,56	0,780	430-432	2,92	1,460	540-542	3,67	1,835
322- 324	1,36	0,680	432-434	3,31	1,655	542-544	3,78	1,890
324- 326	1,39	0,695	434-436	3,54	1,770	544-546	3,81	1,905
326- 328	2,12	1,060	436-438	3,70	1,850	546-548	3,76	1,880
328- 330	1,99	0,995	438-440	3,38	1,690	548-550	3,84	1,920
330- 332	2,00	1,000	440-442	3,64	1,820	550-552	3,77	1,885
332- 334	1,90	0,950	442-444	3,90	1,950	552-554	3,77	1,885
334- 336	1,87	0,935	444-446	3,78	1,890	554-556	3,82	1,910
336- 338	1,61	0,805	446-448	3,92	1,960	556-558	3,67	1,835
338- 340	1,90	0,950	448-450	4,03	2,015	558-560	3,67	1,835
340- 342	1,94	0,970	450-452	4,37	2,185	560-562	3,69	1,845
342- 344	1,89	0,945	452-454	3,89	1,945	562-564	3,76	1,880
344- 346	1,66	0,830	454-456	4,06	2,030	564-566	3,67	1,835
346- 348	1,81	0,905	456-458	4,13	2,065	566-568	3,70	1,850
348- 350	1,80	0,900	458-460	3,98	1,990	568-570	3,70	1,850
350- 352	2,02	1,010	460-462	4,15	2,075	570-572	3,57	1,785
352- 354	2,08	1,040	462-464	4,17	2,085	572-574	3,74	1,870
354- 356	2,27	1,135	464-466	3,96	1,980	574-576	3,72	1,860
356- 358	1,70	0,850	466-468	3,91	1,955	576-578	3,61	1,805

Table 1: Spectral irradiance of extraterrestrial solar radiation.

λ	$\Delta\bar{I}_0$	$\bar{I}_{0\lambda}$	λ	$\Delta\bar{I}_0$	$\bar{I}_{0\lambda}$	λ	$\Delta\bar{I}_0$	$\bar{I}_{0\lambda}$
578-580	3,64	1,820	688-690	2,80	1,400	798- 800	2,29	1,145
580-582	3,72	1,860	690-692	2,86	1,430			
582-584	3,69	1,845	692-694	2,85	1,425	800- 805	5,72	1,144
584-586	3,62	1,810	694-696	2,92	1,460	805- 810	5,61	1,122
586-588	3,61	1,805	696-698	2,92	1,460	810- 815	5,52	1,104
588-590	3,35	1,675	698-700	2,87	1,435	815- 820	5,59	1,118
590-592	3,62	1,810	700-702	2,84	1,420	820- 825	5,31	1,062
592-594	3,54	1,770	702-704	2,81	1,405	825- 830	5,26	1,052
594-596	3,57	1,785	704-706	2,83	1,415	830- 835	5,15	1,030
596-598	3,60	1,800	706-708	2,76	1,380	835- 840	5,12	1,024
598-600	3,44	1,720	708-710	2,77	1,375	840- 845	5,08	1,016
600-602	3,44	1,720	710-712	2,84	1,420	845- 850	5,11	1,022
602-604	3,55	1,775	712-714	2,78	1,390	850- 855	4,83	0,966
604-606	3,47	1,735	714-716	2,75	1,375	855- 860	5,06	1,012
606-608	3,52	1,760	716-718	2,74	1,370	860- 865	4,96	0,992
608-610	3,39	1,695	718-720	2,71	1,355	865- 870	4,80	0,960
610-612	3,47	1,735	720-722	2,71	1,355	870- 875	4,92	0,984
612-614	3,42	1,710	722-724	2,81	1,405	875- 880	4,86	0,972
614-616	3,42	1,710	724-726	2,79	1,395	880- 885	4,80	0,960
616-618	3,36	1,680	726-728	2,73	1,365	885- 890	4,75	0,950
618-620	3,49	1,745	728-730	2,70	1,350	890- 895	4,70	0,940
620-622	3,37	1,680	730-732	2,72	1,360	895- 900	4,62	0,924
622-624	3,30	1,650	732-734	2,71	1,355	900- 905	4,51	0,902
624-626	3,24	1,620	734-736	2,71	1,355	905- 910	4,39	0,878
626-628	3,40	1,700	736-738	2,69	1,345	910- 915	4,37	0,874
628-630	3,32	1,660	738-740	2,61	1,305	915- 920	4,28	0,856
630-632	3,23	1,615	740-742	2,59	1,295	920- 925	4,13	0,826
632-634	3,33	1,665	742-744	2,57	1,285	925- 930	4,16	0,832
634-636	3,32	1,660	744-746	2,58	1,290	930- 935	4,14	0,828
636-638	3,32	1,660	746-748	2,58	1,290	935- 940	4,03	0,806
638-640	3,31	1,655	748-750	2,57	1,285	940- 945	3,98	0,796
640-642	3,18	1,590	750-752	2,52	1,260	945- 950	3,93	0,786
642-644	3,28	1,640	752-754	2,50	1,250	950- 955	3,85	0,770
644-646	3,26	1,630	754-756	2,50	1,250	955- 960	3,87	0,774
646-648	3,20	1,600	756-758	2,45	1,225	960- 965	3,84	0,768
648-650	3,14	1,570	758-760	2,45	1,225	965- 970	3,84	0,768
650-652	3,25	1,625	760-762	2,44	1,220	970- 975	3,80	0,760
652-654	3,21	1,605	762-764	2,44	1,220	975- 980	3,84	0,768
654-656	2,97	1,485	764-766	2,41	1,205	980- 985	3,85	0,770
656-658	2,81	1,405	766-768	2,41	1,205	985- 990	3,82	0,764
658-660	3,07	1,535	768-770	2,38	1,190	990- 995	3,80	0,760
660-662	3,15	1,575	770-772	2,37	1,185	995-1000	3,73	0,746
662-664	3,16	1,580	772-774	2,36	1,180	1000-1005	3,71	0,742
664-666	3,13	1,565	774-776	2,35	1,175	1005-1010	3,69	0,738
666-668	3,07	1,535	776-778	2,34	1,170	1010-1015	3,65	0,730
668-670	3,04	1,520	778-780	2,40	1,200	1015-1020	3,57	0,714
670-672	2,90	1,450	780-782	2,38	1,190	1020-1025	3,54	0,708
672-674	2,93	1,465	782-784	2,35	1,175	1025-1030	3,50	0,700
674-676	3,03	1,515	784-786	2,35	1,175	1030-1035	3,46	0,692
676-678	2,97	1,485	786-788	2,33	1,165	1035-1040	3,44	0,688
678-680	2,94	1,470	788-790	2,30	1,150	1040-1045	3,42	0,684
680-682	2,95	1,475	790-792	2,27	1,135	1045-1050	3,37	0,674
682-684	2,93	1,465	792-794	2,26	1,130	1050-1055	3,29	0,658
684-686	2,89	1,445	794-796	2,33	1,165	1055-1060	3,24	0,648
686-688	2,79	1,395	796-798	2,31	1,155	1060-1065	3,22	0,644

Table 1 (continued): Spectral irradiance of extraterrestrial solar radiation.

λ	$\Delta\bar{I}_0$	$\bar{I}_{0\lambda}$	λ	$\Delta\bar{I}_0$	$\bar{I}_{0\lambda}$	λ	$\Delta\bar{I}_0$	$\bar{I}_{0\lambda}$
1065-1070	3,21	0,642	1340-1345	1,99	0,398	1615-1620	1,21	0,242
1070-1075	3,18	0,636	1345-1350	1,96	0,392	1620-1625	1,22	0,244
1075-1080	3,14	0,628	1350-1355	1,93	0,386	1625-1630	1,22	0,244
1080-1085	3,09	0,618	1355-1360	1,91	0,382	1630-1635	1,20	0,240
1085-1090	3,07	0,614	1360-1365	1,88	0,376	1635-1640	1,18	0,236
1090-1095	3,04	0,608	1365-1370	1,85	0,370	1640-1645	1,17	0,234
1095-1100	3,03	0,606	1370-1375	1,85	0,370	1645-1650	1,17	0,234
1100-1105	3,03	0,606	1375-1380	1,84	0,368	1650-1655	1,17	0,234
1105-1110	3,02	0,604	1380-1385	1,82	0,364	1655-1660	1,17	0,234
1110-1115	2,99	0,598	1385-1390	1,81	0,362	1660-1665	1,16	0,232
1115-1120	2,93	0,586	1390-1395	1,79	0,358	1665-1670	1,14	0,228
1120-1125	2,88	0,576	1395-1400	1,78	0,356	1670-1675	1,12	0,224
1125-1130	2,85	0,570	1400-1405	1,76	0,352	1675-1680	1,10	0,220
1130-1135	2,83	0,566	1405-1410	1,74	0,348	1680-1685	1,10	0,220
1135-1140	2,81	0,562	1410-1415	1,73	0,346	1685-1690	1,10	0,220
1140-1145	2,79	0,558	1415-1420	1,72	0,344	1690-1695	1,08	0,216
1145-1150	2,76	0,552	1420-1425	1,73	0,346	1695-1700	1,08	0,216
1150-1155	2,75	0,550	1425-1430	1,72	0,344	1700-1705	1,07	0,214
1155-1160	2,74	0,548	1430-1435	1,67	0,334	1705-1710	1,04	0,208
1160-1165	2,68	0,536	1435-1440	1,65	0,330	1710-1715	1,04	0,208
1165-1170	2,67	0,534	1440-1445	1,62	0,324	1715-1720	1,05	0,210
1170-1175	2,65	0,530	1445-1450	1,61	0,322	1720-1725	1,01	0,202
1175-1180	2,61	0,522	1450-1455	1,58	0,316	1725-1730	0,97	0,194
1180-1185	2,57	0,514	1455-1460	1,56	0,312	1730-1735	0,95	0,190
1185-1190	2,56	0,512	1460-1465	1,57	0,314	1735-1740	0,95	0,190
1190-1195	2,54	0,508	1465-1470	1,56	0,312	1740-1745	0,94	0,188
1195-1200	2,50	0,500	1470-1475	1,55	0,310	1745-1750	0,93	0,186
1200-1205	2,48	0,496	1475-1480	1,53	0,306	1750-1755	0,94	0,188
1205-1210	2,47	0,494	1480-1485	1,51	0,302	1755-1760	0,94	0,188
1210-1215	2,48	0,496	1485-1490	1,51	0,302	1760-1765	0,92	0,184
1215-1220	2,46	0,492	1490-1495	1,51	0,302	1765-1770	0,90	0,180
1220-1225	2,41	0,482	1495-1500	1,49	0,298	1770-1775	0,88	0,176
1225-1230	2,42	0,484	1500-1505	1,48	0,296	1775-1780	0,86	0,172
1230-1235	2,41	0,482	1505-1510	1,47	0,294	1780-1785	0,85	0,170
1235-1240	2,39	0,478	1510-1515	1,46	0,292	1785-1790	0,85	0,170
1240-1245	2,36	0,472	1515-1520	1,44	0,288	1790-1795	0,86	0,172
1245-1250	2,35	0,470	1520-1525	1,44	0,288	1795-1800	0,86	0,172
1250-1255	2,35	0,470	1525-1530	1,43	0,286	1800-1805	0,84	0,168
1255-1260	2,27	0,454	1530-1535	1,39	0,278	1805-1810	0,82	0,164
1260-1265	2,21	0,442	1535-1540	1,38	0,276	1810-1815	0,80	0,160
1265-1270	2,20	0,440	1540-1545	1,38	0,276	1815-1820	0,80	0,160
1270-1275	2,23	0,446	1545-1550	1,37	0,274	1820-1825	0,79	0,158
1275-1280	2,23	0,446	1550-1555	1,37	0,274	1825-1830	0,78	0,156
1280-1285	2,19	0,438	1555-1560	1,36	0,272	1830-1835	0,77	0,154
1285-1290	2,21	0,442	1560-1565	1,34	0,268	1835-1840	0,76	0,152
1290-1295	2,21	0,442	1565-1570	1,31	0,262	1840-1845	0,76	0,152
1295-1300	2,20	0,440	1570-1575	1,30	0,260	1845-1850	0,75	0,150
1300-1305	2,17	0,434	1575-1580	1,29	0,258	1850-1855	0,74	0,148
1305-1310	2,12	0,424	1580-1585	1,27	0,254	1855-1860	0,72	0,144
1310-1315	2,09	0,418	1585-1590	1,25	0,250	1860-1865	0,72	0,144
1315-1320	2,09	0,418	1590-1595	1,23	0,246	1865-1870	0,70	0,140
1320-1325	2,07	0,414	1595-1600	1,24	0,248	1870-1875	0,68	0,136
1325-1330	2,04	0,408	1600-1605	1,23	0,246	1875-1880	0,69	0,138
1330-1335	2,02	0,404	1605-1610	1,22	0,244	1880-1885	0,70	0,140
1335-1340	2,00	0,400	1610-1615	1,22	0,244	1885-1890	0,69	0,138

Table 1 (continued): Spectral irradiance of extraterrestrial solar radiation.

λ	$\Delta\overline{I}_0$	$\overline{I}_{0\lambda}$	λ	$\Delta\overline{I}_0$	$\overline{I}_{0\lambda}$	λ	$\Delta\overline{I}_0$	$\overline{I}_{0\lambda}$
1890-1895	0,69	0,136	2360-2380	1,19	0,060	3180-3200	0,40	0,020
1895-1900	0,68	0,136	2380-2400	1,16	0,058	3200-3220	0,39	0,020
1900-1905	0,68	0,136	2400-2420	1,10	0,055	3220-3240	0,38	0,019
1905-1910	0,69	0,138	2420-2440	1,10	0,055	3240-3260	0,37	0,019
1910-1915	0,69	0,138	2440-2460	1,00	0,050	3260-3280	0,36	0,018
1915-1920	0,68	0,136	2460-2480	1,02	0,051	3280-3300	0,35	0,018
1920-1925	0,67	0,134	2480-2500	0,98	0,049	3300-3320	0,34	0,017
1925-1930	0,66	0,132	2500-2520	0,95	0,048	3320-3340	0,34	0,017
1930-1935	0,66	0,132	2520-2540	0,93	0,047	3340-3360	0,33	0,017
1935-1940	0,65	0,130	2540-2560	0,90	0,045	3360-3380	0,32	0,016
1940-1945	0,64	0,128	2560-2580	0,87	0,044	3380-3400	0,32	0,016
1945-1950	0,64	0,128	2580-2600	0,85	0,043	3400-3420	0,31	0,016
1950-1955	0,62	0,124	2600-2620	0,83	0,042	3420-3440	0,30	0,015
1955-1960	0,62	0,124	2620-2640	0,80	0,040	3440-3460	0,30	0,015
1960-1965	0,63	0,126	2640-2660	0,78	0,039	3460-3480	0,29	0,015
1965-1970	0,63	0,126	2660-2680	0,76	0,038	3480-3500	0,28	0,014
1970-1975	0,64	0,128	2680-2700	0,74	0,037	3500-3520	0,28	0,014
1975-1980	0,64	0,128	2700-2720	0,72	0,036	3520-3540	0,27	0,014
1980-1985	0,62	0,124	2720-2740	0,70	0,035	3540-3560	0,27	0,014
1985-1990	0,61	0,122	2740-2760	0,68	0,034	3560-3580	0,26	0,013
1990-1995	0,61	0,122	2760-2780	0,66	0,033	3580-3600	0,26	0,013
1995-2000	0,60	0,120	2780-2800	0,65	0,033	3600-3620	0,25	0,013
			2800-2820	0,63	0,032	3620-3640	0,25	0,013
2000-2020	2,25	0,113	2820-2840	0,61	0,031	3640-3660	0,24	0,012
2020-2040	2,20	0,110	2840-2860	0,60	0,030	3660-3680	0,24	0,012
2040-2060	2,09	0,105	2860-2880	0,58	0,029	3680-3700	0,23	0,012
2060-2080	1,97	0,099	2880-2900	0,57	0,029	3700-3720	0,23	0,012
2080-2100	1,91	0,096	2900-2920	0,55	0,028	3720-3740	0,22	0,011
2100-2120	1,81	0,091	2920-2940	0,54	0,027	3740-3760	0,22	0,011
2120-2140	1,75	0,088	2940-2960	0,53	0,027	3760-3780	0,21	0,011
2140-2160	1,65	0,083	2960-2980	0,51	0,026	3780-3800	0,21	0,011
2160-2180	1,60	0,080	2980-3000	0,50	0,025	3800-3820	0,21	0,011
2180-2200	1,47	0,074	3000-3020	0,49	0,025	3820-3840	0,20	0,010
2200-2220	1,52	0,076	3020-3040	0,48	0,024	3840-3860	0,20	0,010
2220-2240	1,42	0,071	3040-3060	0,47	0,024	3860-3880	0,19	0,010
2240-2260	1,44	0,072	3060-3080	0,45	0,023	3880-3900	0,19	0,010
2260-2280	1,35	0,068	3080-3100	0,44	0,022	3900-3920	0,19	0,010
2280-2300	1,32	0,066	3100-3120	0,43	0,022	3920-3940	0,18	0,009
2300-2320	1,21	0,061	3120-3140	0,42	0,021	3940-3960	0,18	0,009
2320-2340	1,12	0,056	3140-3160	0,41	0,021	3960-3980	0,18	0,009
2340-2360	1,23	0,062	3160-3180	0,40	0,020	3980-4000	0,17	0,009
						4000-25000	11,49	-

Table 1 (continued): Spectral irradiance of extraterrestrial solar radiation, $\overline{I}_{0\lambda}$ (based on WMO 1981).

λ   Wavelength interval, in nm;

$\Delta\overline{I}_0$ Irradiance within that interval, in W m^{-2};

$\overline{I}_{0\lambda}$ Mean spectral irradiance within that interval, in in W m^{-2} nm^{-1}.

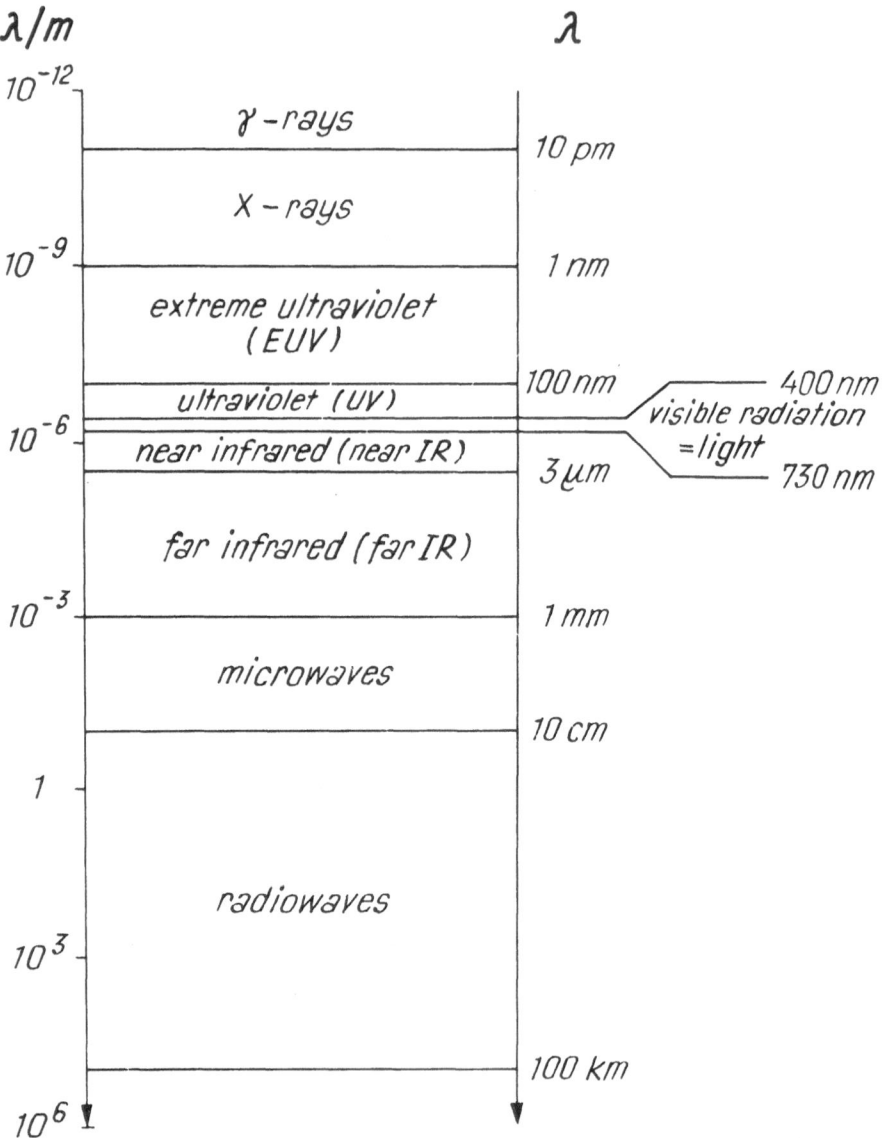

Fig. 1: Spectral ranges of electromagnetic radiation;
λ = wavelength.

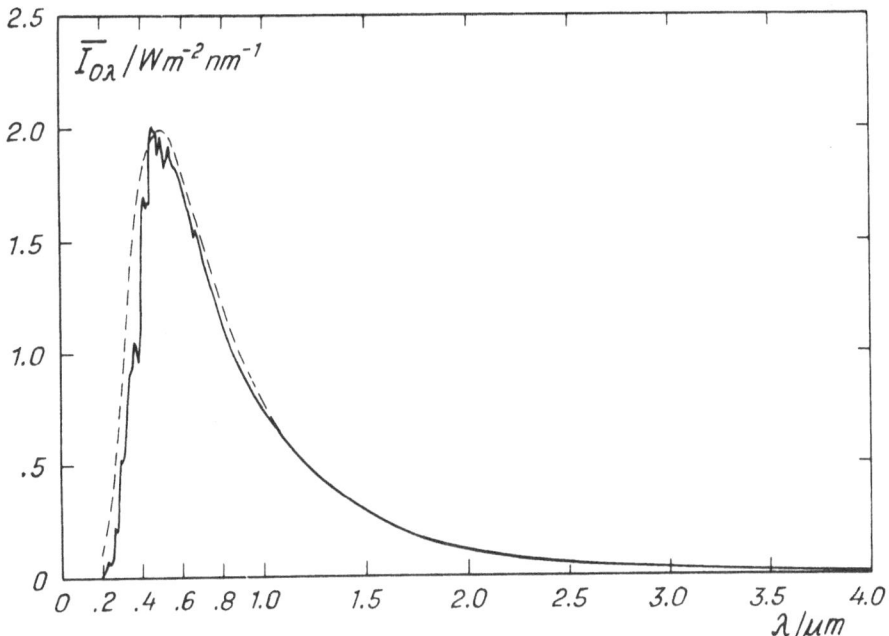

Fig. 2: Spectral distribution of extraterrestrial solar radiation (solid line, smoothed) compared with that of a black body of 6 000 K (dashed line).

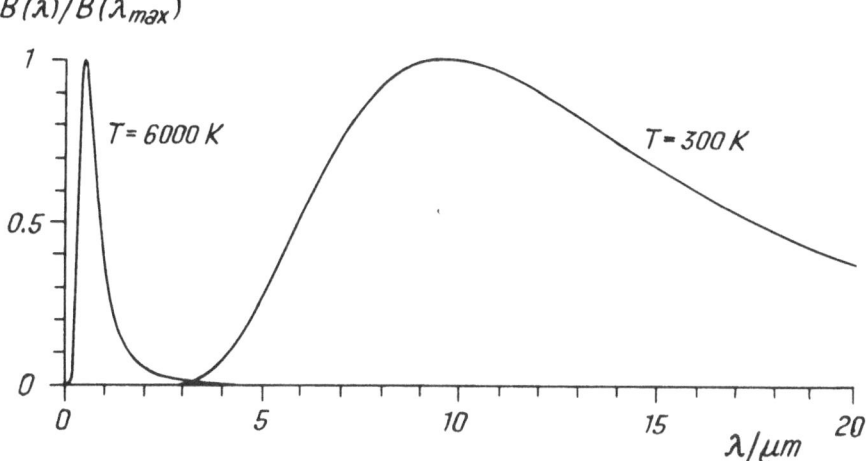

Fig. 3: Relative spectral distribution $B(\lambda)/B(\lambda_{max})$ of a black body of T = 6 000 K (λ_{max} = 0.48 μm) and of T = 300 K (λ_{max} = 9.6 μm).

the solar spectrum is divided into two halves of equal
energy; thus, one half of the extraterrestrial solar power
falls into the infrared spectral range.

A diagram of the spectral distribution of extraterrestrial solar radiation according to Table 1 is shown on
Fig.2. The maximum of the curve lies around λ_{max} = 0.48 µm.
If we consider the sun as a radiating black body following
Planck's spectral distribution law, the so-called color
temperature T of the sun may be calculated from λ_{max} with
the help of Wien's displacement law,

$$\lambda_{max} \cdot T = 2897.8 \text{ µm K} \tag{1}$$

giving T ≈ 6 000 K. On Fig.2, the spectral distribution
curve of a 6 000 K black body is plotted for comparison. As
can be seen, the curve is only a rough approximation to the
extraterrestrial solar spectrum. It may be worthwhile to
mention that in the spectral range between 100 and 200 nm
(UV-C), the sun behaves like a black body of T = 4 500 K
whereas between 4 and 10 nm (E-UV), the solar spectrum corresponds to that of a 500 000 K black body.

The temperature of the earth surface and of the lower
atmosphere is of the order of 300 K. According to Wien's
displacement law, Eq. 1, this terrestrial temperature corresponds to a black body with a spectral distribution peaking
at λ_{max} = 9.6 µm. On Fig.3, the spectral distributions of
a 6000 K radiatior (= sun) and a 300 K radiatior (= earth)
are plotted. Since there is only a minor overlap, solar
radiation on the one hand and terrestrial radiation on the
other hand can most often be treated separately in measurements and computations. Solar radiation is sometimes called
shortwave, and terrestrial radiation is called longwave
radiation in meteorology.

If the spectral irradiance $\overline{I_{0\lambda}}$ in Fig.2 is integrated
over all wavelengths λ, the total irradiance of extraterrestrial solar radiation, in W m^{-2}, is obtained:

$$\int_{0}^{\infty} \overline{I_{0\lambda}} \, d\lambda = \overline{I_0}. \tag{2}$$

The subscript 0 means "extraterrestrial", and the bar
indicates "at mean earth-sun distance". The quantity $\overline{I_0}$ is
called the solar constant. Its value was recently
(see WMO 1981) determined with an accuracy of ± 7 W m^{-2} as

$$\overline{I_0} = 1367 \ W \ m^{-2}. \qquad (2a)$$

If there should exist short time fluctuations of the solar constant as occasionally supposed or asserted, they are to be expected within that uncertainty of 7 W m^{-2}. Secular or even longer time variations of the solar constant, on the other hand, can not firmly be excluded on the basis of the present state of knowledge.

Since the earth surrounds the sun on an elliptical orbit during one year, the extraterrestrial solar irradiance varies with the time of year. Its actual value I_0 at a given day may be computed from the solar constant $\overline{I_0}$ by the relationship

$$I_0 = \overline{I_0} \cdot (\overline{r}/r)^2 \qquad (3)$$

where r = actual earth-sun distance,
$\overline{r}$ = mean earth-sun distance
= 1 astronomical unit = 149.598 $\cdot$ 10^6 km.
The ratio $(\overline{r}/r)^2$ is known from astronomical tables or can be calculated with sufficient accuracy by the help of the equation

$$(\overline{r}/r)^2 = 1 + 2e \cdot \cos[\overline{\Omega}(t-1) - \Psi] \qquad (4)$$

where 2e = 0.03344 = twice the numerical eccentricity of the ecliptic,
Ω = 360°/365.25d = 0.9856°/d = mean angular velocity of the revolution of the earth around the sun,
t = time in days (d) counted from 1 $\hat{=}$ 1 Jan 1200 true local time (TLT),
Ψ = 1.7346° = difference of the geocentric longitudes of the sun at perigee and at 1 Jan 1200 TLT, respectively.
Numerically, Eq. 4 reads

$$(\overline{r}/r)^2 = 1 + 0.03344 \cos(0.9856^\circ \cdot t - 2.720^\circ). \qquad (4a)$$

The "distance correction" $(\overline{r}/r)^2$ varies from 1.03344 at perigee (3 January) to 0.96656 at apogee (3 July) that is by about $\pm$ 3.3 % during the year.

2. RADIATION FLUXES AT THE GROUND

The energy balance of a horizontal surface at the ground, or of a solid body near the ground, is given by

$$Q + K + H + L + (W) + (P) = 0. \qquad (5)$$

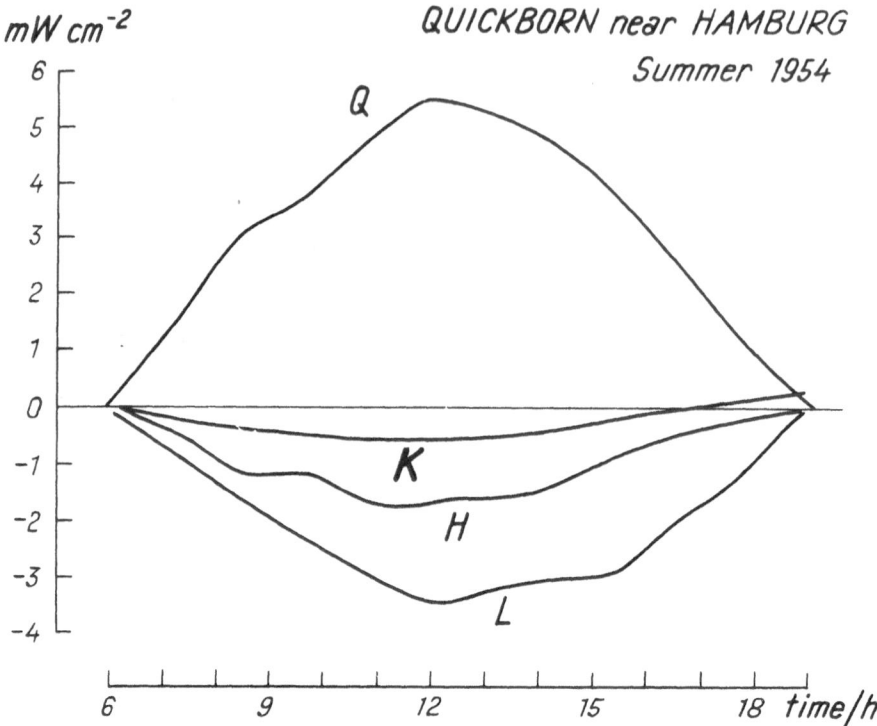

Fig. 4: Mean daily course of the different terms of the energy balance of the earth surface on clear days with low winds (after Geiger 1961, based on measured data of Franken-berger 1955).

Each term in this equation stands for an energy flux density or power density, in $W\ m^{-2}$. The fluxes in Eq. 5 are counted positive when they are directed towards the surface from above or from below. Since the surface, as an infinitely thin layer, can not store or deliver energy, the sum of all energy fluxes to and from the surface is zero. The symbols have the following meaning:

Q = net total radiation = sum of all positive and negative radiation fluxes to the surface;

K = heat flux from the interior of the body (ground) to its surface;

H = sensible heat flux from the atmosphere to the surface due to molecular and convective heat conduction (diffusion and turbulence);

L = latent heat flux due to condensation or evaporation at the surface;

W = heat flux due to advection i.e. heat transported by horizontal air currents. W is set zero if
   a) the measuring surface is located at a horizontal and homogeneous plane of sufficient extension so that the so-called katabatic flow is negligible, and
   b) the measuring time is small compared to the time of an air mass exchange (approach of a cold front, for instance);

P = heat flux brought to the surface by falling precipitation. P is often not taken into consideration because the measurements are confined to times without precipitation.

As an example, Fig. 4 shows the daily course of the energy fluxes Q, K, H and L. The net total radiation Q is, at daytime, to be compensated by the heat fluxes K, H and L. The net total radiation Q in Eq.5 is by itself composed of

$$Q = (G - R) + (A - E). \qquad (6)$$

Q is also called the total radiation balance. The other radiation flux densities in Eq.6 are defined as follows:

G = global radiation = sum of direct and diffuse solar radiation on the horizontal surface;

R = reflected global radiation = that fraction of G which is reflected by the body (ground);

A = atmospheric radiation = downward thermal radiation of the atmosphere (from atmospheric gases, mainly water vapor, and from clouds);

E = terrestrial surface radiation = upward thermal radiation of the body (ground).

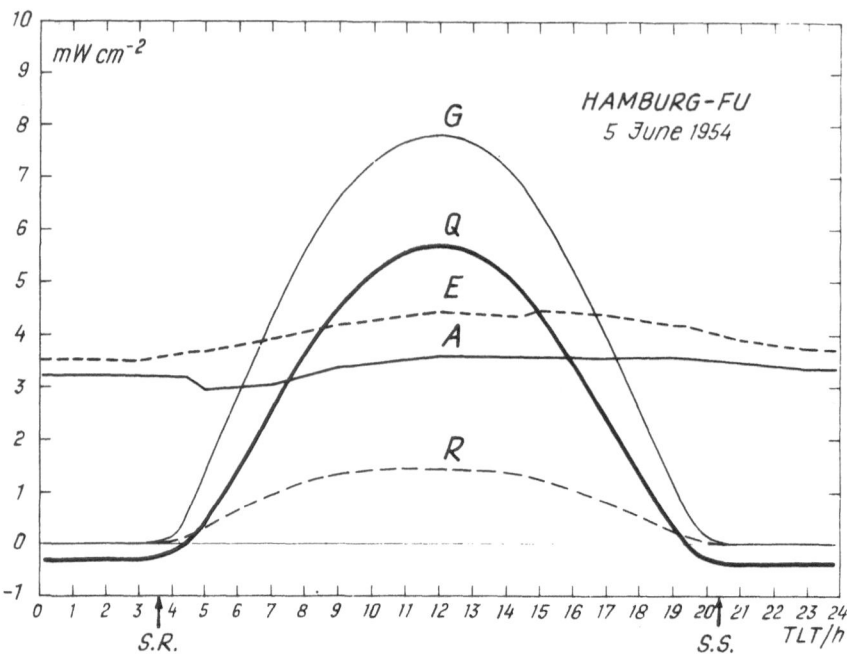

Fig. 5: Daily course of the different terms of the radiation balance of the earth surface on a clear day.
S.R. = sunrise, S.S. = sunset. (After Fleischer and Gräfe 1955/56).

G and R are solar or shortwave radiation fluxes therefore

$$Q_s = G - R \qquad\qquad (7)$$

is called net solar or net global radiation, or shortwave radiation balance. A and E are terrestrial or longwave radiation fluxes so that

$$Q_t = A - E \qquad\qquad (8)$$

is called the longwave radiation balance, and

$$- Q_t = E - A \qquad\qquad (9)$$

the (upward) net terrestrial surface radiation.

An example of the daily courses of Q, G, R, A and E is given on Fig.5. Whereas the shortwave radiation fluxes exhibit a pronounced variation during daylight hours, the longwave radiation fluxes vary but slightly because the temperatures of atmosphere and ground vary by but a few K during the day.

The ratio
$$\varrho_s = R/G \qquad\qquad (10)$$

is called the (shortwave) albedo of the body (ground). ϱ_s mainly depends on the optical properties of the body that is on its spectral reflectance $\varrho_s(\lambda)$. But there is also a slight dependence on the spectral distribution of the incoming global radiation, G_λ, as is demonstrated by

$$\varrho_s = \frac{R}{G} = \frac{\int\limits_{\lambda=0}^{\infty} \varrho_s(\lambda)\, G_\lambda\, d\lambda}{\int\limits_{\lambda=0}^{\infty} G_\lambda\, d\lambda}. \qquad\qquad (11)$$

Some representative albedo values are tabulated on Table 2.

Strictly speaking, the terrestrial surface radiation E is composed of two terms: (1) the thermal radiation of the body (ground)

$$E_1 = \alpha_t \cdot \sigma T^4 \qquad\qquad (12)$$

where α_t = effective longwave absorptance of the surface, slightly depending on ints temperature T;

artificial surfaces		natural surfaces	
velvet, black	0.04	water surfaces	0.06–0.12
paper, black	0.05–0.06	earth, black, moist	0.08
road, black top	0.08	earth, grey, moist	0.10–0.12
tiles, limestone	0.25	forest, coniferous	0.10–0.14
road, concrete	0.25–0.35	forest, deciduous	0.12–0.20
cardboard, yellow	0.30	earth, black, dry	0.14
wood, pine	0.40	grass land	0.15–0.35
paper, white	0.60–0.80	sand, grey	0.18–0.23
enamel	0.72–0.79	earth, grey, dry	0.25–0.30
aluminum	0.77–0.81	sand, white	0.34–0.40
magnesium oxide	0.98	snow, fresh	0.74–0.93

Table 2: Shortwave albedo Q_s of several surfaces

artificial surfaces		natural surfaces	
metal, polished	0.04–0.06	bright sand	0.89
brass, dead	0.22	bright limestone	0.92
paper	0.80–0.90	coarse gravel	0.92
roofing felt	0.93	plant leaves	0.96
glass	0.94	soil with lawn	0.98
brick wall	0.94	water	0.98
"Rubens'black"	0.96	snow	0.996

Table 3: Effective longwave absorptance α_t of several surfaces (for T = 300 K)

σ = Stefan–Boltzmann–constant
= $5.6697 \cdot 10^{-8}$ W m^{-2} K^{-4};

and (2) the reflected atmospheric radiation

$$E_2 = Q_t \cdot A \tag{13}$$

where $Q_t = 1 - \alpha_t$ = effective longwave reflectance of the surface. Thus, E is strictly given by

$$E = E_1 + E_2 = \alpha_t \cdot \sigma T^4 + (1-\alpha_t) \cdot A.$$

A table of α_t for different surfaces is presented in Table 3. Experience shows that in most cases, Eq.14 can be approximated by

$$E = \sigma T^4 \tag{15}$$

with an error of the order of 2 %.

3. DIRECT AND DIFFUSE SOLAR RADIATION

In the radiation balance Q, it is the global radiation G which is to be used as a natural energy source for engineering applications. G is composed of two terms:

$$G = B + D \tag{16}$$

where B = direct solar radiation on the horizontal surface, D = diffuse solar or sky radiation on the horizontal surface. B is connected with I, the direct solar radiation on a receiving surface being normal to the direction of incidence, by the cosine-projection

$$B = I \cdot \cos\zeta = I \cdot \sin\gamma \tag{17}$$

where ζ = zenith angle of the sun,
$\gamma = 90^\circ - \zeta$ = elevation angle of the sun; a formula for $\sin\gamma$ will be given in a subsequent section.

I is that part of the extraterrestrial solar radiation I_0 which reaches the earth surface after extinction by scattering by the molecules of the air, by scattering and absorption by the aerosol particles suspended in the air ("turbidity") and by absorption by the molecules of ozone and water vapor. Fig.6 gives an example of how the extraterrestrial solar spectrum is modified by stepwise introducing the four extinction processes just mentioned. The

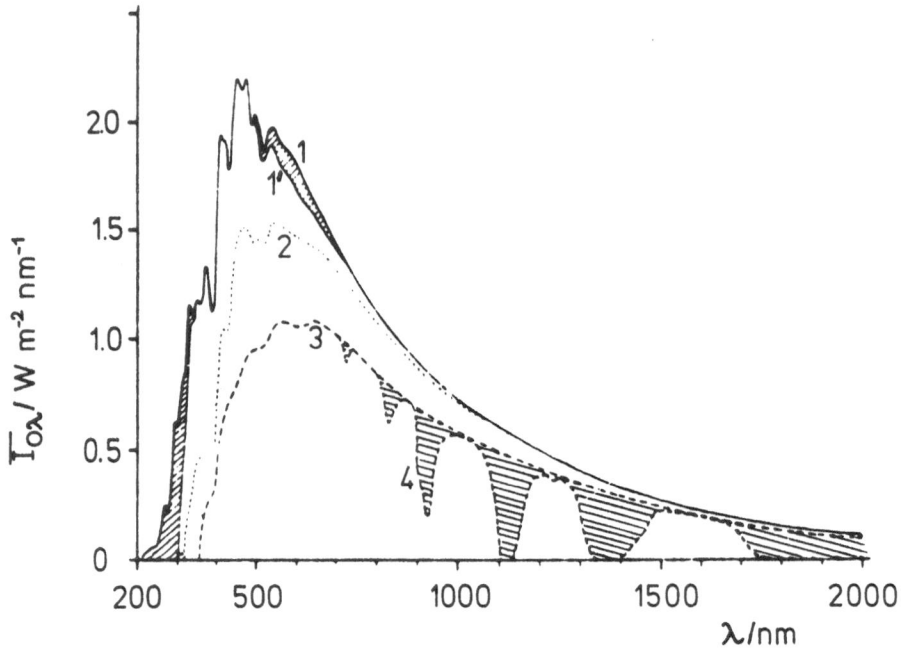

Fig. 6: Solar spectrum: 1) extraterrestrial; 1') below the ozone layer; 2) after additional scattering by the molecules of the air; 3) after additional extinction by aerosol particles; 4) after additional absorption by water vapor. (After Möller 1973).

direct solar radiation reaching the ground, I, is represented by the area under curve 4.

If the sun is covered by clouds, the direct solar radiation is zero so that G = D. The diffuse solar radiation
or sky radiation is produced by scattering of direct solar
radiation by the air molecules, aerosol particles and cloud
droplets and crystals. The diffuse solar radiation received
by a horizontal surface at the ground, D, depends on the
position of the sun at the sky, on the turbidity of the atmosphere, and on the amount and thickness of clouds and
their spatial distribution over the sky hemisphere. Fig.7
serves to introduce the spherical coordinates θ = zenith
angle and α = azimuth angle of the unit hemisphere as well
as the solid angle element

$$d\Omega = \sin\theta \ d\theta \ d\alpha \ \Omega_0 \tag{18}$$

where Ω_0 = unit solid angle = 1 steradian (sr).

If $L(\theta,\alpha)$ is the radiance received at the ground from
the solid angle element with the spherical coordinates θ
and α of the sky, the diffuse solar radiation D at the
ground is given by

$$D = \int_\Omega L(\theta,\alpha) \cos\theta \ d\Omega$$

$$= \int_{\alpha=0}^{2\pi} \int_{\theta=0}^{\pi/2} L(\theta,\alpha) \cos\theta \sin\theta \ d\theta \ d\alpha \ \Omega_0. \tag{19}$$

As may be seen from Eq.19, D heavily depends on the spatial
distribution of sky radiance, $L(\theta,\alpha)$. Examples of sky radiance distributions for cloudless and for partly clouded sky
are given on Fig.8.

4. SOLAR COORDINATES AND TIME

By astronomical considerations which can not be discussed extensively here, the coordinates of the sun in the
horizontal system are given by

$$\sin\gamma = \sin\varphi \sin\delta + \cos\varphi \cos\delta \cos\omega; \tag{20}$$

$$\cos\psi = \frac{\sin\varphi \sin\gamma - \sin\delta}{\cos\varphi \cos\gamma} \tag{21}$$

where γ = solar elevation angle;

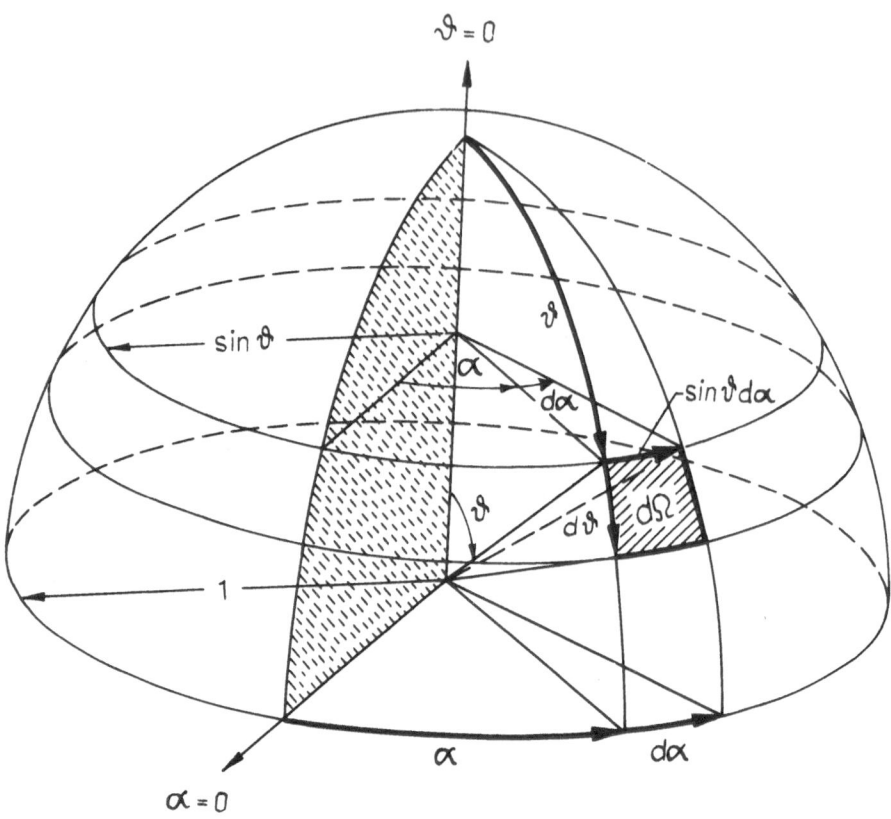

Fig. 7: Definition of zenith angle θ and azimuth angle α and of the solid angle element $d\Omega = \sin \theta \, d\theta \, d\alpha \cdot \Omega_0$ (after Kasten and Raschke 1974).

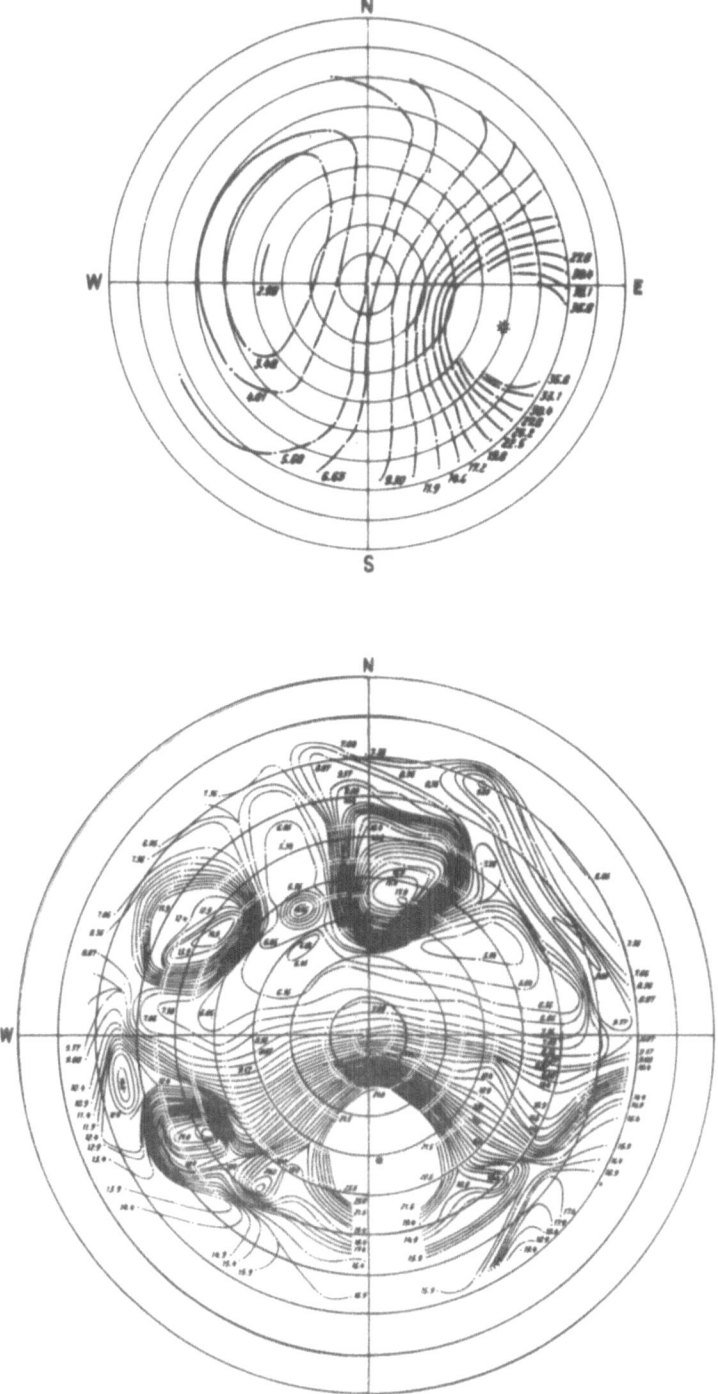

Fig. 8: Spectral sky radiance distributions $L_\lambda(\vartheta,\alpha)$, in $\mu W\ cm^{-2}\ nm^{-1}\ sr^{-1}$, of cloudless sky (left) and of partly clouded sky (right) for wavelength $\lambda = 561$ nm, presented as isophots on a polar diagram. Zenith is in the center. (After Dehne 1974).

ψ = solar azimuth angle, counted from south positive to west, negative to east;

φ = geographical latitude of the point of observation;

δ = solar declination = elevation of the sun above the equator at the time of observation;

ω = solar hour angle = angle between the hour circle (geographical direction of the sun) and the meridian (geographical south direction), at the time of observation.

The variation of δ during 1 day is small so that for practical purposes, δ is considered to be independent of ω. δ may be found in astronomical tables or may be calculated by means of

$$\sin\delta = \sin\varepsilon \cdot \sin\Lambda \qquad (21a)$$

where $\varepsilon = 23.44^{\circ}$ = obliquity of the ecliptic, Λ = geocentric longitude of the sun. Λ is given by

$$\Lambda = [\bar{\Omega}(t-1)-\Psi]+2e\cdot(180^{\circ}/\pi)\cdot\sin[\bar{\Omega}(t-1)-\Psi] +\Lambda_p \qquad (21b)$$

where $\bar{\Omega}$, t, Ψ, $2e$ habe already been explained after Eq.4, $\Lambda_p = -77.51^{\circ}$ = geocentric longitude of the sun at perigee. Numerically, Eq.(21a) reads

$$\sin\delta = 0.3978\cdot\sin[0.9856^{\circ}\cdot t + 1.92^{\circ}\cdot\sin(0.9856^{\circ}\cdot t - 2.72^{\circ})$$
$$- 80.23^{\circ}] \qquad (21c)$$

where t = time in days (d) counted from $1 \hat{=} 1$ Jan 1200 TLT. δ varies from -23.44° at winter solstice through 0° at the equinoxes to $+23.44^{\circ}$ at the summer solstice.

Whereas the solar declination δ describes the dependence of solar elevation angle γ on the time of year (date), the hour angle ω defines the dependence of γ on the time of day (clock time). ω is counted from the meridian (south direction) as positive in the afternoon (westerly directions of the sun), as negative on forenoon (easterly sun):

ω/deg	-180	-90	0	+90	+180
TLT/h	0	6	12	18	24
position of the sun	N	E	S	W	N

The term TLT = true local time will be explained later on.

Special cases of Eq.20 are:

a) Sunrise or sunset: $\gamma = \gamma_0 = 0$.

$0 = \sin\gamma_0 = \sin\varphi \sin\delta + \cos\varphi \cos\delta \cos\omega_0$.

$$\cos\omega_0 = - \frac{\sin\varphi \sin\delta}{\cos\varphi \cos\delta} = - \tan\varphi \tan\delta. \tag{22}$$

Example: Equinoxes, $\delta = 0$. Then $\cos\omega_0 = 0$; $\omega_0 = \pm 90$;

$$TLT = \begin{cases} 18 \text{ h (sunset)} \\ 6 \text{ h (sunrise).} \end{cases}$$

b) Culmination at noon (TLT = 12 h):

Then $\omega_N = 0$; $\cos\omega_N = 1$.

$\sin\gamma_N = \sin\varphi \sin\delta + \cos\varphi \cos\delta = \cos(\varphi-\delta)$.

$$\gamma_N = 90^0 - (\varphi-\delta). \tag{23}$$

The conversion of the hour angle ω of the sun to time defines the so-called true solar or true local time. During 1 day (1d = 1 440 min), the hour angle ω runs from -180^0 at TLT = 0 h through 0^0 at TLT = 12 h (noon) to $+180^0$ at TLT = 24 h. Therefore the following proportion holds:

$$\frac{TLT - 12 \text{ h}}{\omega} = \frac{1 \text{ d}}{360 \text{ deg}} = 4 \frac{\text{min}}{\text{deg}},$$

$$TLT = \omega \cdot 4 \text{ min/deg} + 12 \text{ h}. \tag{24}$$

At TLT = 12 h = true local noon, the sun culminates at south. In order to make solar radiation measurements at different locations of the earth comparable, a convention of the World Meteorological Organization (WMO) recommends to perform all meteorological radiation measurements according to TLT. Then all radiation records have the natural symmetry with respect to true noon.

True local time TLT is a time running unevenly because the hour angle ω runs unevenly with time, for two reasons:
a) The apparent orbit of the sun, the so-called ecliptic, is not a circle but an ellipse;
b) the time is not measured on the ecliptic but because of the definition of time by the rotation of the earth around its axis, it is measured on the equator.

For practical purposes, a mean hour angle $\bar{\omega}$ is derived by astronomical considerations, and a corresponding mean local time MLT is defined. The difference of both time scales is called the equation of time, Z:

$$Z = TLT - MLT. \tag{25}$$

The variation of Z during the year results from the overlap of a sine curve due to the effect a) mentioned above, and a double sine curve from the effect b):

$$Z = - 2e \cdot \sin[\bar{\Omega}(t-1)-\Psi] + \tan^2(\varepsilon/2) \cdot \sin 2\Lambda \tag{25a}$$

where the symbols have already been explained when the declination δ was introduced (Eq. 21a). Neglecting the second order terms in the expression (21b) for Λ and converting Z to minutes, Eq.25a takes the following numerical form:

$$Z = - 7.7 \text{ min} \cdot \sin(0.9856^\circ \cdot t - 2.72^\circ)$$

$$- 9.9 \text{ min} \cdot \sin(1.9712^\circ \cdot t + 19.55^\circ). \tag{25b}$$

As in the case of solar declination δ, the small variations of Z during one day can be neglected for most applications. Extreme values are $Z \approx - 14$ min in February and $Z \approx + 16$ min in October/November.

As mentioned, the hour angle ω is measured from the meridian of the point of observation. Consequently, true local time TLT and also mean local time MLT are only valid for the point of observation. Civil life requires the use of an uniform time scale within a certain region, the so-called standard time ST. It is defined by the MLT of a certain selected standard meridian of geographical longitude λ_s. The MLT of the 0-meridian which passes through Greenwich in England is called Greenwich Mean Time GMT:

$$GMT = MLT(\lambda_s = 0). \tag{26}$$

The standard time of another meridian λ_s east of Greenwich is related to GMT by

$$ST(\lambda_s) = GMT + \lambda_s \cdot 4 \text{ min/deg.} \tag{27}$$

Example: Central European Time CET is the MLT of the standard meridian $\lambda_s = 15^\circ$E. Thus,

$$CET = GMT + 1 \text{ h.} \tag{28}$$

Analogously, the mean local time of any point located on a meridian λ is connected to GMT by

$$MLT(\lambda) = GMT + \lambda \cdot 4 \text{ min/deg.} \qquad (29)$$

By combining Eq. 27 and Eq. 29, we obtain

$$MLT(\lambda) = ST(\lambda_s) + (\lambda-\lambda_s) \cdot 4 \text{ min/deg} \qquad (30)$$

as the relation between MLT and ST. Finally, we combine Eq.30 and Eq.25 and obtain an equation which allows to compute true local time from standard time:

$$TLT = ST(\lambda_s) + (\lambda-\lambda_s) \cdot 4 \text{ min/deg} + Z. \qquad (31)$$

Note: If locations west of the Greenwich meridian are considered, the longitudes λ, λ_s in the above equations are to be taken negative.

5. GLOBAL RADIATION ON A TILTED PLANE

A plane surface tilted by an angle β against the horizontal and with its normal directed to the azimuth α receives radiation $G(\beta,\alpha)$ which is basically composed of three radiation fluxes:

$$G(\beta,\alpha) = I(\beta,\alpha) + D(\beta,\alpha) + R(\beta,\alpha). \qquad (32)$$

$I(\beta,\alpha)$ is the direct solar radiation on the tilted plane and is given by

$$I(\beta,\alpha) = I \cdot \cos\eta \qquad (33)$$

where I is the direct solar radiation at normal incidence, and η is the angle between the direction of the sun and the normal of the tilted plane. By spherical geometry, η is given by

$$\cos\eta = \sin\gamma \cdot \cos\beta + \cos\gamma \cdot \sin\beta \cdot \cos(\alpha-\psi) \qquad (34)$$

where γ = solar elevation angle,
ψ = solar azimuth angle,
β = tilt angle of the plane against the horizontal,
α = azimuth angle of the normal of the plane,

cf. Fig.9. The direct normal radiation I can be expressed by G and D with the help of Eqs. 17 and 16:

$$I = B/\sin\gamma = (G-D)/\sin\gamma. \qquad (35)$$

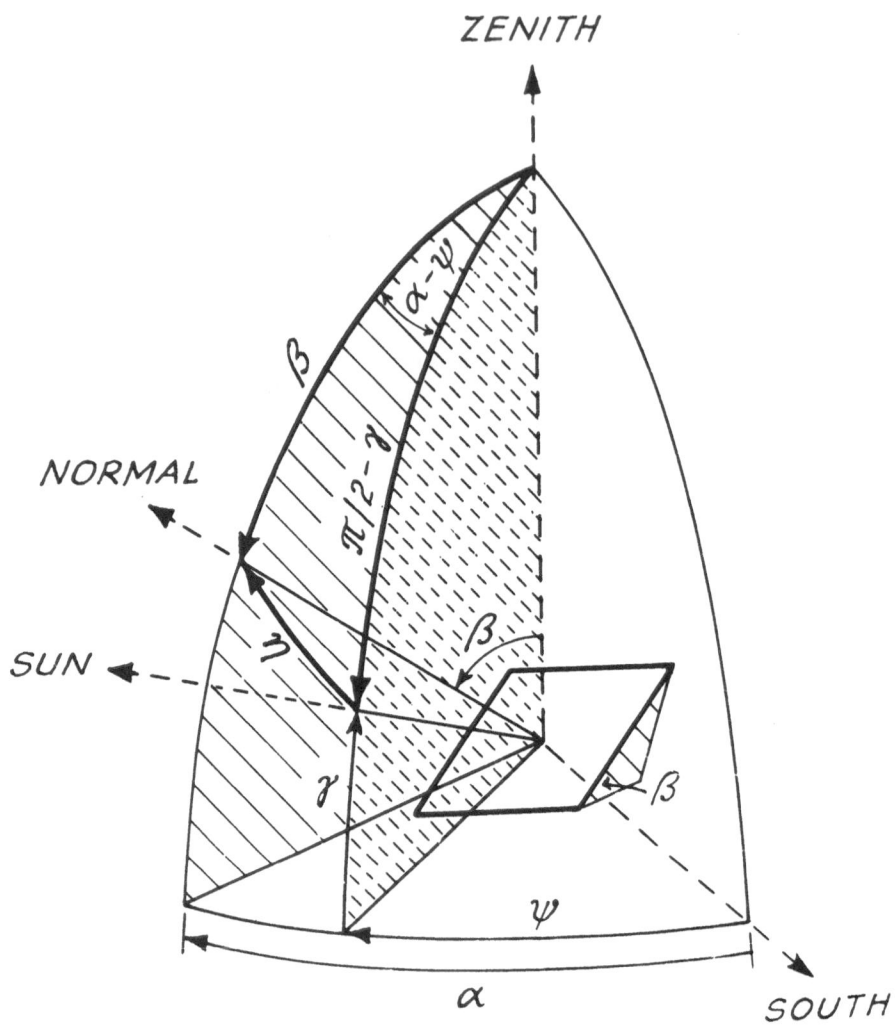

Fig. 9: Definition of the angle η between the direction of
the sun and the normal of the tilted plane.
γ = solar elevation angle,
ψ = solar azimuth angle,
ß = tilt angle of the plane against the horizontal,
α = azimuth angle of the normal of the plane.

$D(\beta,\alpha)$ is the diffuse solar radiation, and $R(\beta,\alpha)$ is the reflected global radiation from those parts of the sky and the earth surface, respectively, which are visible from the tilted plane. Assuming isotropic sky radiation and isotropic reflection from the ground, $D(\beta,\alpha)$ and $R(\beta,\alpha)$ are proportional to the solid angles covered by those parts of the sky and the earth surface, respectively. Spherical geometry then yields

$$D(\beta,\alpha) = D \cdot \cos^2(\beta/2) \qquad (36)$$

$$R(\beta,\alpha) = R \cdot [1 - \cos^2(\beta/2)] = R \cdot \sin^2(\beta/2) \qquad (37)$$

where D and R are the corresponding radiation fluxes on the horizontal plane. According to Eq. 10, R may be derived from G when the short-wave albedo of the earth surface, Q_s, is given

$$R = Q_s \cdot G. \qquad (10)$$

The formulae 36 and 37 may hold for overcast sky. On the other hand, the radiance distribution of cloudless sky is strongly anisotropic. For this case, Temps and Coulson (1977) recommend to modify Eqs. 36 and 37 in the following manner:

$$D(\beta,\alpha) = D.\cos^2(\beta/2).[1+\sin^3(\beta/2)] \cdot (1+\cos^2\eta \cdot \cos^3\gamma), \qquad (36a)$$

$$R(\beta,\alpha) = R \cdot \sin^2(\beta/2) \cdot [1+\sin^2\{(90^o-\gamma)/2\} \cdot \cos(\alpha-\psi)]. \qquad (37a)$$

The general case of a partly clouded sky may be described by introducing the weighting or modulating function (Klucher 1979)

$$F = 1 - (D/G)^2 \qquad (38)$$

which is zero for overcast sky ($D/G = 1$) and close to one for cloudless sky ($D/G \approx 0.3$ so that $(D/G)^2 \approx 0.1$). F is applied to Eqs. 36a and 37a in the following manner:

$$D(\beta,\alpha) = D \cdot \cos^2(\beta/2) \cdot [1+F \cdot \sin^3(\beta/2)] \cdot (1+F \cdot \cos^2\eta \cdot \cos^3\gamma), (39)$$

$$R(\beta,\alpha) = R \cdot \sin^2(\beta/2) \cdot [1+F \cdot \sin^2\{(90^o-\gamma)/2\} \cdot \cos(\alpha-\psi)]. \qquad (40)$$

Summarizing, global radiation on a tilted plane is composed of three terms (Eq.32). These terms are given by Eqs. 33, 39, 40 which may be reduced to functions of G and D by Eqs. 35, 38 and 10. The other variables are angular quantities of the plane and of the sun, and the albedo of the earth surface.

Fig. 10: Moll-Gorczynski pyranometer = solarimeter
(Kipp & Zonen, Delft, The Netherlands).
Responsivity: 0.11 mV/mW·cm^{-2};
temperature dependence: -0.15 %/K;
internal resistance: 10 ohm;
time constant: 4 s;
cosine error: within ± 5 % for angles
                         of incidence up to 80°.

6. RADIATION INSTRUMENTS

The instruments used for meteorological radiation measurements may be classified into two main groups: those for measuring solar (shortwave) radiation and those for measuring total that is solar plus terrestrial (shortwave plus longwave) radiation. If terrestrial (longwave) radiation fluxes are to be determined, a combination of both types of instruments is to be used. The sensor of all instruments described in the following sections is a black surface connected with a thermopile the thermovoltage of which is the electrical signal to be recorded or processed.

Global radiation G is measured by pyranometers. On Fig.10, a Moll-Gorczynski pyranometer also called solarimeter (Kipp & Zonen, Delft, The Netherlands) is shown. The ring-shaped shield is to prevent the thermopile in the socket of the instrument from being heated up by direct solar radiation. The two concentric glass domes are for weather protection and, since the transmission of glass is limited to the spectral range 300 nm $< \lambda <$ 2 500 nm, to confine the spectral response of the instrument to short-wave radiation. For the same reason, the inner dome shields the sensor from the longwave thermal radiation of the weth-er-exposed outer dome. To reduce deposition of dust, dew and rime, it is recommended to ventilate the outer dome through special nozzles by a blower. For measuring reflected global radiation R, a pyranometer facing downward is applied.

Diffuse solar radiation D is also measured by a pyranometer but additionally equipped with a sun-shading ring on an equatorial mount, see Fig.11. The specifications for the shading ring are given by CSAGI (1958) and WMO (1971). The shading ring shields the horizontal surface of the sensor from direct solar radiation so that the system measures

$$G - B = D. \tag{41}$$

The ring has to be reset once every few days according to the declination δ of the sun. Since the ring shields not only from direct solar radiation but also from part of diffuse solar radiation, accurate determination of D by this measuring system requires to apply some correction for the loss of diffuse radiation caused by the ring. On the basis of field experiments at the Meteorological Observatory Hamburg, correction factors for the daily sums of D were determined and tabulated as function of solar elevation at noon, γ_N, and of relative daily sunshine duration S/S_0

Fig. 11: Solarimeter with
sun-shading ring.

Fig. 12: Eppley Normal Incidence
Pyrheliometer (Eplab Inc., Newport,
R.I., USA).
Responsivity: 0.08 mV/mW·cm^{-2};
temperature dependence: compensated
    for $\pm$ 1 % of responsivity in the
    temperature range -30 °C to
    + 40 °C;
internal resistance: 200 ohm;
time constant: 1 s;
angle of view: 5.7°.

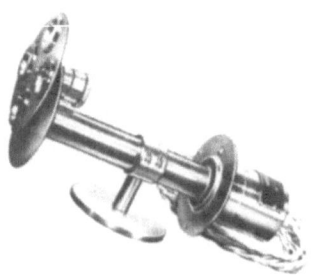

which is a rough measure for the cloud coverage of the sky;
see Table 4. The correction factors range from 1.005
(21 Dec., cloudy) to 1.15 (21 June, cloudless) in Hamburg.

The overall accuracy of pyranometers mainly depends on
solar elevation angle γ. For $\gamma > 10^{\circ}$, it is considered to
be better than 5 % in case of G and R, and between 5 and
10 % in case of D.

An instrument for measuring direct solar radiation on
a receiving surface being normal to the direction of incidence, I, is called pyrheliometer. As an example, Fig.12
shows the Eppley Normal Incidence Pyrheliometer (Eplab Inc.,
Newport, R.I., USA). A long tube and several diaphragms
allow radiation from only a narrow solid angle to reach the
sensor at the end of the tube. A filter wheel permits to
select certain spectral ranges of the incident radiation.

Pyrheliometers are mainly used either as primary reference standard or as secondary calibration standard instruments. With the help of the first type, the secondary standard pyrheliometers are calibrated which in turn serve as
standards for calibrating pyranometers. Secondary standard
pyrheliometers are to have an accuracy of $\pm$ 1 % or better.

For continuously recording direct normal incidence
solar radiation I, the pyrheliometer has to be installed
on a motordriven equatorial mount which allows the pyrheliometer axis to follow the sun. Since this device requires
some care and service it is only used at selected observa-
tories. For routine radiation stations, it is recommended
to determine I from B, the direct radiation on a horizontal
receiving surface. B may be computed from measurements of
G and D with the help of

$$B = G - D; \tag{42}$$

and from that, direct solar radiation on the normal surface
by

$$I = B/\sin\gamma. \tag{35}$$

It can be shown that Eq.35 also holds for hourly sums of I
and B with sufficient accuracy if instead of γ the solar
elevation angle γ_m of the middle of the respective hour is
taken. γ_m is given by an equation being analogous to Eq.20:

$$\sin\gamma_m = \sin\varphi \sin\delta + \cos\varphi \cos\delta \cos\omega_m \tag{43}$$

with ω_m = hour angle of the middle of the respective hour.

S/S_0	0.0	0.1	0.2	0.3	0.4	0.5	0.6	0.7	0.8	0.9	1.0
Y_N											
0	1,000	1,000	1,000	1,000	1,000	1,000	1,000	1,000	1,000	1,000	1,000
2	1,001	1,001	1,001	1,001	1,001	1,001	1,001	1,001	1,002	1,002	1,002
4	1,001	1,001	1,002	1,002	1,002	1,003	1,003	1,003	1,003	1,004	1,004
6	1,002	1,002	1,003	1,003	1,004	1,004	1,004	1,005	1,005	1,006	1,006
8	1,003	1,004	1,004	1,005	1,005	1,006	1,007	1,007	1,008	1,008	1,009
10	1,004	1,005	1,006	1,006	1,007	1,008	1,009	1,010	1,010	1,011	1,012
12	1,005	1,006	1,007	1,009	1,010	1,011	1,012	1,013	1,014	1,015	1,016
14	1,007	1,008	1,009	1,011	1,012	1,013	1,014	1,015	1,017	1,018	1,019
16	1,008	1,010	1,011	1,013	1,014	1,016	1,017	1,019	1,020	1,022	1,023
18	1,010	1,012	1,014	1,015	1,017	1,019	1,021	1,023	1,024	1,026	1,028
20	1,011	1,013	1,015	1,018	1,020	1,022	1,024	1,026	1,028	1,030	1,032
22	1,013	1,015	1,018	1,020	1,023	1,025	1,027	1,030	1,032	1,035	1,037
24	1,014	1,017	1,020	1,023	1,026	1,029	1,031	1,034	1,037	1,040	1,043
26	1,016	1,019	1,022	1,026	1,029	1,032	1,035	1,038	1,042	1,045	1,048
28	1,018	1,022	1,025	1,029	1,032	1,036	1,040	1,043	1,047	1,050	1,054
30	1,019	1,023	1,027	1,031	1,035	1,040	1,044	1,048	1,052	1,056	1,060
32	1,021	1,025	1,030	1,034	1,039	1,043	1,048	1,052	1,057	1,061	1,066
34	1,023	1,028	1,033	1,038	1,043	1,048	1,052	1,057	1,062	1,067	1,072
36	1,025	1,030	1,036	1,041	1,046	1,052	1,057	1,062	1,067	1,073	1,078
38	1,027	1,033	1,038	1,044	1,050	1,056	1,061	1,067	1,073	1,078	1,084
40	1,029	1,035	1,041	1,047	1,053	1,060	1,066	1,072	1,078	1,084	1,090
42	1,031	1,038	1,044	1,051	1,057	1,064	1,071	1,077	1,084	1,090	1,097
44	1,033	1,040	1,047	1,054	1,061	1,068	1,075	1,082	1,089	1,096	1,103
46	1,035	1,042	1,049	1,057	1,065	1,072	1,079	1,087	1,094	1,102	1,109
48	1,037	1,045	1,052	1,060	1,068	1,076	1,083	1,091	1,099	1,106	1,114
50	1,039	1,047	1,055	1,063	1,071	1,079	1,087	1,095	1,104	1,112	1,120
52	1,041	1,049	1,058	1,066	1,075	1,083	1,091	1,100	1,108	1,117	1,125
54	1,042	1,051	1,060	1,068	1,077	1,086	1,095	1,104	1,113	1,122	1,131
56	1,044	1,053	1,062	1,072	1,081	1,090	1,099	1,108	1,118	1,127	1,136
58	1,046	1,056	1,065	1,075	1,084	1,094	1,103	1,113	1,122	1,132	1,141
60	1,048	1,058	1,068	1,077	1,087	1,097	1,107	1,117	1,126	1,136	1,146

Table 4: Correction factors for diffuse solar radiation D measured by a solarimeter with shading ring of 5 cm width and 60 cm diameter, as used by the Meteorological Observatory Hamburg.

Y_N = solar elevation angle at noon; S/S_0 = relative daily sunshine duration.

Next to global radiation G, the total radiation balance or net total radiation Q is the most important radiation quantity in meteorology. Q is measured by net pyrradiometers also called radiation balance meters. Fig.13 shows the Schulze-Däke radiation balance meter (B. Lange, Berlin (West)). The instrument is basically similar to a pair of pyranometers; but instead of glass domes it possesses hemispherical covers of polyethylene which is transparent to wavelengths up to more than 50 µm so that solar (shortwave) plus terrestrial (longwave) radiation is received by the sensors. By connecting the thermopiles of the upward and downward facing sensors against each other, the electrical output of the instrument is directly proportional to the net total radiation provided both sensors have the same calibration constants in both the shortwave and longwave spectral range.

In the general case, the different magnitude of the calibration factors of the upward facing sensor for shortwave and longwave radiation, f_{us} and f_{ul}, and of the corresponding calibration factors of the downward facing sensor, f_{ds} und f_{dl}, have to be accounted for. The instrument temperature T_i is also to be known for what reason a resistance thermometer is provided in the instrument. The upward facing sensor records the difference between incoming total radiation G + A and the thermal radiation σT_i^4 emitted by the instrument: $G + A - \sigma T_i^4$; the downward facing sensor measures the difference between incoming R + E and emitted σT_i^4: $R + E - \sigma T_i^4$. The radiation fluxes G and R are to be known from independent measurements by pyranometers.

By this way, atmospheric radiation A and terrestrial surface radiation E are determined. The net total radiation is then easily computed from its components G, R, A, E:

$$Q = (G - R) + (A - E). \tag{6}$$

The overall accuracy of the quantities A, E and Q determined by the net pyrradiometer is around 10 %.

For further details on radiation instruments, their calibration, installation and maintenance, the reader is refered to the surveys given by Robinson (1966), WMO (1971), Latimer (1972), Coulson (1975) and Dehne (1977).

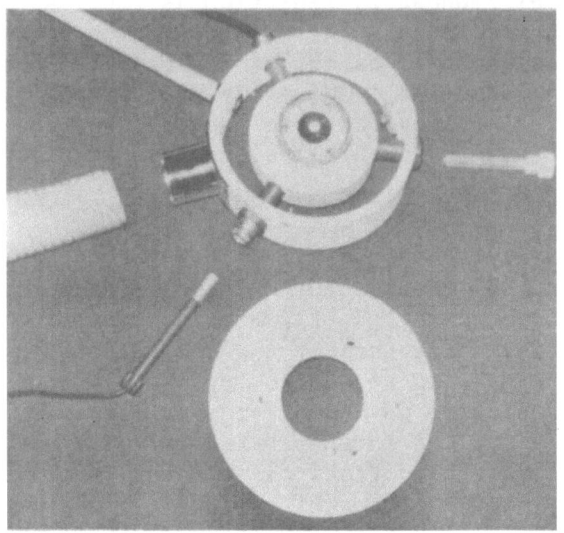

Fig. 13: Schulze-Däke radiation balance meter = net pyrradiometer (B. Lange, Berlin (West)).

Responsivity:	0.9 mV/mW·cm^{-2};
temperature dependence:	$\pm$ 6 % in the temperature range from −10 ^{o}C to + 30 ^{o}C;
internal resistance:	2.5 ohm;
time constant:	150 s for 99 % value;
cosine error:	within $\pm$ 2.5 % for angles of incidence up to 80°.

7. DAILY COURSE OF HOURLY RADIATION

In order to get a feeling for the variation of solar radiation, a few examples of original records will be presented. Fig.14 is a reproduction of the records of global radiation G and diffuse solar radiation D on two selected days with no clouds and low turbidity, one day close to winter solstice, the other day close to summer solstice. Global radiation on these two extreme days differs by a factor 4.1 in maximum, and by 8.4 in the daily sums which are given by the areas below the respective curves. To the contrary, diffuse solar radiation D in winter reaches the same maximum as in summer, and the ratio of the daily sums is 2.6.

Fig.15 illustrates the large variation of solar radiation due to changing cloudiness during one day. Bright towering cumulus of a thunderstorm approaching the site of observation make, by reflection, G momentaneously to increase up to 1000 W m^{-2}; after the front has reached the site, the sky is completely covered by thick dark cloudiness and G is reduced to about 10 W m^{-2}. Later on, the cloud deck is varying in thickness so that G undergoes relatively smooth variations.

This example of a clouded day exhibiting large radiation fluctuations within short time intervals demonstrates that evaluation of radiation records on a routine basis requires some kind of smoothing or averaging. In meteorology, hourly sums of the radiation fluxes are considered to give representative radiation values even in cases of large variation in cloudiness and consequently of widely scattered points on the chart records. On the other hand, the time resolution of one hour seems to be short enough to permit correlation with other meteorological quantities such as actual cloud observations, for instance.

The hourly sum is a time integral of the irradiance or energy flux density or power density of radiation, and is called (hourly) irradiation. Whereas the basic unit of irradiance is W m^{-2}, the basic unit of irradiation is Ws m^{-2} = J m^{-2}. Alternate units of irradiance are mW cm^{-2} or kW m^{-2}, of irradiation J cm^{-2}, MJ m^{-2}, and Wh cm^{-2} or kWh m^{-2}.

For demonstrating typical mean daily courses of the different radiation quantities, the results of a 10 year period of measurements at the Meteorological Observatory Hamburg are presented. On Fig.16, mean hourly sums of global radiation G, in J cm^{-2}, are plotted versus true local time

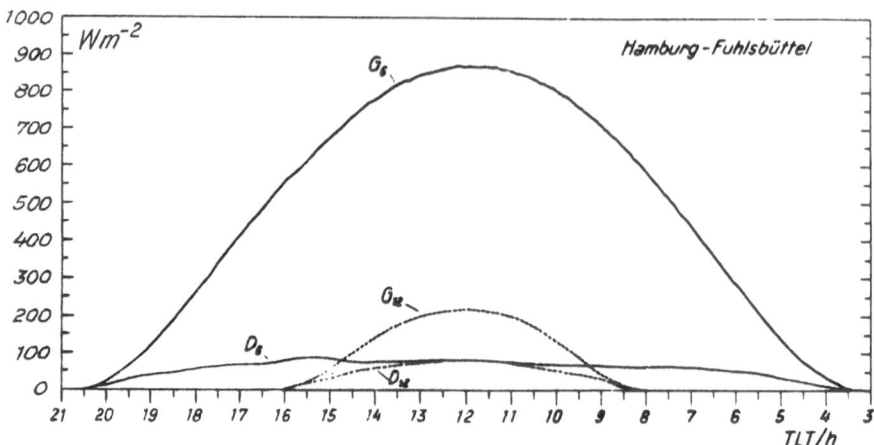

Fig. 14: Global radiation G and diffuse solar radiation D on two cloudless days in June and December, resp. G_6, D_6: 29 June 1976; G_{12}, D_{12}: 31 Dec 1969. Note: Time runs from right to left on this diagram. (After Dehne 1977a).

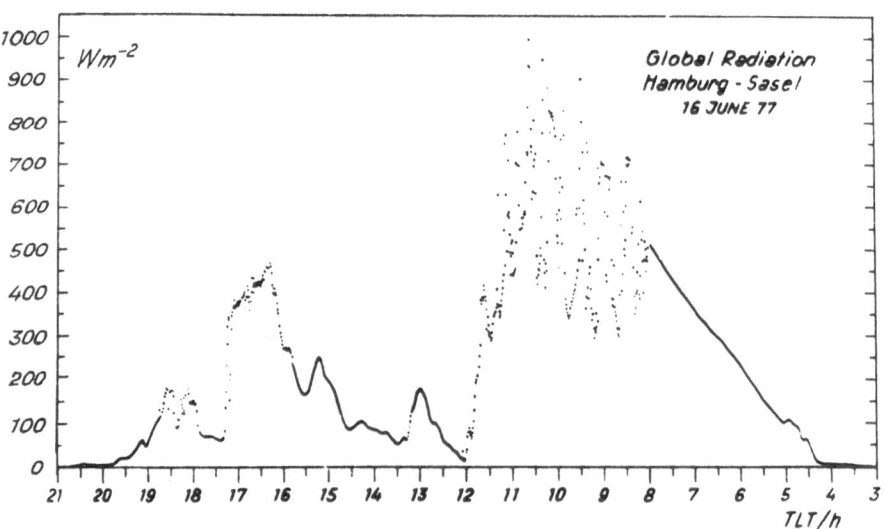

Fig. 15: Global radiation G on a day with varying cloudiness. Note: Time is running from right to left on this diagram. (After Dehne 1977a).

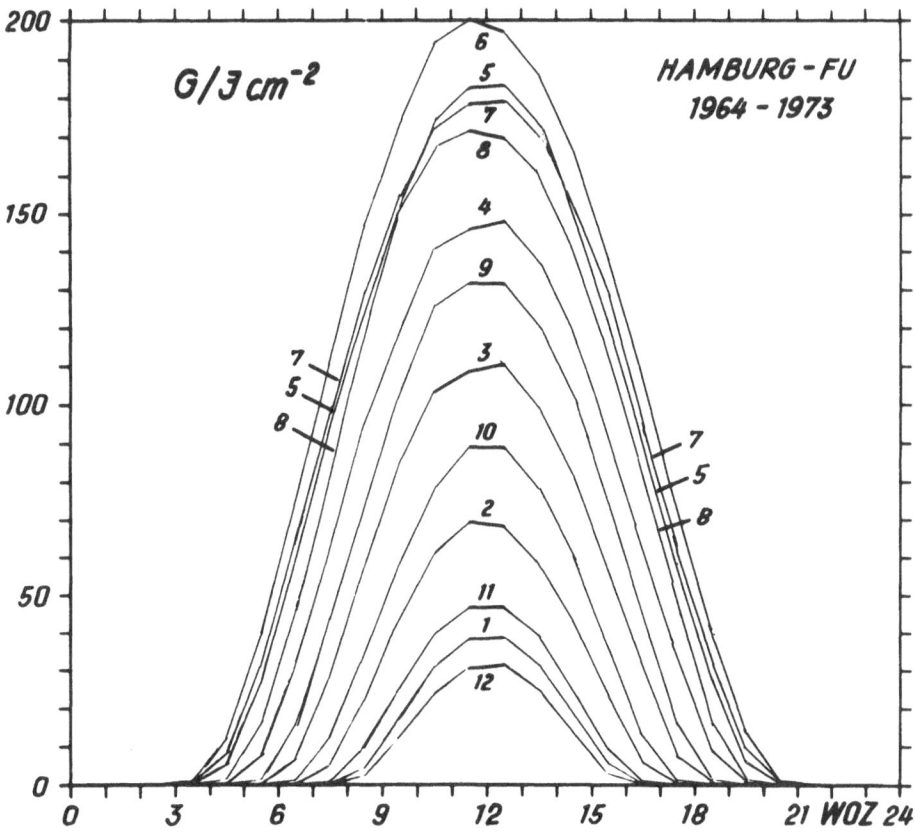

Fig. 16: Global radiation G: 10 year means of the hourly
sums, in J cm^{-2}, for each of the twelve months. Time of day
is given as true local time Figures at the curves indicate
number of the months. (After Kasten 1977).

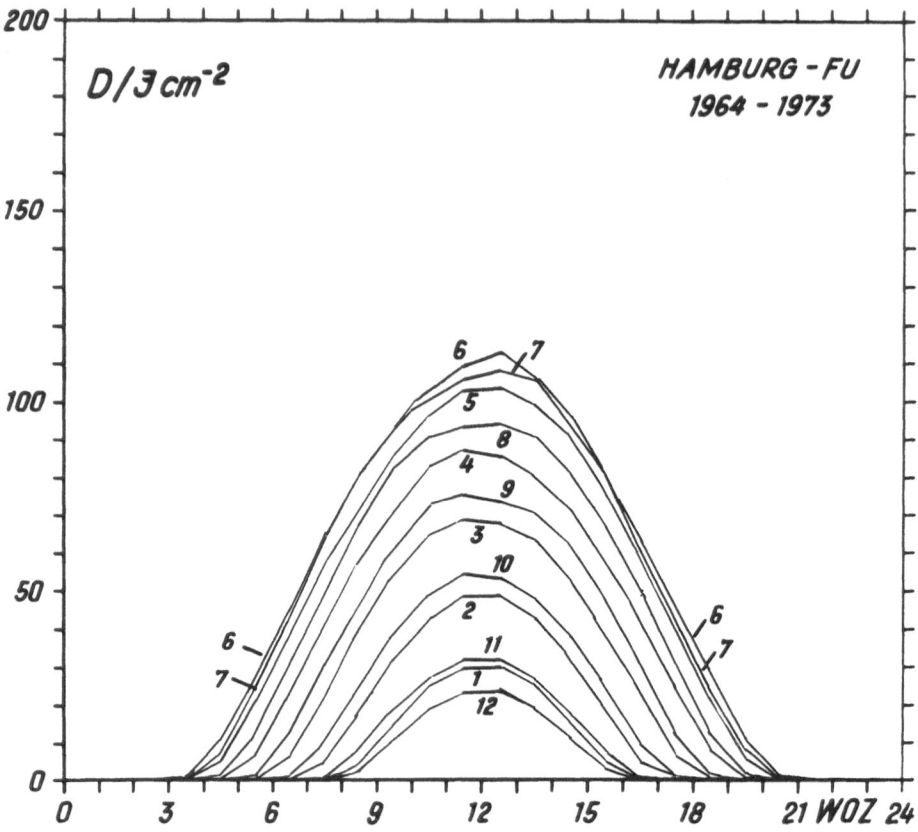

Fig. 17: Diffuse solar radiation D. See legend to Fig. 16.

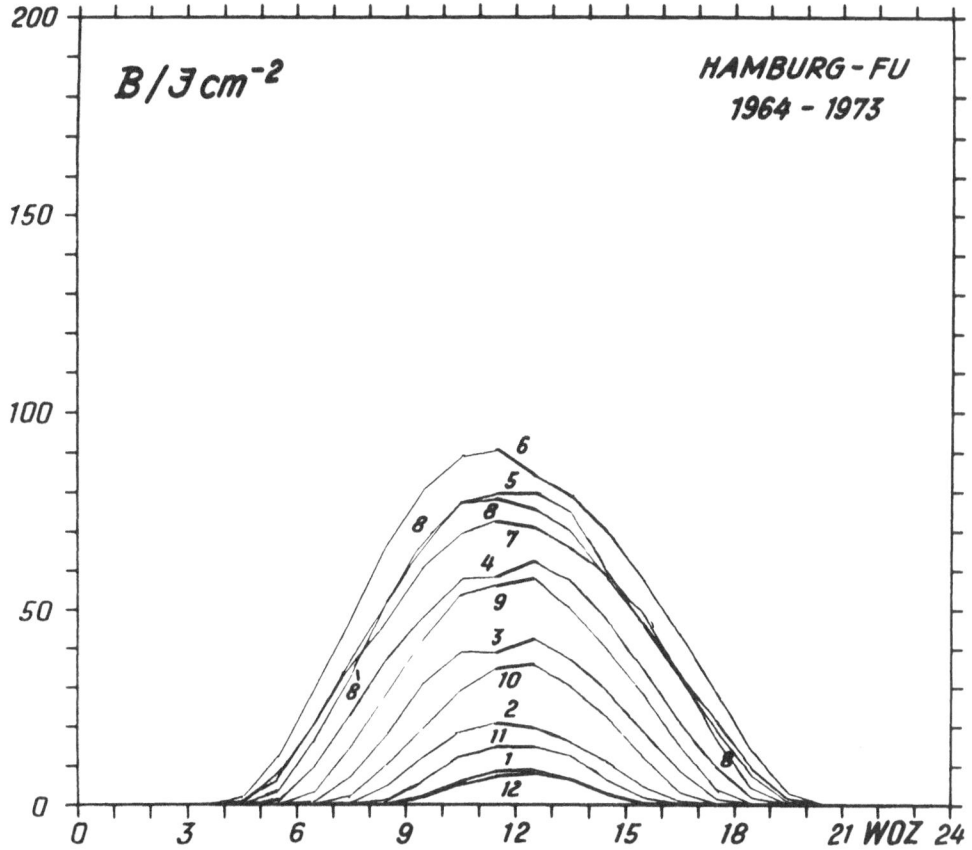

Fig. 18: Direct solar radiation B. See legend to Fig. 16.

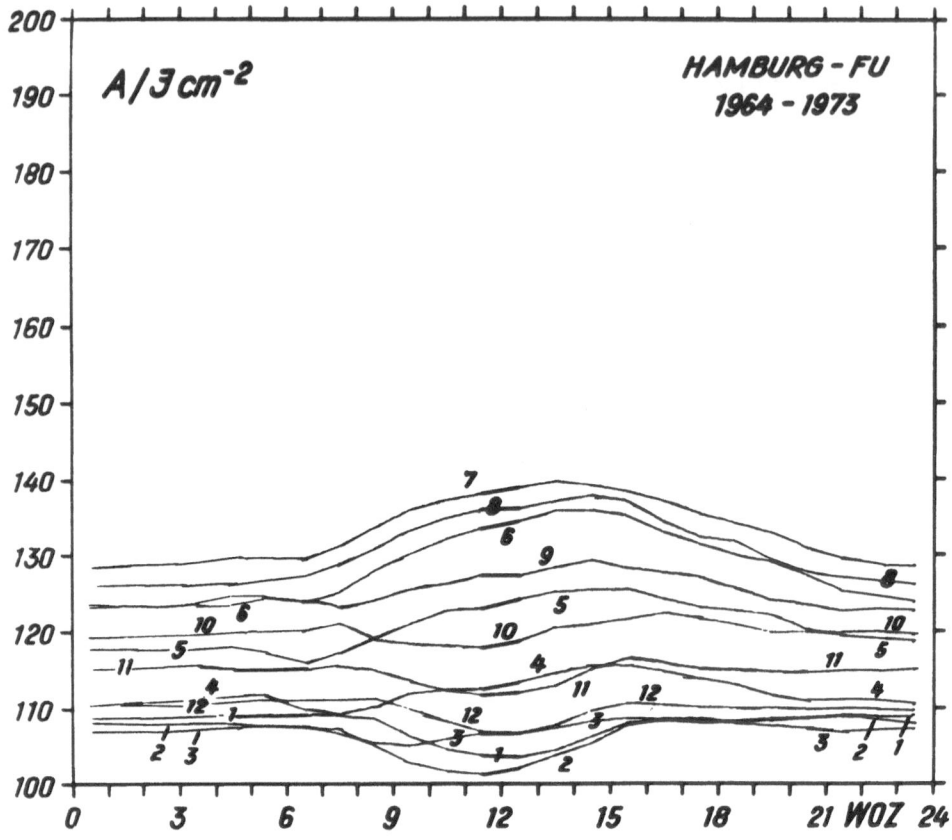

Fig. 19: Atmospheric radiation A. See legend to Fig. 16.

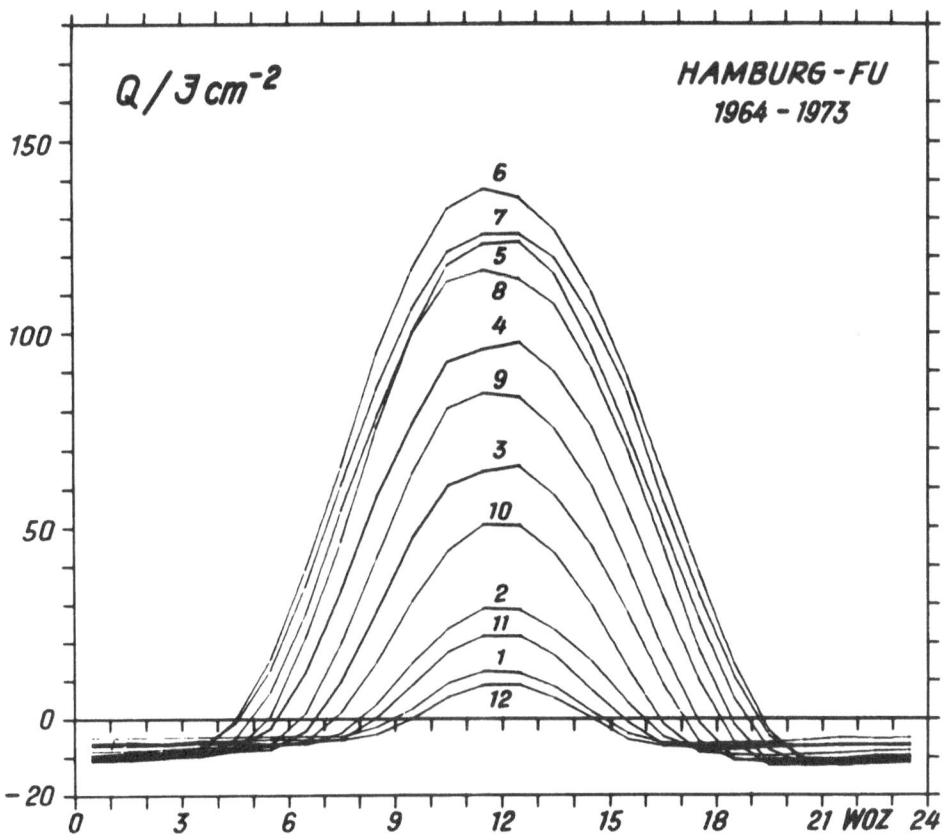

Fig. 20: Net total radiation (total radiation balance)
Q = (G-R) - (E-A).
See legend to Fig. 16.

for each of the twelve months indicated by the figures at
the curves. G reaches its maximum at true solar noon but
a slight depression is noticed, in the spring months March
and April, just before noon due to convective cloudiness
which disappears in the afternoon.

Diffuse solar radiation, Fig.17, exhibits values being
about half of G in summer but of almost equal magnitude in
winter. Direct solar radiation on the horizontal surface,
B (Fig.18), has a distinct minimum before noon in March and
April which is caused by convective cloudiness just mentioned.

Atmospheric radiation A, Fig.19, has a daily course
which is quite different from those of the preceding solar
radiation sums. Due to delayed warming of the atmosphere,
there is a typical asymmetry with a maximum around 1400 TLT.
For the same reason, the night values in the spring months
March and April are relativey low in contrast to the corresponding months in fall, September and October. Peculiar
are the minima in March, April and May around the time of
sunrise, and the broad minima around noon from October
through February. Rise of radiating inversion layers to
higher, colder levels of the atmosphere and finally their
dissolution are believed to be responsible for these reduced values of downward thermal radiation.

Finally, the curves of the net total radiation or total radiation balance Q, Fig.20, show a course similar to
global radiation G for which reason several parameteriza-
tions of Q by G have been proposed in the literature. Of
course, the negative values of Q at night can not be expressed in terms of G. The nighttime values of Q are more
negative in the summer months than during winter corresponding to the opposite behaviour of the net terrestrial
surface radiation because $Q = - (E-A)$ at nighttime.

8. FREQUENCY DISTRIBUTION OF HOURLY RADIATION

Many applications of solar energy require a certain
minimum of incoming solar radiation in order to operate
efficiently. Therefore, the probabilities for the occurence
of irradiance above given threshold levels are of interest.

Continuous measuring series of all solar and terrestrial radiation fluxes have been performed since 1954 at
Meteorologisches Observatorium Hamburg of Deutscher Wetterdienst (DWD). This institute also operates the radiation
network of DWD, see Fig.21. At present, 24 global radiation

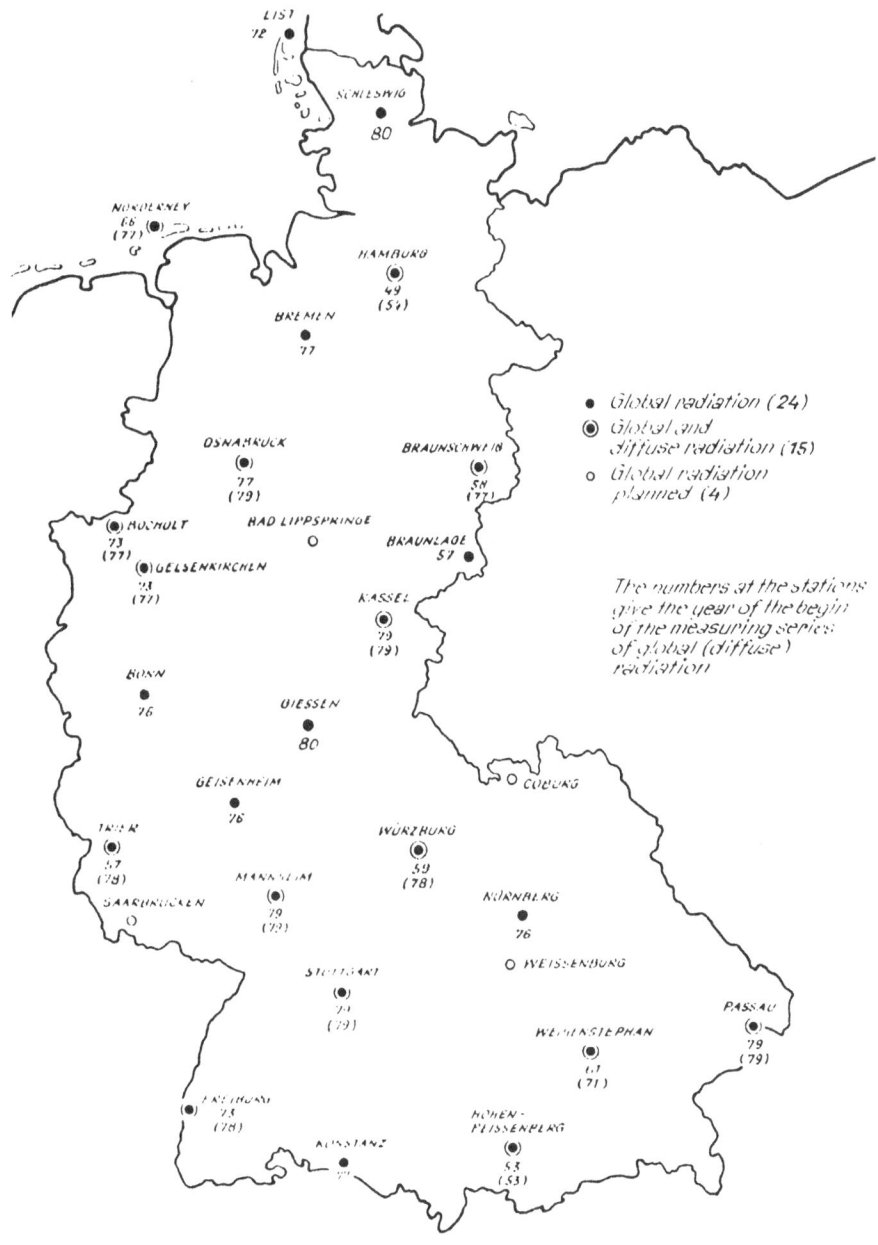

Fig. 21: Radiation network of Deutscher Wetterdienst
(as of 31 Dec 1980).

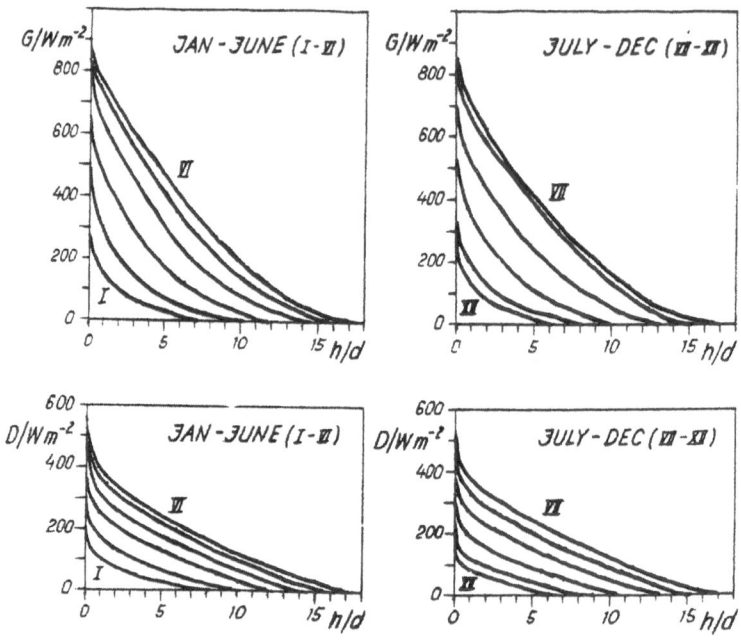

Fig. 22: Global radiation G and diffuse solar radiation D.
Cumulative frequency, in hours per day. Mean hourly irradiance, in W m^{-2}, Hamburg-Fuhlsbüttel 1966-1975.

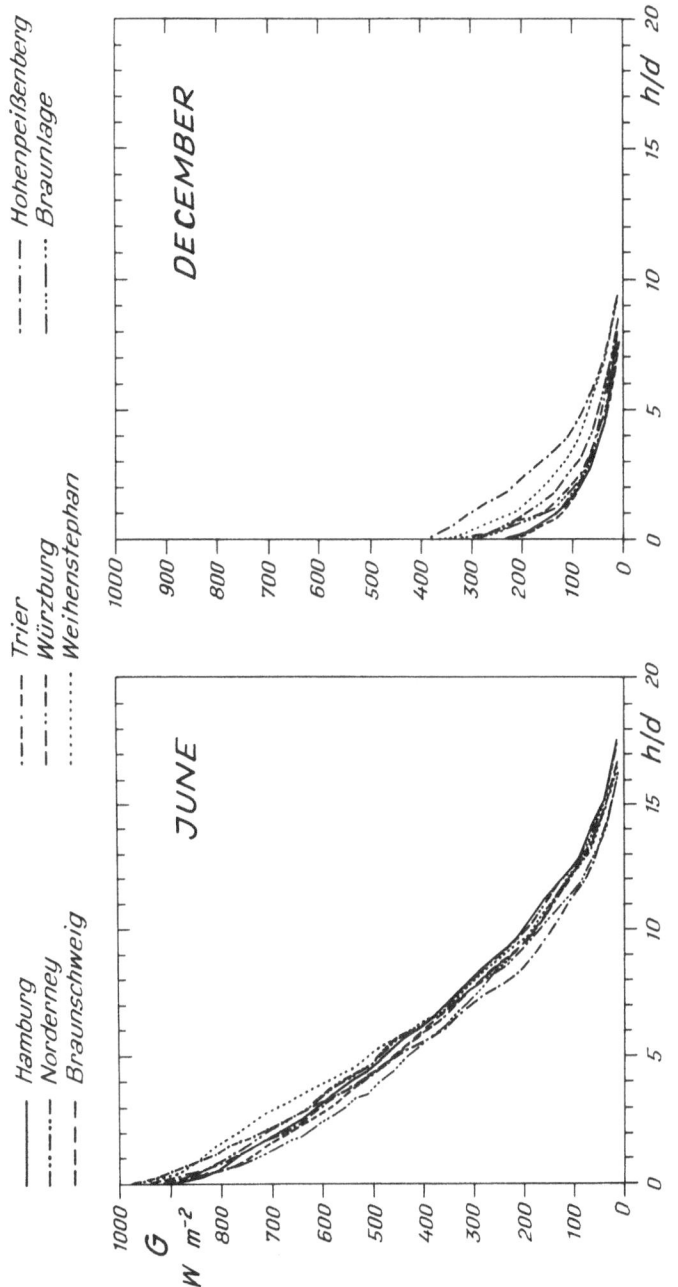

Fig. 23: Global radiation G. Cumulative frequency, in hours per day. Mean hourly irradiance, in W m⁻². 8 stations in Germany 1966-1975.

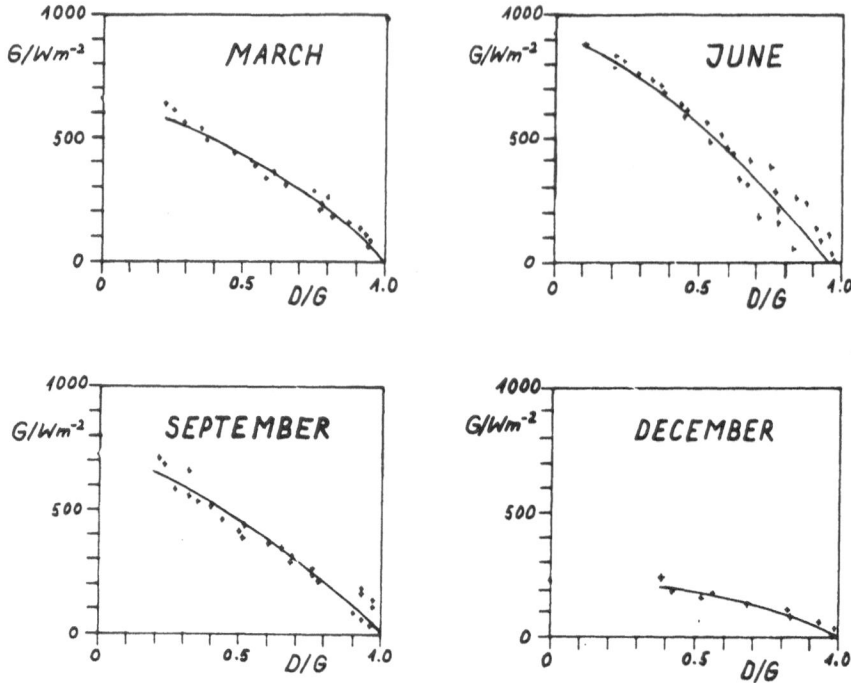

Fig. 24: Ratio of diffuse solar to global radiation, D/G, for given G above specified threshold. Mean monthly means, Hamburg—Fuhlsbüttel, 1966—1975.

stations are in operation, 15 of which additionally record
diffuse solar radiation. The numbers at the stations give
the starting year of operation. From 8 global radiation
stations, continuous records of more than 10 years of hourly
radiation sums are available.

These data were analyzed month by month with regard
to the number of hours when the hourly irradiations were
$> 0, 9, 18, 27....$ J cm^{-2} corresponding to mean hourly irradiances of $> 0, 25, 50, 75....$ W m^{-2}. The results are
plotted as mean monthly means of cumulative frequency, in
hours per day (h/d), of mean hourly irradiance, in W m^{-2},
for each month, see Fig. 22 for the station Hamburg. In the
upper row, the frequency curves of global radiation G for
January to June (I - VI) and July to December (VII - XII)
are plotted, in the lower row the corresponding curves of
diffuse solar radiation D. Global radiation frequencies
for all 8 German stations were summarized on one diagram
for each month, see Fig.23 for June and December as examples.

The frequency of global radiation at Hamburg was also
analyzed with respect to the D/G-ratio in the following
manner: For each month, the hourly sums or mean hourly irradiances, respectively, of G were assorted into the classes
defined above. For each individual G-value within a class,
the corresponding simultaneously measured D-value was taken
irrespectively of any D-threshold. The ratio D/G of individual G,D-pairs within each class were averaged and then
plotted versus G. As examples, the resulting diagrams for
March, June, September and December are presented on Fig.24.

9. YEARLY COURSE OF DAILY RADIATION

To illustrate the yearly course of the different radiation quantities, daily radiation sums seem to be most suitable. Fig.25 presents the 20 year average, 1954 - 1973, of
the five radiation quantities directly measured at the
Meteorological Observatory Hamburg. Diffuse solar radiation
D seems to follow the astronomical seasons with only minor
disturbances. But global radiation G exhibits several peculiar deviations from the astronomically conditioned course.
A depression is noticed from end of June through end of
August which is caused by enhanced cloudiness during the
so-called "European Monsoon". Another depression appears in
May which is known as May coolness to farmers. Several other features which are called singularities in meteorological
literature, can be traced in the yearly course of G.

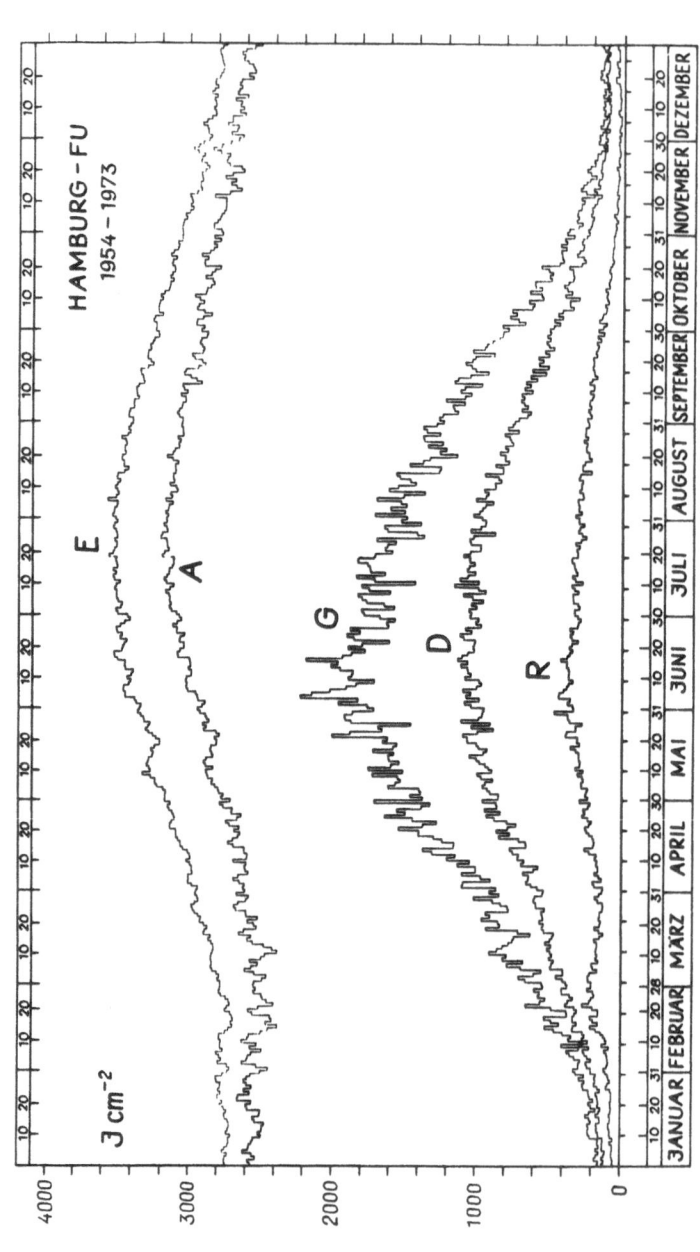

Fig. 25: 20 year means of daily sums, in J cm^{-2}, of global radiation G, diffuse solar radiation D, reflected global radiation R, terrestrial surface radiation E and atmospheric radiation A (after Kasten 1977).

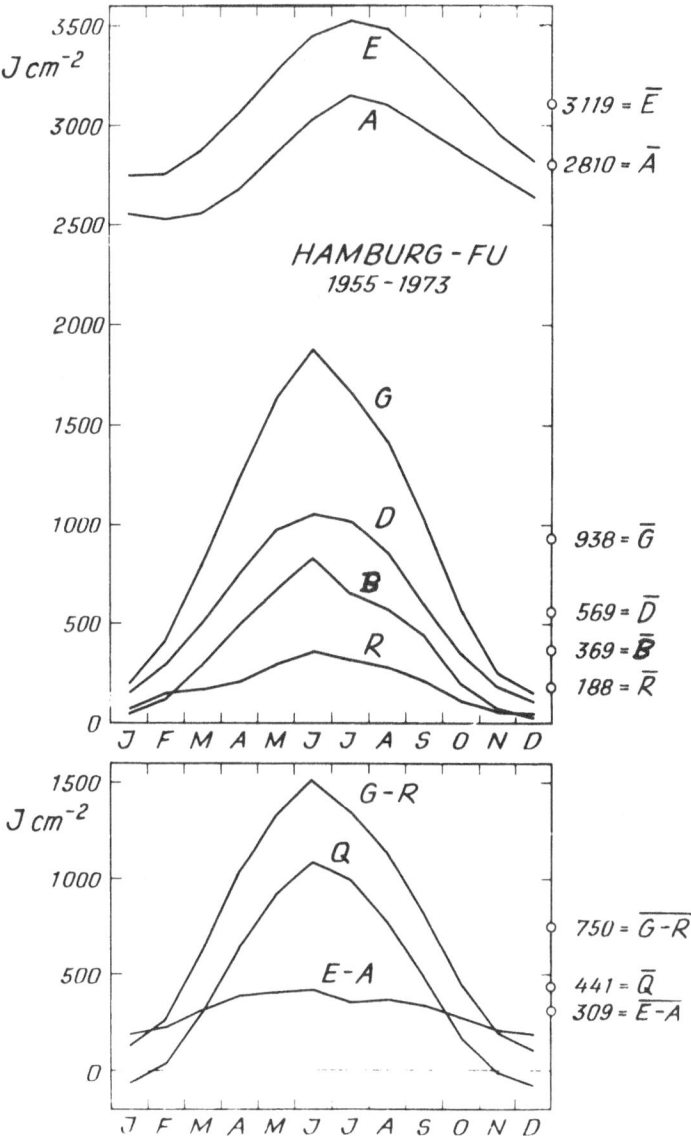

Fig. 26: 19 year means of monthly means of daily sums of radiation fluxes, in J cm^{-2}. Circles and numbers on right hand vertical scale give 19 year means of yearly means of daily sums (after Kasten 1977).

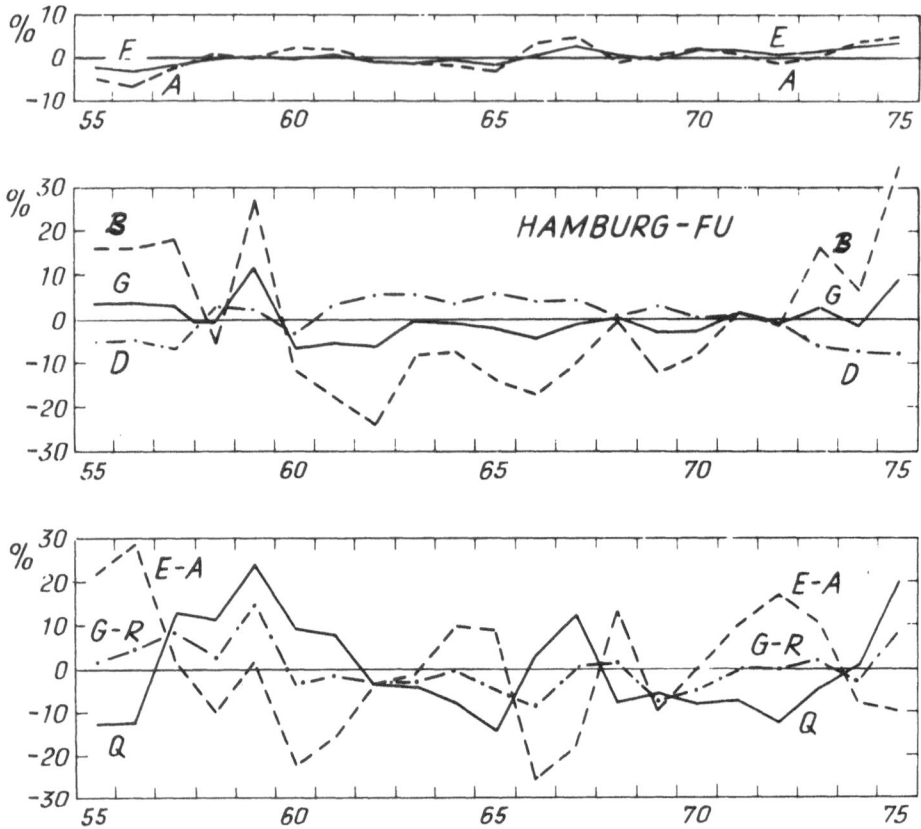

Fig. 27: Deviations of yearly means, 1955 to 1975, of daily sums of radiation fluxes from their 21 year means, in percent (after Kasten 1977).

Direct solar radiation $B = G - D$, not shown here, follows the course of G very closely because the curve of D is
smooth. Reflected global radiation R is relatively high in
the winter months especially in February due to snow cover
on the ground. The longwave radiation sums E and A are influenced by the delayed warming of the ground and the atmosphere so that their maxima appear at the end of July. The
yearly courses of E and A are almost parallel but because
of the inertness of the ground, the terrestrial surface
radiation E comes out smoother than the atmospheric radiation A. Both quantities show a marked minimum during the
May coolness. Several other singularities are observed in
the curve of A, especially the winterly cold breaks.

In order to get a general view of the seasonal variations of the radiation sums, monthly means of the daily
sums were computed and averaged over 19 full years of records, see Fig.26. Since each discrete point is obtained by
averaging about 600 single values, the connecting polygons
are very smooth. Clearly,the time lag of the longwave fluxes
E and A with respect to the shortwave solar fluxes is seen.
The curve of direct solar radiation B exhibits the marked
"monsoon" depression in July mentioned earlier. The deficit
of incoming solar radiation induces a cooling of the ground
which shows up as a minimum of the net terrestrial surface
radiation E-A. On the right edge of the diagram, the mean
yearly means of the daily radiation sums are indicated.

Individual yearly means of the daily radiation sums
for the period 1955 to 1975 are presented as percentage
deviations from their 21 year means on Fig.27. Largest amplitudes are displayed by direct solar radiation B which
reached + 35 % in 1975 and 1976. Cloudiness is the factor
most influencing B. Much of the direct radiation scattered
by clouds reaches the ground as diffuse solar radiation D.
Therefore, the year to year variation of the yearly mean of
global radiation $G = B+D$ is relatively small ($\pm$ 10 % maximum). In the uppermost diagram, the terrestrial radiation fluxes E and A show but little variation (< 7 %).
Large amplitudes are exhibited by the net terrestrial surface radiation E-A which are caused by the yearly warming
or cooling of the ground relative to the atmosphere. Since
the net global radiation G-R has relatively small variations, the net total radiation $Q = (G-R) - (E-A)$ runs opposite to E-A in most years.

10. GEOGRAPHICAL DISTRIBUTION OF DAILY GLOBAL RADIATION

In cooperation with the responsible national institutions, an inventory of existing radiometric and helio-
graphic stations in the European Community and of their
available data records has been screened with respect to
quality of maintenance and to length of continuous recording. On the basis of the available data of the selected
stations, the years from 1966 till 1975 have been defined
as common reference period. Fig.28 gives the geographical
distribution of the 56 selected stations.

Daily sums of global radiation and of sunshine duration
were collected and subjected to quality control procedures.
The screened data were arranged in 1344 tables, one for
each station and each month, of the individual daily sums
from which the 10 year means of the monthly means, maxima
and minima were compiled in another set of 56 tables, one
for each station. On the basis of these summary tables,
preliminary maps of global radiation for each month and the
year were designed. The summary tables and the maps have
been published by the Commission of the European Communities (Palz 1979).

Examples of the preliminary maps are given on the
following Figures. On Fig.29 and 30, the geographical distribution of mean daily global radiation in June and December, respectively, are displayed. Fig. 31 shows the mean
minimum of daily global radiation in June, and Fig.32 the
mean maximum in December.

In the geographical distribution of global radiation,
the different climates in the various regions of the EC are
reflected because solar radiation is the primary factor
governing all other climatic parameters. Further, the interaction of various air masses of either maritime or continental origin, having different optical properties such as
transparency of the air and cloudiness, and of pronounced
orography such as the Alps and the highlands, cause essential differences in the global radiation pattern from country to country.

Principle features of the monthly G-maps are:

The daily sums of G generally decrease with increasing
geographical latitude as particularly evidenced by comparing
the areas south and north of 45°N. For astronomical reasons,
the gradient of the isolines is weaker in the summer half-
year when the greater length of day counteracts the meteor-

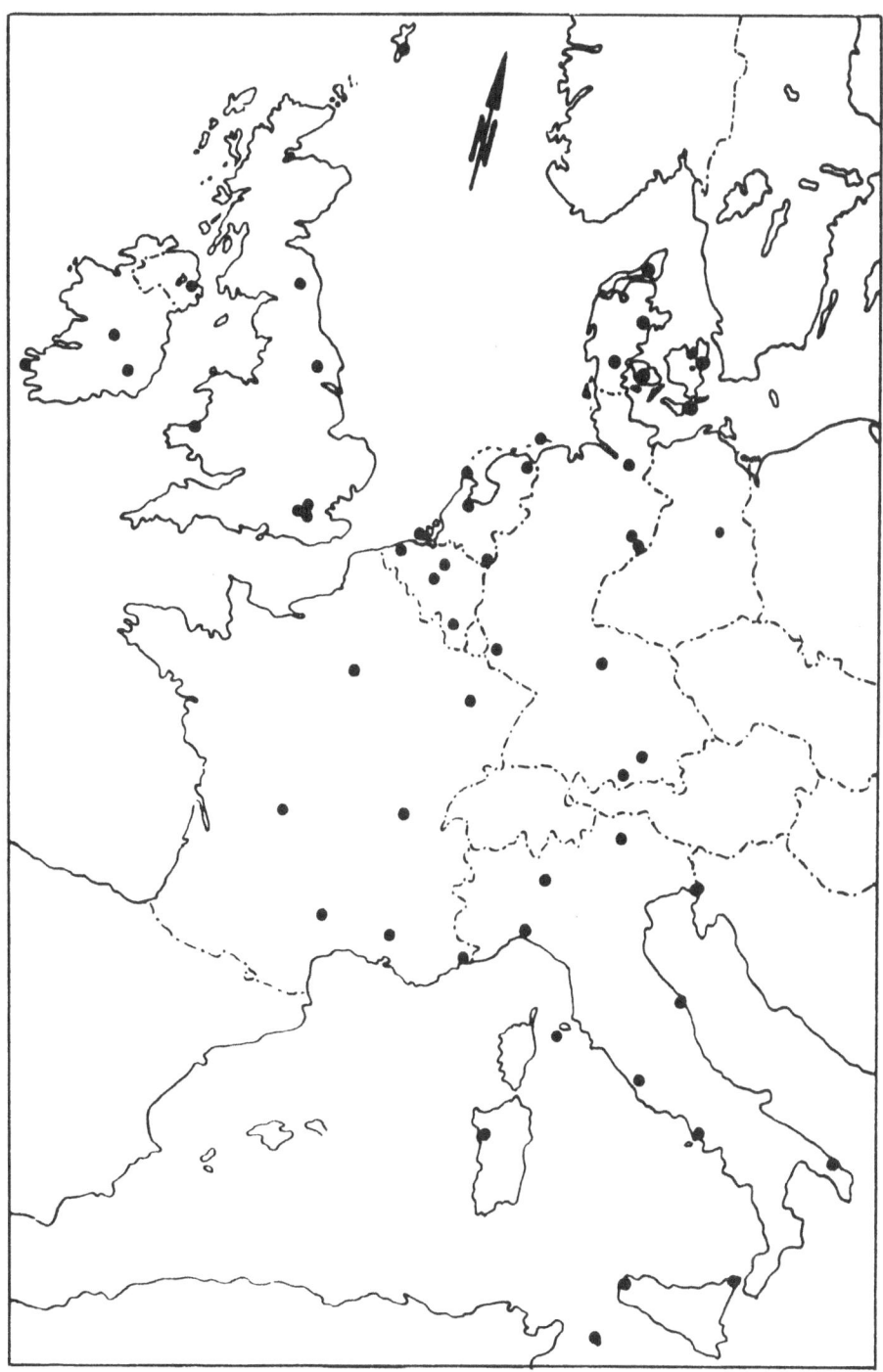

Fig. 28: Stations in the region of the EC selected for the
global radiation atlas (⊚ sunshine duration only).

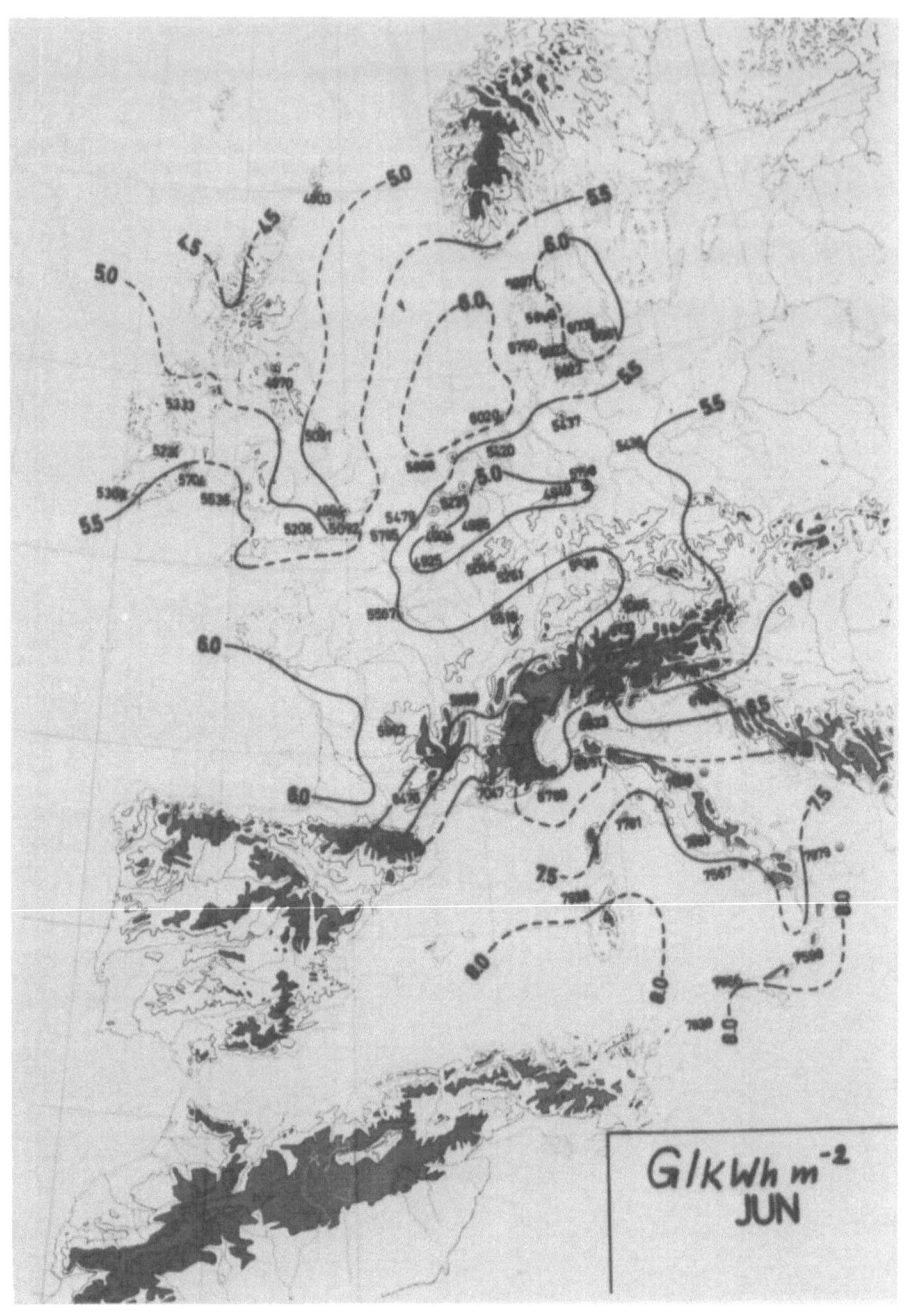

Fig. 29: Geographical distribution of mean daily global
radiation in June, 1966-1975.

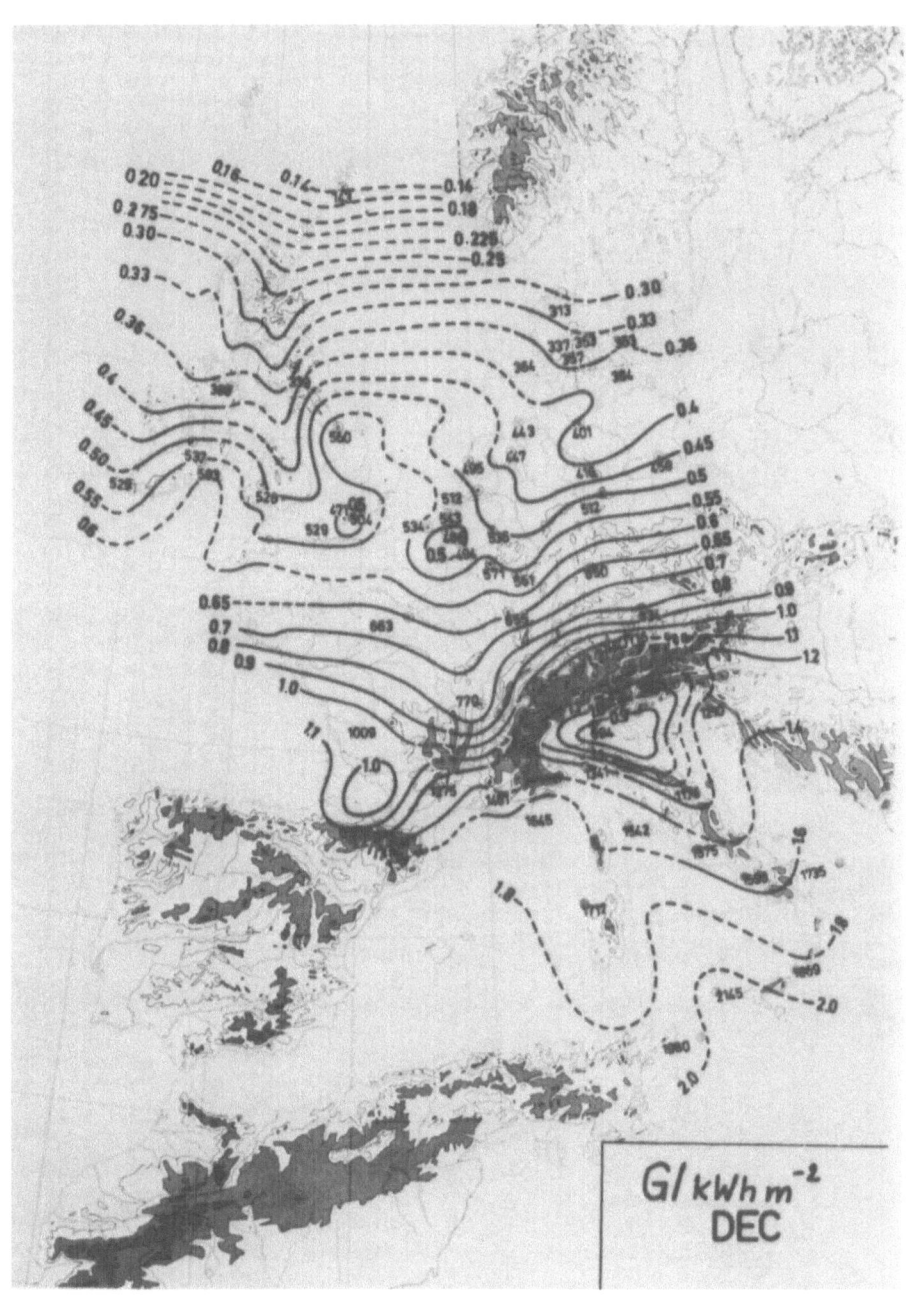

Fig. 30: Geographical distribution of mean daily global
radiation in December, 1966-1975.

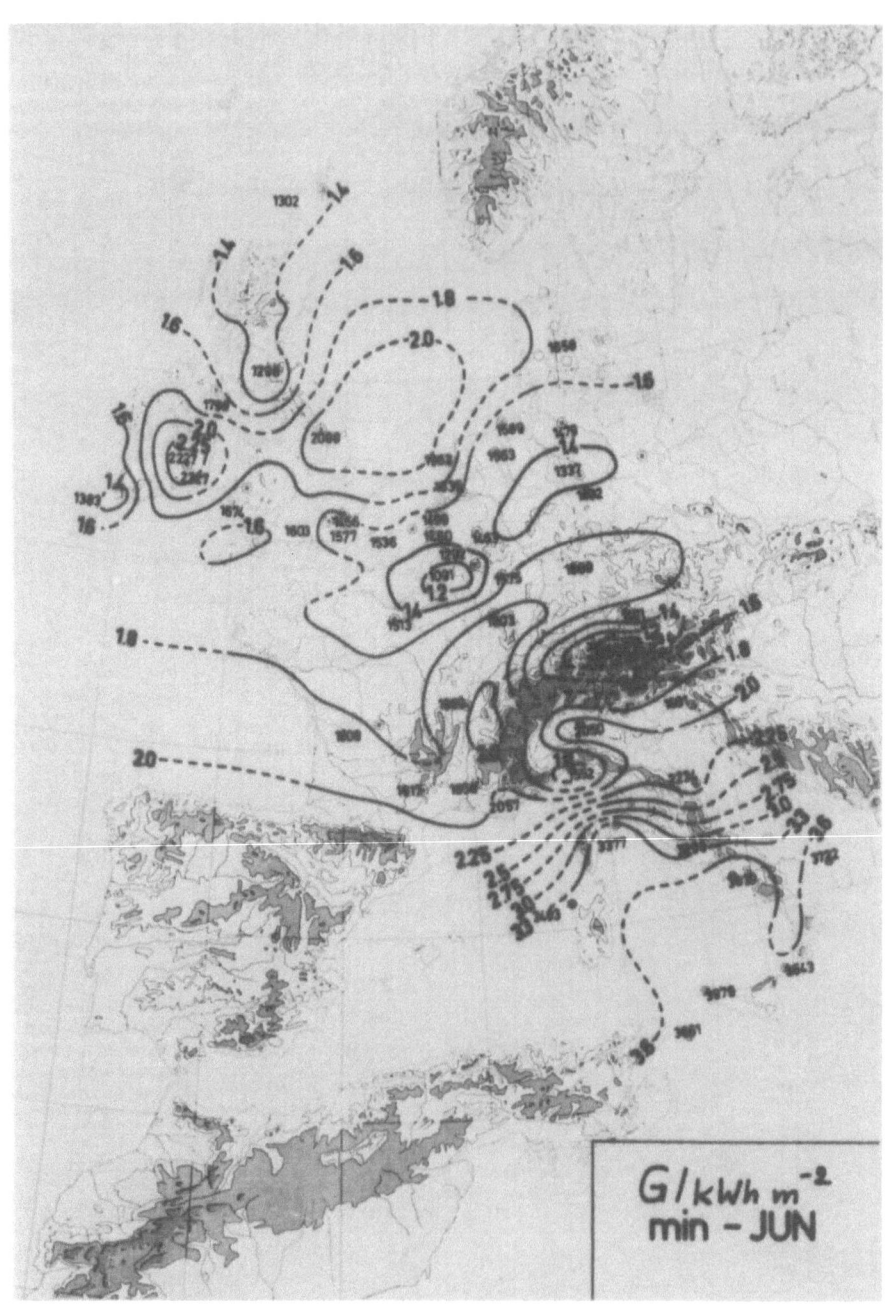

Fig. 31: Geographical distribution of mean minimum of
daily global radiation in June, 1966-1975.

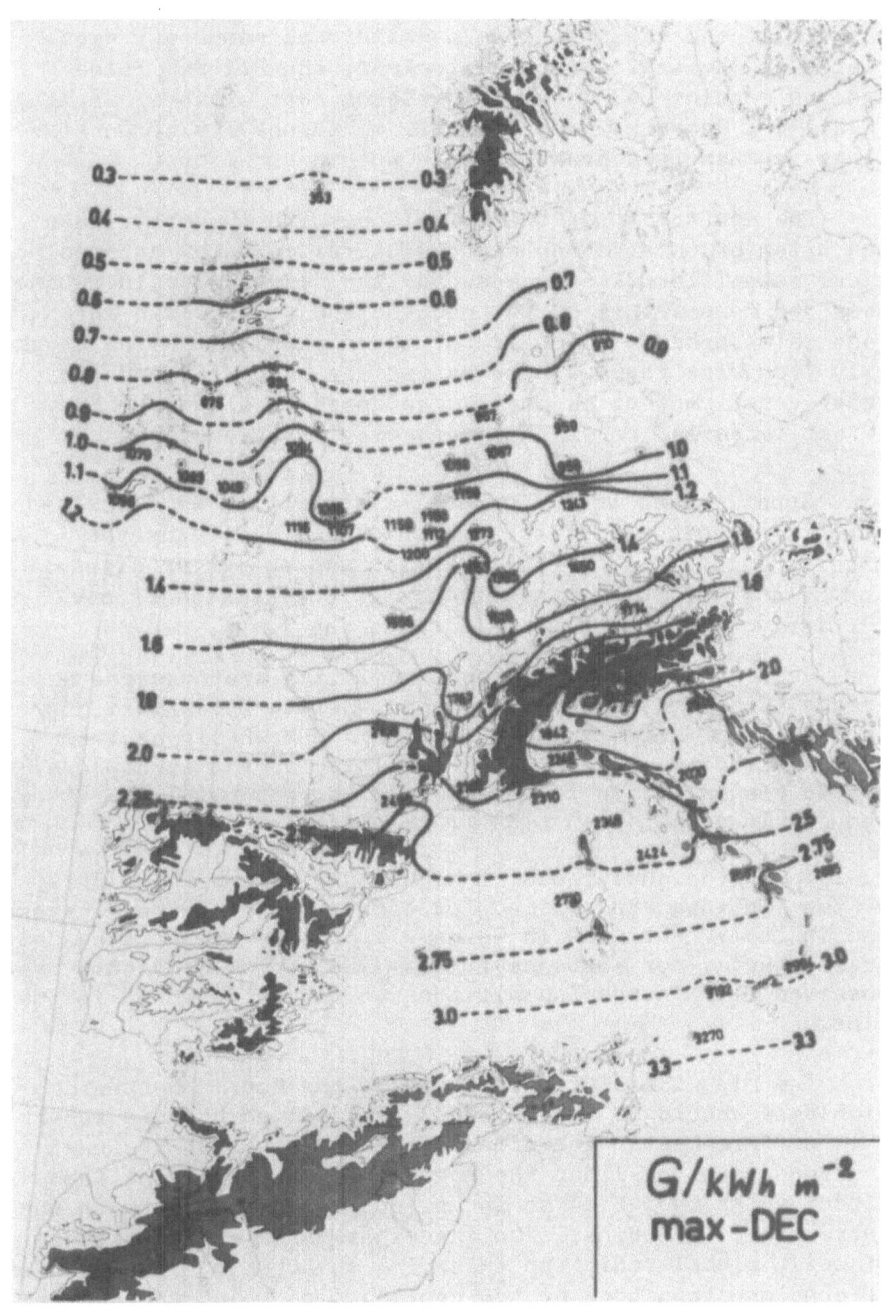

Fig. 32: Geographical distribution of mean maximum of
daily global radiation in December, 1966-1975.

ological influences on the daily radiation sums.

Besides the latitudinal effect just mentioned, the pattern of the isolines show a meridional component especially at the west coasts of Ireland, the British Isles and the continental countries Belgium, Netherlands, Germany (F.R.) and Denmark. However, this meridionality of the isolines is much less pronounced in autumn and winter.

The conformity of the G-isolines with the coastlines can often be observed on cloud pictures taken by meteorological satellites. It is caused by the differences in roughness and temperature of the sea and the solid earth surface which produce an uplift of the air masses moving generally from west to east; the raised air is cooled and its water vapor content is partly condensed to clouds. This effect is called coastal convergence in meteorology.

Superimposed on the general distribution as described above are regional differences which are caused by the orography influencing the formation and extent of clouds, and by the variable transmittance of the atmosphere particularly at higher altitudes above sea level.

Decreased values of global radiation are observed around large cities such as London, Milano and Napoli and/or in industrial areas such as the Rhein-Ruhr-Basin or the Belgian industry center; high turbidity of the atmosphere due to air pollution is believed to be responsible for these reduced G-values.

In the neighbourhood of highlands, steeper gradients of the isolines are noticed, particularly in southern France and in Italy, but also in Germany at the Harz mountains near Braunschweig, for instance. Extremely strong G-gradients are observed with increasing altitude as most pronounced in the Alps.

The global radiation pattern in the Alps and other highlands should be interpreted with caution because remarkable differences in irradiance from one place to the next are found due to either the mountain chain shielding the direct solar radiation or to upslope and lee effects on the formation of cloudiness. On the other hand, an annual mean of daily global radiation exceeding 4 700 Wh m^{-2} is measured at some mountain tops of the central Alps which are above the clouds most of the time.

The isolines on the geographical maps shall only give
a general impression of the large scale features of the
distribution of global radiation. Local details or peculi-
arities con not be read from these maps.

11. FREQUENCY DISTRIBUTION OF DAILY GLOBAL RADIATION

Knowledge of the statistical distribution of measured
radiation data is useful for planning and using solar energy systems. Information is needed on the probability of how
many days per month the daily irradiation will exceed given
thresholds in the different months of the year. Even more
specific is the question on the probability of how many consecutive days the daily irradiation will persist above given
thresholds. The informations on the availability of global
radiation may assist in adapting the design of solar energy
systems to the radiation climate prevailing at a given location.

The statistical analysis (Kasten and Golchert 1980)
was based on the data which were provided by the Member-
states of the European Communities for preparing the so-
called European global radiation atlas (Palz 1979). The data
comprise 10 year (1966-1975) continuous records of daily
sums of global radiation G on horizontal plane at 49 stations in the region of the EC.

These data were analyzed in terms of mean monthly cumulative frequency distributions of daily irradiation above
given thresholds. Furthermore, for each month the mean maximum numbers of consecutive days on which the daily irradiation persists above given thresholds were established. The
results of this statistical analysis were presented on 2
times 49 tables, and for 8 selected stations also on 2 times
8 diagrams.

A few results are presented here. Figs.33 and 34 show
the curves of cumulative frequency and of mean maximum number of consecutive days, respectively, for the station
Lerwick at the far north of the EC. The corresponding diagrams for Trapani at the far south of the EC with exception-
ally sunny climate are given on Figs.35 and 36. Whereas the
northern stations show an almost linear decrease of daily
global irradiation from low to high frequencies, the curves
of the southern stations exhibit a wide plateau up to very
high frequencies from where they abruptly drop. On the basis
of the statistical material gathered in the report, large
scale radiation climate zones and sub-zones within the
region of the EC were attempted to be defined:

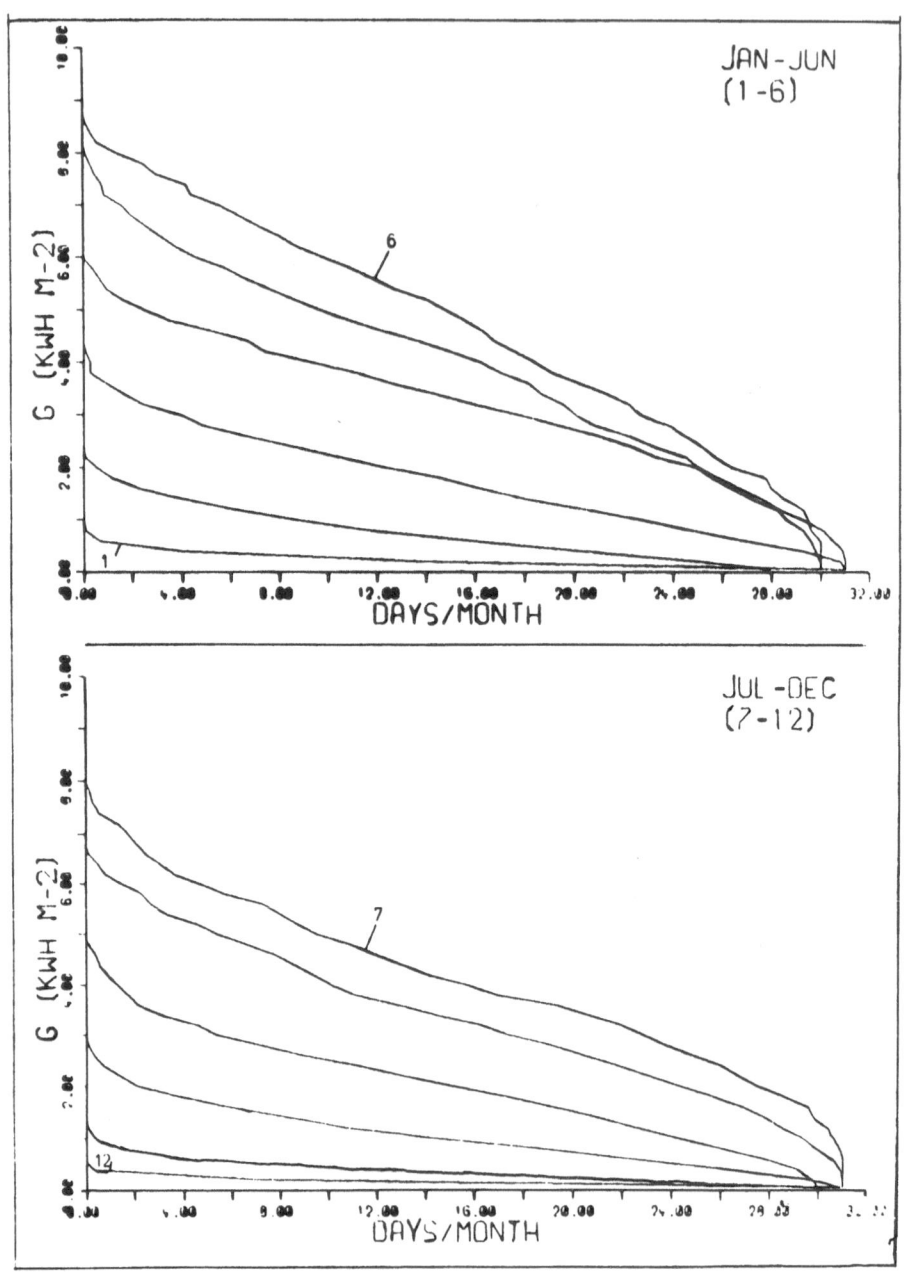

Fig. 33: Cumulative frequency, in days per month, of daily sums of global radiation G, in kWh m^{-2}. Lerwick 1966–1975.

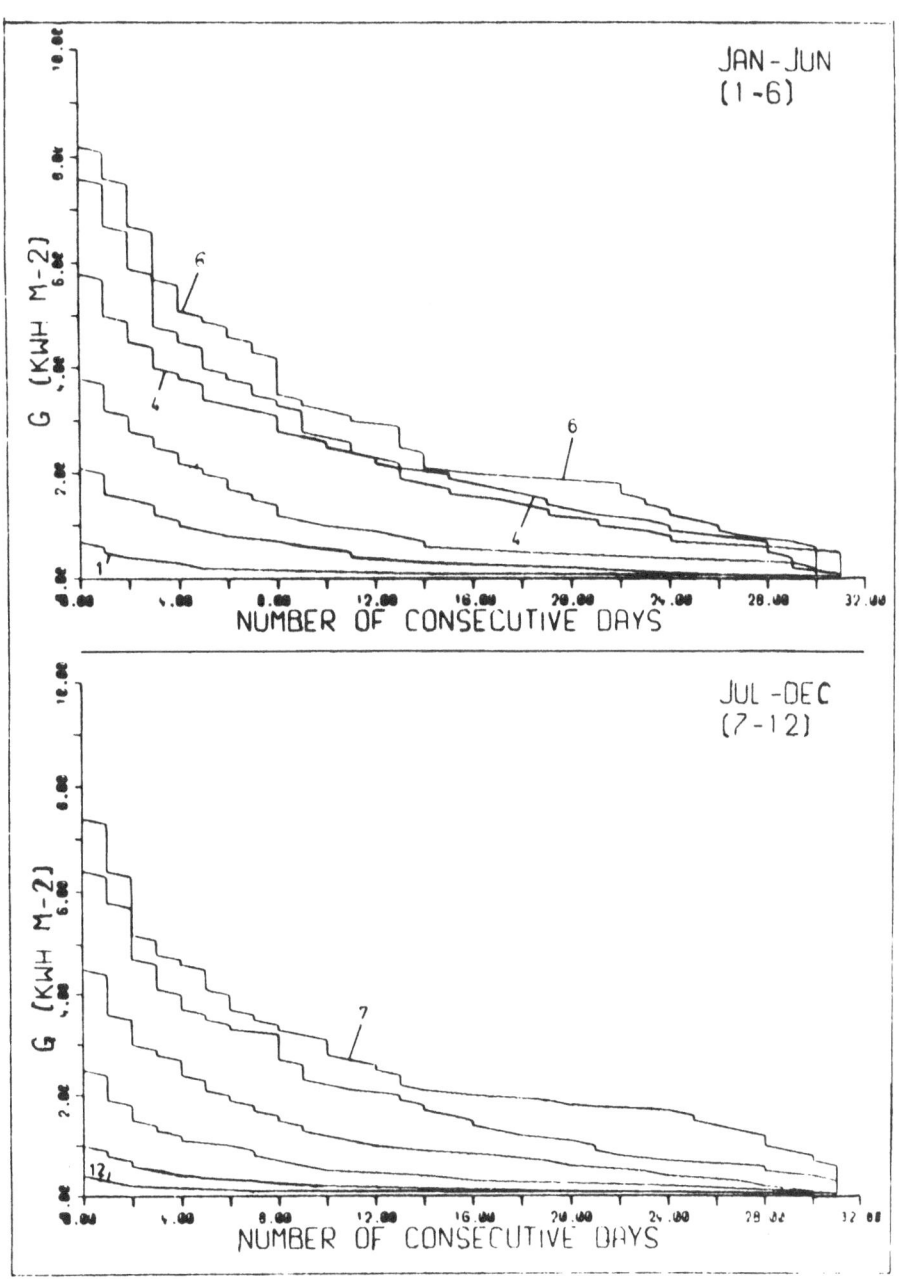

Fig. 34: Mean maximum number of consecutive days with
daily sums of global radiation G above given thresholds.
Lerwick 1966-1975.

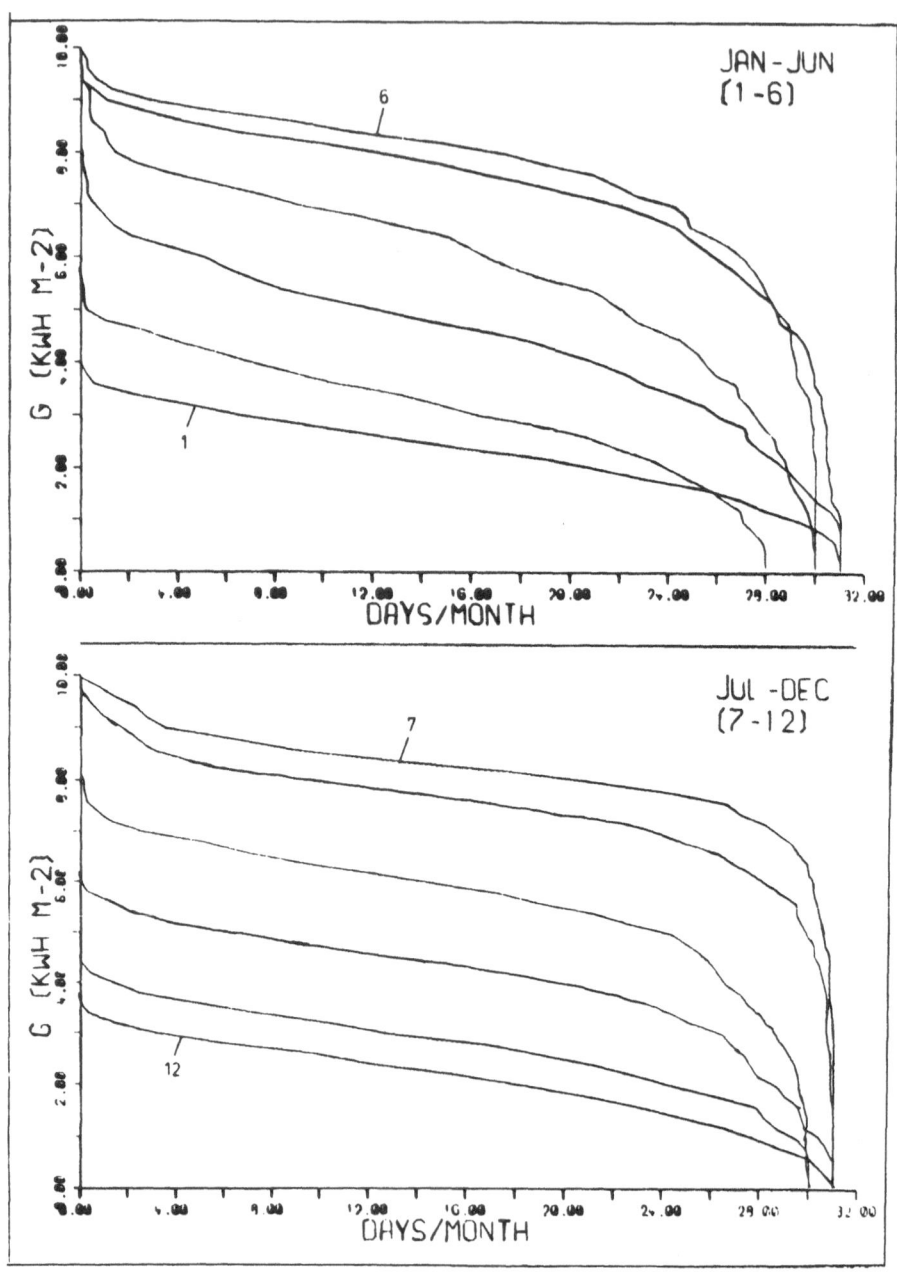

Fig. 35: Cumulative frequency, in days per month, of daily
sums of global radiation G, in kWh m^{-2}. Trapani 1966-1975.

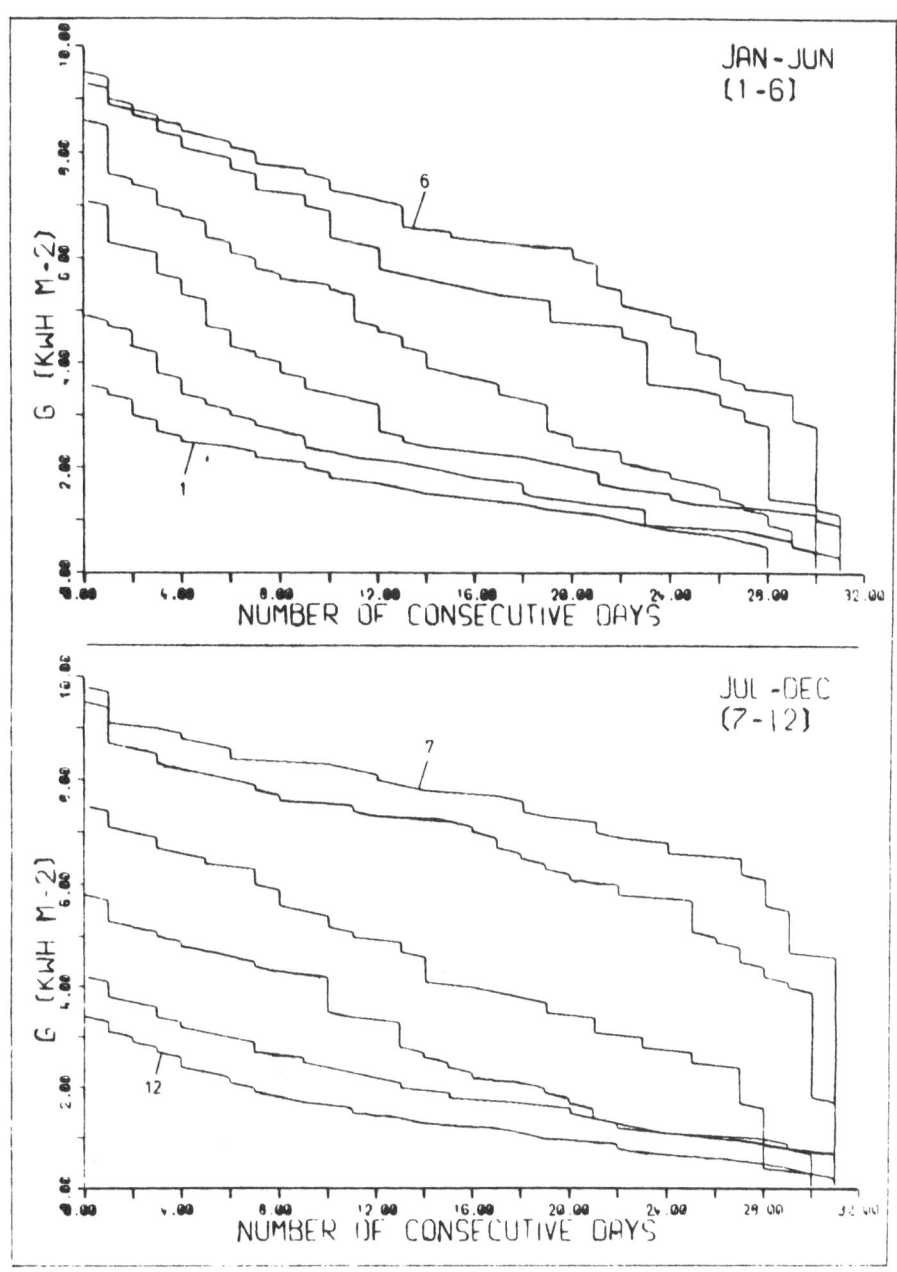

Fig. 36: Mean maximum number of consecutive days with daily sums of global radiation G above given thresholds. Trapani 1966-1975.

Zone 1: Maritime radiation climate (Ireland, United Kingdom, northern France, Benelux countries, northwest Germany north of 50°N).
Sub-zone 1a: Monsoon-like type (lowland and highland region north of 50°N). Characterized by maximum irradiation in June but a decrease in July and August due to monsoon effects and, in the highlands, due to orographic effects; low values in winter.
Sub-zone 1b: Coastal type (coastal and offshore region). Characterized by a pronounced maximum of irradiation in May and June; low values in winter. During most months of the year, this sub-zone can be attributed to sub-zone 1a.

Zone 2: Continental radiation climate (central Europe south of 50°N).
Sub-zone 2a: Medium altitude type. Characterized by maximum irradiation in July and August.
Sub-zone 2b: Alpine type. Characterized by only moderate seasonal variations; maximum irradiation in July; relatively high irradiation in autumn and winter when mountain tops are above clouds.

Zone 3: Mediterranian radiation climate (southern France; Italy).
Sub-zone 3a: River basin type (northern Italy with Po river). Characterized by pronounced decrease of irradiation in winter as a consequence of heavy and very frequent fogs; maximum irradiation in July and August.
Sub-zone 3b: Sunny type (southern France, southern Italy with Sicily and Sardinia). Characterized by very high irradiation from April to October with maximum in July; relatively high irradiation in winter months too.

REFERENCES

1 Coulson, K.L. (1975): "Solar and terrestrial radiation, methods and measurements". Academic Press, New York, San Francisco, London.
2 CSAGI (1958): "Annals of the International Geophysical Year", ed. by Special Committee for the International Geophysical Year (CSAGI), vol. V, part VI: Radiation instruments and measurements. Pergamon Press, London, New York, Paris.
3 Dehne, K. (1974): "Entwicklung eines Sky-Scanners zur schnellen Vermessung der räumlichen Verteilung spektraler Himmelsstrahldichten". Berichte des Deutschen Wetterdienstes 17, No. 134.

4 Dehne, K. (1977): "Messinstrumente zur Beobachtung der Strahlungsintensität". Informationswerk Sonnenenergie, vol.3, pp. 199-217. Udo Pfriemer Verlag, München.

5 Dehne, K. (1977 a): "Solare Strahlungsmessungen im Rahmen der Weltorganisation für Meteorologie (WMO)". 1st German Solar Energy Forum, Proceedings, vol.II, pp. 15-23. Deutsche Gesellschaft für Sonnenenergie e.V. (DGS), München.

6 Fleischer, R. and K. Gräfe (1955/56): "Die Ultrarot-Strahlungsströme aus Registrierungen des Strahlungsbilanz-messers nach Schulze". Annalen der Meteorologie 7, pp.87-95.

7 Frankenberger, E. (1955): "Über vertikale Temperatur-, Feuchte- und Windgradienten in den untersten 7 Dekametern der Atmosphäre, den Vertikalaustausch und den Wärmehaushalt an Wiesenboden bei Quickborn/Holstein 1953/54". Berichte des Deutschen Wetterdienstes 3 No. 20.

8 Geiger, R. (1961): "Das Klima der bodennahen Luftschicht", 4th edition, p. 246. Vieweg, Braunschweig.

9 Kasten, F. (1977): "Daily and yearly time variation of solar and terrestrial radiation fluxes as deduced from many years records at Hamburg". Solar Energy 19, pp.589-593.

10 Kasten, F. and H.J. Golchert (1980): "Frequency distributions of global radiation on horizontal surface for the region of the EC". Special Report on Contract ESF-004-80 D (B), CEC, Brussels.

11 Kasten, F. and E. Raschke (1974): "Reflection and transmission terminology by analogy with scattering". Applied Optics 13, pp. 460-464.

12 Klucher, T.M. (1979): "Evaluation of models to predict insolation on tilted surfaces". Solar Energy 23, pp. 111-114.

13 Latimer, J.R. (1972): "Radiation measurements". International Field Year for the Great Lakes, Technical Manual Series No. 2. Secretariat of the Canadian National Committee for the International Hydrological Decade, Ottawa/Canada.

14 Möller, F. (1973): "Einführung in die Meteorologie", vol.2, p.30. Bibliographisches Institut, Mannheim, Wien, Zürich.

15 Palz, W. (editor) (1979): "European solar radiation atlas, vol.I: Global radiation on horizontal surfaces". Grösschen, Dortmund.

16 Robinson, N. (1966): "Solar radiation". Elsevier Publishing Company, Amsterdam, London, New York.

17 Temps, R.C. and K.L. Coulson (1977): "Solar radiation incident upon slopes of different orientations". Solar Energy 19, pp. 179-184.

18 WMO (1971): "Guide to meteorological instrument and observing practices", 4th edition, WMO-No. 8.TP.3. Secretariat of the World Meteorological Organization (WMO),
   Geneva.
19 WMO (1981): "Report of the Working Group on Radiation
   Measurement". WMO CIMO-VIII/Doc.31 (28 July 1981). Secretariat of the World Meteorological Organization (WMO),
   Geneva.

SOLAR THERMAL COLLECTORS *

E. ARANOVITCH

Commission of the European Communities
Joint Research Centre - Ispra Establishment
I-21020 Ispra (Va), Italy

INTRODUCTION

The solar collector is the main component of a solar system.
It transforms radiant energy from the sun, in the spectral range
0.3 - 3 μm, into usable heat (Fig. 1). The density of fluxes involved is low, at a maximum of the order of 1 kW/m^2 (1000
times less than in the case of nuclear reactors). Heat losses
depend essentially on temperature levels. A black plate, at 70°C,
put in ambience at 10°C, will lose about 1 kW/m^2. The useful

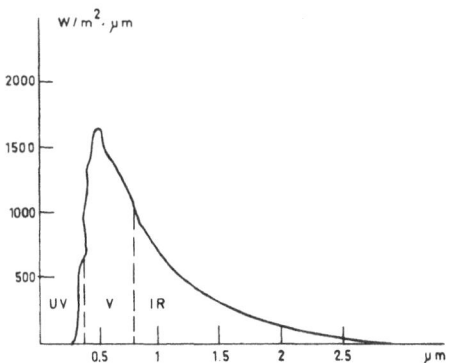

Figure 1. Spectral distribution of solar radiation at sea
level (very similar to a blackbody spectrum at 6000 K).

* From 'Heat Transfer Processes in Solar Collectors', M. E. Aranovitch, first published by Elsevier Sequoia.

G. Beghi (ed.), Performance of Solar Energy Converters: Thermal Collectors and Photovoltaic Cells, 65–102.
Copyright © 1983 ECSC, EEC, EAEC, Brussels and Luxembourg

extracted energy will be the difference between the absorbed
energy and the energy losses:

$$(USEFUL\ ENERGY) = (ABSORBED\ ENERGY) - \\ (ENERGY\ LOSSES).$$

Only carefully designed collectors which minimize the heat losses
will show an acceptable efficiency for practical applications. The
first part of this lecture is devoted to the description and the evaluation of the thermal processes concerned with solar collectors
in order that the user may become familiar, either with the limitations, or the potential improvements which can be expected
from solar systems.

The text deals principally with non-focusing flat-plate collectors,
which are mostly used for housing applications. Such collectors
have the advantage of using both direct and diffuse radiation
(diffuse radiation may account for 50% of the total radiation in
most Central and North-European regions).

BASIC SCHEME OF A FLAT-PLATE COLLECTOR

The essential parts of a flat-plate collector are represented in
Fig. 2. They are:
- the "black" absorber plate which transfers the absorbed energy
  to a fluid;
- the transparent cover, the purpose of which is to limit heat
  losses through the so-called "greenhouse effect";
- the thermal insulation which limits backlosses.

The solar collector has to deal with two distinct processes:
- the absorption of radiant energy, which requires the highest
  possible transmission coefficient, τ, for the transparent cover
  and the highest possible absorption coefficient, α, for the absorber plate. The effective parameter will be the product $(\alpha\tau)$;
- the loss of energy in the infra-red spectrum due to:
a) radiation losses between the absorber plate and the transparent
   cover;
b) natural convection losses between the absorber plate and the
   transparent cover;
c) conduction losses through the back-insulation and the edges.
It can be seen that the three modes of heat transfer: radiation,
natural convection, and conduction are involved here. They will
successively be dealt with in the next sections.

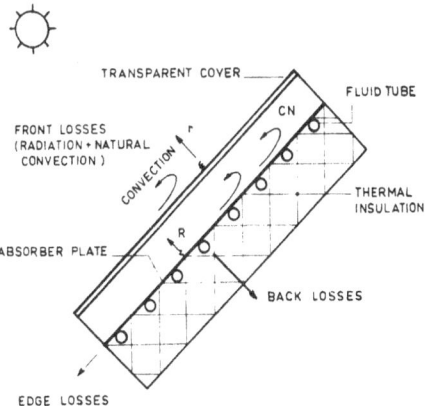

FRONT LOSSES
(RADIATION + NATURAL
CONVECTION)

TRANSPARENT COVER

FLUID TUBE

CN

CONVECTION

r

THERMAL
INSULATION

ABSORBER PLATE

R

BACK LOSSES

EDGE LOSSES

Fig. 2. Scheme of a flat-plate collector. The green-house effect is very important for the efficiency of the system.

HEAT LOSSES DUE TO RADIATION

Thermal radiation is emitted by bodies because of their temperature. By definition a blackbody is a perfect absorber of radiation. It is an ideal concept since all substances will reflect some radiation. A blackbody is also a perfect emitter of thermal radiation.

The wavelength distribution of the radiation for a blackbody is given by Planck´s law:

$$E_\lambda = \frac{C_1}{\lambda^5 (e^{C_2/\lambda T} - 1)} \tag{1}$$

where $C_1 = 3.7405 \times 10^{-16}$ W/m^2
$C_2 = 0.0143879$ mK
E_λ = energy per unit area and per unit time for a given wavelength
λ = wavelength, m
T = absolute temperature, K.

The wavelength corresponding to the maximum intensity of blackbody radiation is given by Wien´s law (Fig. 3):

$$\lambda_{max} = \frac{2897.8}{T} \text{ μm} \qquad (2).$$

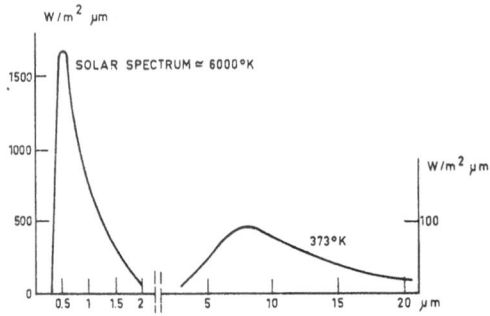

Fig. 3. Spectral distribution of blackbody radiation.

It can be noted that for a temperature of 6000 K the spectral
distribution of blackbody radiation is near the surface temperature of the sun.

By integrating Planck's law over all wavelengths, the total
energy emitted by a blackbody is found to be:

$$E = \int_0^\infty E_\lambda d\lambda = \sigma T^4 \qquad (3),$$

where σ is the Stefan-Boltzmann constant and equal to 5.6697 x
10^{-8} W/m²K.
In the case of a Lambertonian or a diffuse surface, the blackbody's emissive power is related to the blackbody's intensity,
I_b, by:

$$E = \pi I_b \qquad (4).$$

This equation is also valid for monochromatic radiation:

$$E_\lambda = \pi I_{b\lambda} \qquad (5).$$

In the case of a grey surface an emissivity coefficient, ϵ, and
an absorption coefficient, α, are introduced. For a given wavelength λ, Kirchoff's law states that:

- 68 -

$$\alpha_\lambda = \epsilon_\lambda \tag{6}.$$

The energy emitted by a grey surface, defined by ϵ, at a temperature of T K is equal to

$$\epsilon \sigma T^4.$$

The heat transfer between two infinite parallel planes, at temperatures T_1 and T_2, acting as grey surfaces defined by ϵ_1 and ϵ_2 is:

$$q_r = \frac{\sigma(T_2^4 - T_1^4)}{1/\epsilon_1 + 1/\epsilon_2 - 1} \tag{7}.$$

The global emittance, ϵ_g, between the two planes is:

$$\epsilon_g = \frac{1}{1/\epsilon_1 + 1/\epsilon_2 - 1} \tag{8}.$$

Example 1.

What is the global emittance of two parallel planes having respectively emissivity coefficients of 0.96 and 0.87?
What is the heat transfer when these two planes are respectively at 85°C and 40°C?

$$\epsilon_1 = 0.87 \quad T_1 = 273 + 40 = 313 \text{ K}$$
$$\epsilon_2 = 0.96 \quad T_2 = 273 + 85 = 358 \text{ K}.$$

The global emittance is:

$$\epsilon_g = \frac{1}{1/0.87 + 1/0.96 - 1} = 0.84.$$

The heat transfer is:

$$Q = (0.84)(5.67) \times 10^{-8} [(358)^4 - (313)^4] = 325 \text{ W/m}^2.$$

We have a similar situation in a solar collector, considering the radiation transfer between the absorber plate and a transparent cover in glass or in plastic films, because such materials behave as grey bodies for infra-red radiation (Fig. 4). It is only because of these special selective properties that the "green-

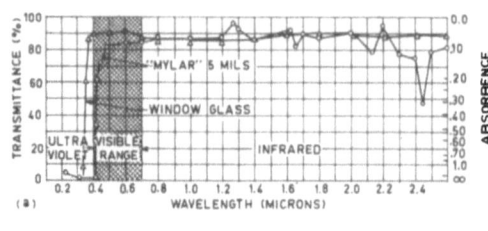

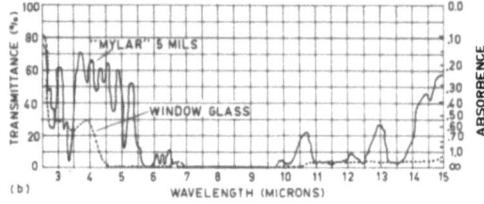

Fig. 4. Absorption spectrum for "Mylar" W-2 film
(a) low range, (b) high range.

house effect" can be achieved and that solar collectors can hope
to attain acceptable efficiencies (for instance polyethylene remains transparent to infra-red radiation).

In conclusion we shall assume that the heat transfer between the
absorber plate and the transparent cover can be represented
with an acceptable approximation by equation (7). The global
emittance is then of the order of 0.85. It will be seen how this
value can be reduced with selective surfaces.

HEAT LOSSES BY NATURAL CONVECTION

Heat losses by natural convection from the absorber plate to the
transparent cover are characterized by a Nusselt number, Nu:

$$Nu = \frac{U_{cn} d}{\lambda_{air}} \tag{9}$$

where U_{cn} = heat transfer coefficient between absorber plate
and cover,
d = plate spacing,
λ_{air} = thermal conductivity of air (Fig. 5).

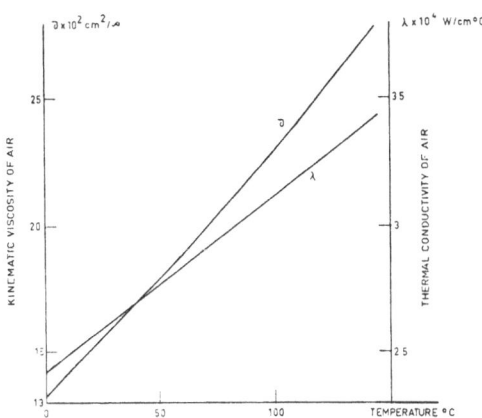

Fig. 5. Kinematic viscosity and thermal conductivity
of air.

This Nusselt number can be expressed as a function of the
Grashof number, Gr, and of the slope s:

$$Gr = \frac{g\beta d^3 \Delta T}{v^2} = \frac{gd^3 \Delta T}{v^2 T} \tag{10}$$

where g = gravitational constant,
β = expansion coefficient; $\beta \simeq \frac{1}{T}$,
ΔT = temperature difference between plates,
v = kinematic viscosity of air (Fig. 5),
T = average absolute temperature of air between plates, K.

(Some authors use the Rayleigh number, Ra, which is equal to
the product of the Grashof number and the Prandtl number. In
the case of gases the Prandtl number can be considered constant
and equal to 0. 7 and there is no necessity to introduce it as a
variable.)

Example 2.

Calculate the Grashof number in the case of an absorber plate
at a temperature of 70°C, a glass cover at a temperature of
40°C with a spacing of 2. 5 cm.

$g = 981 \ cm/s^2$,
$d = 2. 5 \ cm$,

$$\Delta T = 70 - 40 = 30^\circ C,$$

$$T = \frac{70 + 40}{2} + 273 = 328 \text{ K}$$

$$\nu = 17.5 \times 10^{-2} \text{ cm}^2/s$$

$$Gr = \frac{(981)(2.5)^3(30)}{(17.5 \times 10^{-2})^2(328)} = 45670.$$

The heat transfer coefficient U_{cn} between the absorber plate and the cover plate can be calculated if the function which relates the Nusselt number, Nu, to the Grashof number, Gr, and the slope, s, is known. A great number of experimental correlations have been proposed for the calculation of heat transfer by natural convection in closed inclined cells. It can be noted that there exists a dispersion of the order of 20% between the different formulas recommended by authors such as Tabor, de Graaf, van der Held, Dropkin, Somerscales, etc.

In Figure 6 are represented correlations recommended by Tabor and Dropkin-Somerscales for different inclinations. Tabor recommends:

$$Nu = 0.152\,(Gr)^{0.281} \qquad (11)$$
$$\text{for } s = 0 \text{ and } 10^4 < Gr < 10^7$$

$$Nu = 0.093\,(Gr)^{0.310} \qquad (12)$$
$$\text{for } s = 45^\circ C \text{ and } 10^4 < Gr < 10^7$$

$$Nu = 0.033\,(Gr)^{0.381} \qquad (13)$$
$$\text{for } s = 90^\circ \text{ and } 1.5 \times 10^4 < Gr < (1.5)10^5$$

$$Nu = 0.062\,(Gr)^{0.327} \qquad (14)$$
$$\text{for } s = 90^\circ \text{ and } 1.5 \times 10^5 < Gr < 10^7$$

Dropkin-Somerscales recommend:

$$Nu = [0.060 - 0.017_5 (s/90)]\,(Gr)^{1/3} \qquad (15)$$
$$\text{with } Gr > 2.10^5.$$

It can be observed that the Dropkin-Somerscales formula gives somewhat lower values than the Tabor formulas which might be considered more reliable because they were experimentally established with air, whereas the former was established with liquids.
Moreover, the Dropkin-Somerscales formula is recommended for Grashof numbers superior to 1.5×10^5 where, as in solar

collectors, the Grashof number between absorber plate and glass cover is generally inferior to 10^5.

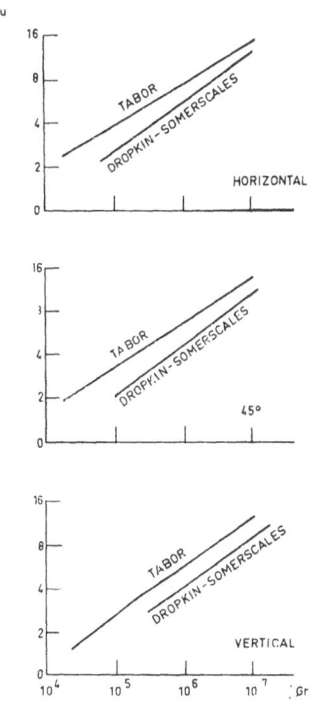

Fig. 6. Natural convection in inclined cells.

On the other hand, the Dropkin-Somerscales formula has a great merit for simplicity and can easily be used in design applications. So, after changing the values of the constants in order to be nearer to the values recommended by Tabor, one obtains:

$$Nu = [0.078 - 0.026 \, (s/90)](Gr)^{1/3} \qquad (16)$$
$$\text{with } 10^4 < Gr < 10^5.$$

Example 3.

Referring to example 1, calculate with equation (15) the Nusselt number and the heat transfer coefficient, U_{cn}, assuming a slope of 45°.

$$Gr = 45670$$
$$Nu = [\, 0.078 - 0.026(45/90)] \, (45670)^{1/3} = 1.67$$
$$s = 45^\circ$$

$$U_{cn} = \frac{Nu.\lambda}{d} = \frac{(1.67)(2.81) \times 10^{-4}}{2.5} = 1.88 \times 10^{-4} \text{ W/cm}^2\text{°C}$$

$$\lambda_{air} = 2.81 \times 10^{-4} \text{ W/cm°C}.$$

One of the other advantages of such a formula is that formally the heat transfer, U_{cn}, is then independent of the spacing, d. Combination of equations (9), (10) and (16) gives:

$$U_{cn} = \left[\frac{\lambda g^{1/3}}{\nu^{2/3} T^{1/3}}\right] [0.078 - 0.026 \, (s/90)] \times (\Delta T)^{1/3} \tag{17}.$$

It can be shown that the first term between brackets is only slightly dependent on temperature in the range of applications for solar collectors, so a further simplification is introduced by assuming that it is a constant and finally the heat transfer coefficient is represented by:

$$U_{cn} \simeq [1 - 0.33 \, (s/90)] \Delta T^{1/3} \tag{18}$$

where U_{cn} is expressed in W/m^2 °C.
The heat transfer per unit surface is then:

$$q_{cn} = U_{cn} \Delta T = [1 - 0.33 \, (s/90)] \Delta T^{4/3} \tag{19}.$$

Example 4.

Assuming a slope of $45°$ and referring to example 1, what is the heat transfer by natural convection between the two planes?

$$q_{cn} = [1 - 0.33(45/90)] (85 - 40)^{4/3} = 96 \text{ W/m}^2.$$

This example shows that for non-selective surfaces, heat losses by radiation are predominant over heat losses by natural convection.

Determination of the spacing between the absorber plate and the glass cover

From a fabrication point of view the manufacturer will be interested in reducing the spacing, d, between the glass cover and the absorber plate. For Grashof numbers inferior to 2000 it can be assumed that air remains stagnant and that heat losses are due only to conduction through the air.

$$q = \frac{(\lambda_{air}) \Delta T}{e} \tag{20}.$$

For Grashof numbers superior to 10,000 we assume that heat losses vary little with the spacing and can be represented by equation (19). For Grashof numbers between 2000 and 10,000, there exists a transition zone where correlations are not clearly defined. The heat losses, expressed as a function of the spacing, d, will be represented by equation (20) for

$$\text{Gr} < 2000 \quad \text{that is d} < \left(\frac{2000 \; T^2}{g \Delta T} \right)^{1/3}$$

and by equation (19) for

$$\text{Gr} > 10,000 \quad \text{that is d} > \left(\frac{10,000 \; T^2}{g \, \Delta T} \right)^{1/3}.$$

An example is shown in Fig. 7. In practical cases a value of d equal to 2.5 cm seems reasonable.

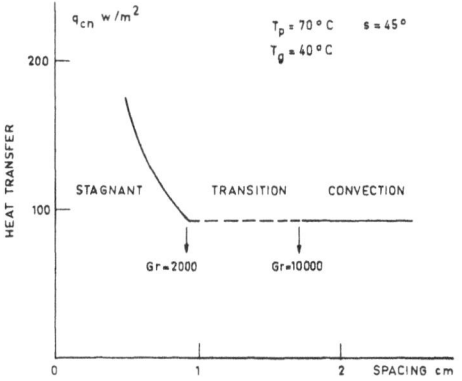

Fig. 7. Heat transfer by natural convection.

BACK LOSSES AND EDGE LOSSES

Back losses and edge losses, for convenience, are generally

characterized by a single heat transfer coefficient, U_b. In a poorly designed solar collector, the edge losses can become of significant importance (in the case, for instance, of a metallic contact between the absorber plate and the external casing). In a properly designed collector the edge losses should be kept inferior to 5% of the total losses and the heat transfer coefficient, U_b, can be represented with an acceptable approximation by:

$$U_b \simeq \frac{\lambda_b}{e_b} \qquad (21)$$

where λ_b = thermal conductivity of the insulating material,
$\quad e_b$ = thickness of the insulating material.

The corresponding losses are then:

$$q_b \simeq h_b(T_p - T_a) \qquad (22).$$

HEAT LOSSES FROM THE TRANSPARENT COVER TO THE AMBIENT AIR

The heat losses from the transparent cover to the ambient air are due to radiative and convective exchanges which are affected by the wind velocity.
Radiative exchanges are not only influenced by the ground and surrounding conditions (snow, reflective windows, etc.), but also by long wave radiation from the sky; especially in the case of a very clear sky when the "sky temperature" can be significantly lower than the ambient air temperature.

Swinbank relates sky temperature to the local air temperature by:

$$T_{sky} = 0.0552\ T_{air}^{1.5}\ (K) \qquad (23).$$

Whillier recommends a simpler formula:

$$T_{sky} = T_{air} - 6 \qquad (24).$$

Some conclusive experimental work on this subject is still lack-

ing. Except in very special cases (involving snow for instance)
it is recommended, for design purposes, to use a single formula
which accounts globally for radiative and convective losses.

Two correlations proposed by Tabor and MacAdams are represented in Fig. 8. The linear equation for MacAdams:

$$U_a = 5.7 + 3.8 \ V \qquad (25)$$

where U_a = heat transfer coefficient from cover to the ambient
air $(W/m^2 {}^o C)$,
V = wind velocity (m/s),

fits well with the values recommended by Tabor, when the wind
velocity is in the range of 2 - 5 m/s.

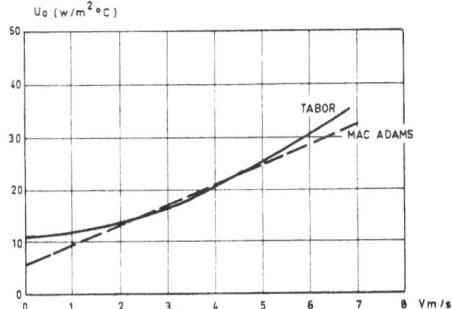

Fig. 8. Heat transfer coefficient from cover to
ambient air.

FIN AND BOND EFFECT

An important number of solar collectors are designed on the
principle of an absorber plate, clamped or bonded to fluid tubes
as seen in Fig. 9. In such a structure there will be a temperature difference between the average plate temperature, T_p, and
the fluid temperature, T_m, leading to a loss of efficiency.

Considering an element dx, an energy balance gives the following differential equation:

$$e\lambda \frac{d^2T}{dx^2} + \phi_s - U_1(T - T_a) = 0 \qquad (26)$$

with the following boundary conditions:

$$-\lambda \frac{dT}{dx} = 0 \qquad\qquad \text{when } x = 0 \qquad (27)$$

$$-\lambda \frac{dT}{dx} = U_{bd}(T - T_m) \quad \text{when } x = L \qquad (28).$$

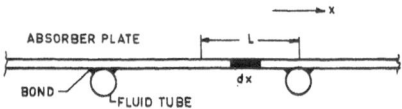

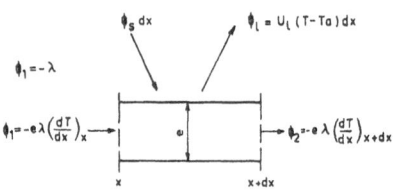

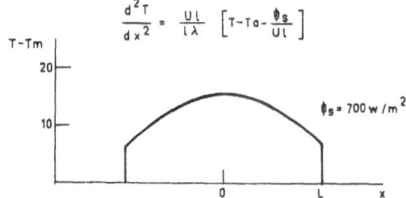

Fig. 9. Fin and bond effect.

L   = length of fin,
e   = thickness of fin,
x   = abscissa,
T   = temperature of fin, function of x,
T_p = average temperature of fin,
T_m = temperature of fluid,
T_a = ambient temperature,

U_1 = heat transfer coefficient between absorber plate and ambient air,

U_{bd} = heat transfer coefficient for the bond between absorber plate and fluid tube,

ϕ_s = absorber solar radiation per unit surface,

λ = thermal conductivity of the fin.

Integration of equation (26) gives:

$$\frac{\phi_s - U_1(T - T_a)}{\phi_s - U_1(T_m - T_a)} = \frac{\cosh mx}{\left(\dfrac{\lambda m}{U_{bd}}\right) \sinh mL + \cosh mL} \tag{29}$$

with

$$m^2 = \frac{U_1}{\lambda e} \tag{30}.$$

Introducing the average value T_p of T

$$T_p = \frac{1}{L} \int_0^L T \, dx \tag{31}$$

one obtains finally:

$$F' = \frac{\phi_s - U_1(T_p - T_a)}{\phi_s - U_1(T_m - T_a)} = \frac{\tanh mL}{mL\left[1 + \left(\dfrac{\lambda m}{U_{bd}}\right)\tanh mL\right]} \tag{32}.$$

It should be noted that $\phi_s - U_1(T_p - T_a)$ represents the useful energy extracted per unit surface and $\phi_s - U_1(T_m - T_a)$ represents the useful energy which would have been extracted if the absorber plate were at the temperature T_m of the fluid. So the ratio of these two terms is equal to an efficiency factor F', related to the fin and bond effect. The standard fin efficiency factor for straight fins is given by:

$$F = \frac{\tanh mL}{mL} \tag{33}.$$

Combination with equation (32) gives:

$$F' = \frac{F}{1 + \beta F} \tag{34}$$

with
$$\beta = \frac{LU_1}{eU_{bd}} \tag{35}.$$

If $U_{bd} = \infty$, the thermal resistance due to the bond is null and the efficiency factor F′is equal to the fin efficiency F. Experience shows that a simple clamping or wiring of the fluid tubes to the absorber plate can result in an important loss of performance.

Example 5.

Calculate the efficiency factor, F, and the factor F′for an aluminium fin with the following characteristics:

$L = 10$ cm, $e = 0.1$ cm, $U_1 = 6$ W/m^2°C, $\lambda = 2$ W/cm°C, $U_{bd} = 6000$ W/m^2°C.

$$mL = \left(\frac{U_1}{e\lambda}\right)^{1/2} L = \left[\frac{6 \times 10^{-4}}{(0.1)(2)}\right]^{1/2} \times 10 = 0.548.$$

$$F = \frac{\tanh 0.548}{0.548} = 0.91,$$

$$\beta = \frac{LU_1}{eU_{bd}} = \frac{10 \times 6}{0.1 \times 6000} = 0.1,$$

$$F′ = \frac{F}{1 + \beta F} = \frac{0.91}{1 + (0.1)(0.91)} = 0.83.$$

The temperature profile in the fin for a value of $\phi_s = 700$ W/m^2 is represented in Fig. 9.

CALCULATION OF TOTAL LOSSES AND OVERALL HEAT TRANSFER COEFFICIENT

The total heat losses, $\dot{q}_1$, per unit surface, can be divided into the front losses, $\dot{q}_f$, consisting of radiative losses, $\dot{q}_r$, and convective losses, $\dot{q}_{cn}$, from the absorber plate to the transparent cover, and of the back losses, $\dot{q}_b$, through the thermal insulation. The front losses are in turn transmitted from the transparent cover to the ambient air.

From the results obtained in the preceding sections, one can write:

$$\dot{q}_f = U_a (T_c - T_a) \tag{36}$$

$$\dot{q}_f = \epsilon_g \sigma (T_p^4 - T_c^4) + [1 - 0.33(s/90)] \times [T_p - T_c]^{4/3} =$$

$$\dot{q}_r + \dot{q}_{cn} \tag{37}$$

$$\dot{q}_b = U_b (T_p - T_a) \tag{38}$$

$$\dot{q}_1 = \dot{q}_f + \dot{q}_b \tag{39}$$

$$\epsilon_g = \frac{1}{1/\epsilon_p + 1/\epsilon_c - 1} \tag{40}$$

$$U_b \simeq \frac{\lambda_b}{e_b} \tag{41}$$

$$U_a = 5.7 + 3.8 \, V \tag{42}.$$

From these equations the front losses, $\dot{q}_f$, the back losses, $\dot{q}_b$, and the total losses, $\dot{q}_1$, can be calculated as a function of the plate temperature T_p, assuming that the other parameters $(\epsilon_p, \epsilon_c, T_a, V, \lambda_b, e_b, s)$ are known. The cover plate temperature, T_c, has to be calculated as an intermediate step.

Practically, one can use the following procedure:
Fix a given value of $\dot{q}_f$, then calculate T_c from equation (32) and then T_p from equation (33); the value of $\dot{q}_b$ can then be determined from equation (34) and $\dot{q}_1$ from equation (35). The calculation can be renewed for other values of q_f and the functions $\dot{q}_1$, $\dot{q}_r$, $\dot{q}_{cn}$, $\dot{q}_b$ of the variable T_p can be obtained.

In Figure 10 the different heat losses have been represented in a numerical example, showing the following approximate repartition:

- radiative losses: 59%,
- convective losses: 26%,
- back losses: 15%.

The overall heat transfer coefficient, U_1, which is an important characteristic of a solar collector is defined by:

$$U_1 = \frac{\dot{q}_1}{T_p - T_a}$$
(43).

In a similar manner one can calculate the heat transfer coefficient related to the convective and radiative losses:

$$U_r = \frac{\dot{q}_r}{T_p - T_c} = \frac{\epsilon_g \sigma(T_p^4 - T_c^4)}{T_p - T_c}$$
(44)

$$U_{cn} = \frac{\dot{q}_{cn}}{T_p - T_c} = \frac{[1 - 0.33(s/90)] (T_p - T_c)^{4/3}}{T_p - T_c}$$
(45).

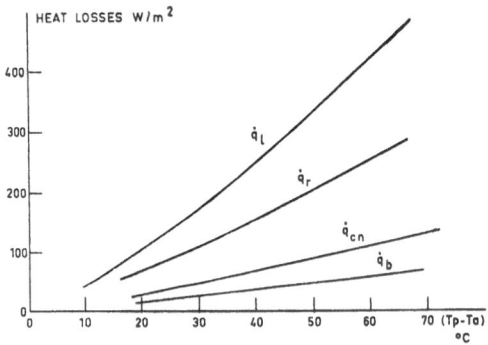

Fig. 10. Heat losses.

It can easily be seen that the various heat transfer coefficients, U_1, U_r, U_{cn}, U_a, U_b are related by the following equation:

$$U_1 = \frac{1}{\dfrac{1}{U_r + U_{cn}} + \dfrac{1}{U_a}} + U_b$$
(46).

In Figure 11 the overall heat transfer coefficient, derived from the preceding example has been represented as a function of $(T_p - T_a)$. Although it varies slightly with temperature, for the characterization of a collector, a constant average value will be assumed.

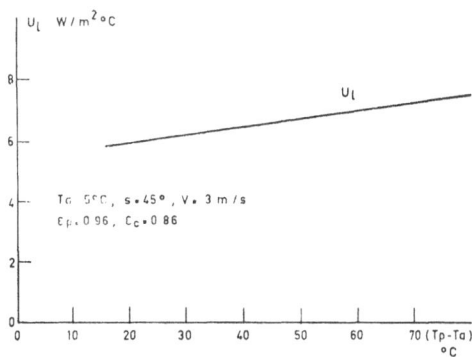

Fig. 11. Overall heat transfer coefficient: single-glass black paint collector.

TRANSMISSION OF RADIATION THROUGH TRANSPARENT COVERS

The transparent cover (or covers) plays an important role in the solar collector because of the so-called "greenhouse effect". Because of its selective properties, it transmits the solar radiation in the spectrum of sunlight and acts as a grey or black body, in the infrared spectrum, for the re-emitted radiation from the absorber plate, "trapping" sunlight.

Not all the incident radiation reaches the absorber plate. Due to the Fresnel reflections and the absorption within the cover material, only a portion of the incident energy is transmitted.

Fresnel reflections at interfaces

Let us consider an incident beam, equal to unity, defined by the angle, θ_1 (Fig. 12). The fraction, ρ, is reflected at the first interface. The fraction, $(1 - \rho)$, is refracted at angle θ_2, according to Snell's law:

$$\sin \theta_1 = n \sin \theta_2 \qquad (47)$$

where n is the refractive index of the transparent medium (the refractive index of air is 1).

At the second interface, the quantity $(1 - \rho)\rho$ is reflected and the quantity $(1 - \rho)^2$ is transmitted. Summing up the transmitted

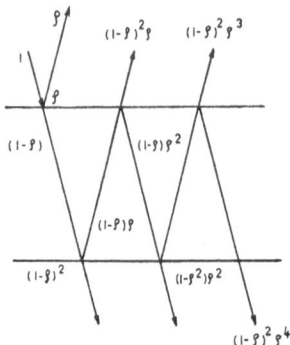

Fig. 12. Transmission through one cover.

terms after the multiple reflections, the transmission coefficient, τ_F, becomes:

$$\tau_F = (1 - \rho)^2 + (1 - \rho)^2 . \rho^2 + (1 - \rho)^2 . \rho^4 + \dots = \frac{1 - \rho}{1 + \rho}$$

(48).

With N covers, the transmission coefficient, τ_F, is:

$$\tau_{F,N} = \frac{1 - \rho}{1 + (2N - 1)\rho}$$

(49).

For non-polarized radiation, the reflection coefficients, ρ_1 and ρ_2, of the two components of polarization are given by Fresnel's formula:

$$\rho_1 = \frac{\sin^2(\theta_2 - \theta_1)}{\sin^2(\theta_2 + \theta_1)}$$

(50)

$$\rho_2 = \frac{\tan^2(\theta_2 - \theta_1)}{\tan^2(\theta_2 + \theta_1)}$$

(51).

For angles of incidence inferior to 45°, it is sufficient to consider an average value for o:

$$\rho = \frac{1}{2}(\rho_1 + \rho_2)$$

(52).

- 84 -

For radiation at normal incidence:

$$\theta_1 \simeq n\theta_2 \to 0 \tag{53}$$

and combination of equations (50) - (53) gives:

$$\rho(0) = \frac{(n-1)^2}{(n+1)^2} \tag{54}$$

Combining equation (54) with equations (48) and (49), the Fresnel transmission coefficients, at normal incidence, are:

$$\tau_F(0) = \frac{2n}{n^2 + 1} \tag{55}$$

$$\tau_{F,N}(0) = \frac{2n}{2n + N(n - 1)^2} \tag{56}$$

Example 6.

Calculate the Fresnel transmission coefficients in the case of 1, 2 and 3 glass covers, at normal incidence, assuming that the refractive index of glass is 1.53. From equation (56):

$$N = 1 \quad \tau_{F,1}(0) = \frac{2 \times 1.53}{2 \times 1.53 + 1(1.53 - 1)^2} = 0.92$$

$$N = 2 \quad \tau_{F,2}(0) = \frac{2 \times 1.53}{2 \times 1.53 + 2(1.53 - 1)^2} = 0.84$$

$$N = 3 \quad \tau_{F,3}(0) = \frac{2 \times 1.53}{2 \times 1.53 + 3(1.53 - 1)^2} = 0.78$$

Example 7.

Calculate the Fresnel transmission in the case of 1 glass cover, at an incidence of 70°, assuming that the refractive index of glass is 1.53. Equation (42) gives:

$$\theta_2 = \text{arc sin} \frac{\sin 70}{1.53} = 37.89^\circ$$

Equations (50) and (51) give:

$$\rho_1 = \frac{\sin^2(37.89 - 70)}{\sin^2(37.89 + 70)} = 0.312$$

$$\rho_2 = \frac{\tan^2(37.89 - 70)}{\tan^2(37.89 + 70)} = 0.041 \ .$$

Equation (48) gives:

$$(\tau_r)_1 = \frac{1 - 0.312}{1 + 0.312} = 0.524$$

$$(\tau_r)_2 = \frac{1 - 0.041}{1 + 0.041} = 0.921 \ .$$

The average transmission is:

$$\tau = \frac{(\tau_r)_1 + (\tau_r)_2}{2} = \frac{0.524 + 0.921}{2} = 0.72.$$

Absorption of radiation

The coefficient of transmission, τ_a, within the material is represented by Bouger's law:

$$\tau_a = e^{-KL} \tag{57},$$

where K is the extinction coefficient which can vary from 0.04/ cm for an excellent glass to 0.32/cm for a poor glass; L is the actual path of radiation through the medium.

Example 8.

Calculate the absorption in the case of 1, 2 and 3 glass covers, at normal incidence, assuming that the extinction coefficient is equal to 0.18/cm and that the thickness of each glass cover is 4 mm. At normal incidence the actual path is equal to the thickness:

$$N = 1; \ L = 0.4 \text{ cm}; \ \tau_a = e^{-(0.18)(0.4)} = 0.93$$

$$N = 2; \ L = 0.8 \text{ cm}; \ \tau_a = e^{-(0.18)(0.8)} = 0.87$$

$$N = 3; \ L = 1.2 \ cm; \ \tau_a = e^{-(0.18)(1.2)} = 0.81 \ .$$

Example 9.

Calculate the absorption coefficient, τ_a, in the case of a glass cover of 4 mm thickness with an extinction coefficient of 0.18/cm for an incident radiation at 70°.

In example 7, the refracted angle was found equal to 37.89^o. The path length is then:

$$L = \frac{0.4}{\cos 37.89} = 0.51 \ cm$$

and $\qquad \tau_a = e^{-(0.18)(0.51)} = 0.91.$

Combined transmission due to reflections and absorption

The combined transmission coefficient, τ, taking into account both reflection and absorption, is given by:

$$\tau = \tau_a \cdot \tau_r \qquad\qquad\qquad (58).$$

Example 10.

Calculate the combined transmission coefficient for 1, 2 and 3 glass covers of 4 mm thickness, with a refractive index equal to 1.53 at normal incidence.

Taking the results from previous examples:

$$N = 1; \ \tau = \tau_r \cdot \tau_a = (0.92)(0.93) = 0.86$$
$$N = 2; \ \tau = \tau_r \cdot \tau_a = (0.84)(0.87) = 0.73$$
$$N = 3; \ \tau = \tau_r \cdot \tau_a = (0.78)(0.81) = 0.66.$$

The combined solar transmittance can be calculated for various incident angles. In Figure 13 it can be seen that this coefficient presents a rather constant value for incident angles inferior to 45°.

Remark:
It has been assumed that the transmittance is independent of the

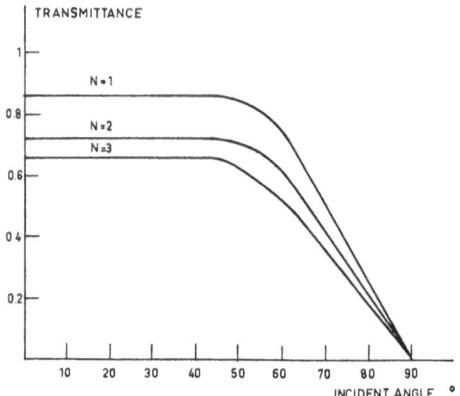

Fig. 13. Transmittance of glass covers.

wavelength. In the case of a strong dependence, the total transmittance can be calculated by:

$$\tau = \frac{\int_0^\infty \tau_\lambda I_\lambda \, d\lambda}{\int_0^\infty I_\lambda \, d\lambda}$$ (59).

In Figure 13 the transmittance curves for 1, 2 and 3 glass covers
are presented as a function of incident angle and in Figure 14 the
spectral transmittance of glass is shown for various iron oxide
contents.

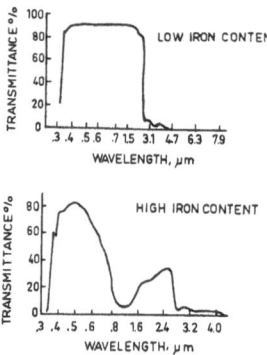

Fig. 14. Spectral transmittance of glass.

THE TRANSMITTANCE-ABSORPTANCE PRODUCT

The effective energy absorbed is determined by the transmittance-absorptance product of the system represented by the transparent cover and absorber plate.

At a given incident angle, if τ is the transmittance of the cover and α the absorptance by the absorber plate $(\tau\alpha)$ is absorbed and $(1-\alpha)\tau$ is reflected back to the cover in the form of mainly diffuse radiation (Fig. 15). The quantity $(1-\alpha)\tau\rho_d$ is reflected back to the absorber plate, where ρ_d is the diffuse reflectance of the cover.

The energy ultimately absorbed is defined by the effective transmittance-absorptance coefficient $(\tau\alpha)_e$:

$$(\tau\alpha)_e = (\tau\alpha) \sum_{n=0}^{n=\infty} [(1-\alpha)\rho_d]^n = \frac{\tau\alpha}{1-(1-\alpha)\rho_d} \qquad (60).$$

With an acceptable approximation the diffuse reflectance, ρ_d, can be taken equal to the specular reflection of the cover at $60°$, as calculated in example 7.

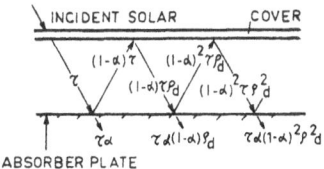

Fig. 15. Absorption of solar radiation by absorber plate.

USEFUL ENERGY ABSORBED AND INSTANTANEOUS EFFICIENCY OF A SOLAR COLLECTOR

If I is the incident energy on the cover, the useful energy, $\dot{q}_u$, absorbed per unit surface, will be the difference between the absorbed energy $(\tau\alpha)_e \cdot I$ and the heat losses $U_1(T_p - T_a)$:

$$\dot{q}_u = (\tau\alpha)_e I - U_1(T_p - T_a) \qquad (61).$$

If A is the surface of the collector exposed to the incident radiation, the useful energy becomes (assuming steady state conditions):

$$\dot{q}_u = A(\tau\alpha)_e I - AU_1(T_p - T_a) \tag{62}.$$

The instantaneous efficiency, η, is defined as the section of the useful energy, $\dot{q}_u$, over the incident energy AI:

$$\eta = \frac{\dot{q}_u}{AI} = (\tau\alpha)_e - U_1 \frac{(T_p - T_a)}{I} \tag{63}.$$

The equation which defines the instantaneous efficiency, is known as the Hottel-Whillier equation. It characterizes a solar collector by only two coefficients $(\tau\alpha)_e$ and U_1. The first coefficient $(\tau\alpha)_e$ relates to the process of absorption of energy and the second coefficient to the thermal loss. It is the basic equation which is used in collector testing for the determination of the thermal performance of a collector. It is often assumed that the two coefficients are constant, even though it has been seen in the preceding paragraphs how they are influenced by various parameters. The approximations and the errors which can result from this hypothesis are compatible with experimental precision.

In Figure 16 the instantaneous efficiency curve of the collector defined previously, is shown.

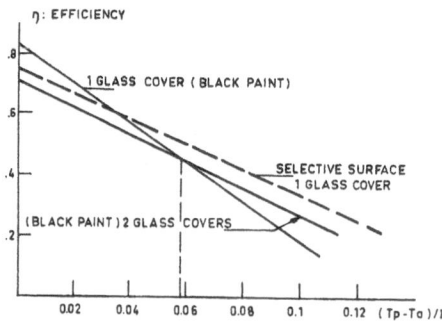

Fig. 16. Instantaneous efficiency curves.

THE TEMPERATURE PROFILE IN THE FLOW DIRECTION

Considering an element, dy, in Figure 17 and introducing the efficiency factor, F', defined previously, the increase of enthalpy of the fluid is equal to the absorbed radiation minus the heat losses:

$$\dot{m}_f C_p dT_f = F'[(\tau\alpha)_e I - U_1(T_f - T_a)] A \frac{dy}{1} \qquad (64).$$

Integration gives the value of the fluid temperature, T_f, at level y:

$$\frac{(\tau\alpha)_e I - U_1(T_f - T_a)}{(\tau\alpha)_e I - U_1(T_{fi} - T_a)} = e^{-F'\beta(y/1)} \qquad (65)$$

where

$$\beta = \frac{AU_1}{\dot{m}_f C_p}, \qquad (66)$$

$\dot{m}_f$ = mass flow,

C_p = specific heat of fluid,

A = surface of collector,

$(\tau\alpha)_e$ = effective transmittance - absorptance coefficient,

I = incident radiation,

U_1 = overall heat transfer coefficient of the collector,

T_a = ambient temperature,

T_{fi} = fluid inlet temperature.

It can be seen from equation (65) that the temperature profile in the flow direction is not linear except for small values of the coefficient β. Introducing the average fluid temperature, T_m, and the outlet fluid temperature, T_{fe}, one obtains:

$$\frac{(\tau\alpha)_e I - U_1(T_m - T_a)}{(\tau\alpha)_e I - U_1(T_{fi} - T_a)} = \frac{1}{F'\beta}[1 - e^{-F'\beta}] \qquad (67)$$

$$\frac{(\tau\alpha)_e I - U_1(T_{fe} - T_a)}{(\tau\alpha)_e I - U_1(T_{fi} - T_a)} = e^{-F'\beta} \qquad (68).$$

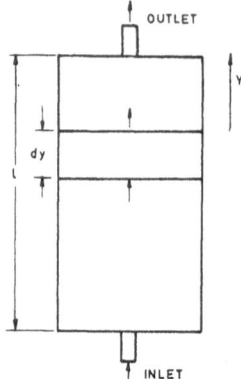

OUTLET

dy

INLET

Fig. 17. Temperature distribution in flow direction.

These last two equations can be rearranged to express the useful energy, q_u:

$$F'[(\tau\alpha)_e I - U_1 (T_m - T_a)] =$$

$$= F'\left[(\tau\alpha)_e I - U_1 \left(\frac{T_{fi} + T_{fe}}{2} - T_a\right)\right] x$$

$$x \left[\frac{2}{F'\beta} \frac{(1 - e^{-F'\beta})}{(1 + e^{-F'\beta})}\right] \qquad (69).$$

As long as the expression

$$\frac{2}{F'\beta} x \frac{(1 - e^{-F'\beta})}{(1 + e^{-F'\beta})}$$

is near unity, as is the case in practical applications, it can be seen that:

$$T_m \simeq \frac{T_{fi} + T_{fe}}{2}.$$

The mean fluid temperature, T_m, is very near the average temperature derived from the inlet and outlet temperature, except for very small mass flows.

In the preceding sections, the heat losses and efficiency of a "normal" collector have been determined. The next sections will be devoted to various ameliorations which can be considered to increase the efficiency of a solar collector.

As it was said earlier, two distinct processes are involved: the absorption of energy and the inevitable losses of energy which increase with temperature. As far as the absorption of energy is concerned, the only improvements which can be expected concern the transmittance of the transparent covers. Some new types of glasses, still very expensive, with better transmission properties, are becoming available. The absorption coefficient of black paint absorber plates is of the order of 0.96 and there is little to be gained in that direction. The increase of efficiency can be expected essentially by the reduction of the heat losses and various solutions are presented here. It should be noted that these solutions generally imply a certain deterioration in the absorption process and one should be careful to verify that the overall effect is positive.

DOUBLE-GLASS SOLAR COLLECTORS

Adding a transparent cover is one of the first possibilities to reduce the heat losses due to radiation and natural convection. On the other hand, the transmittance-absorptance coefficient will be diminished.

For the calculation of the heat losses, a procedure similar to the one described previously can be used leading to the following set of equations (assuming two identical glass covers, Fig. 18):

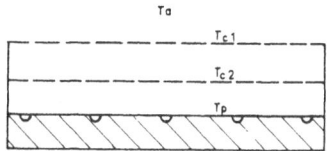

Fig. 18. Schema of double glass collector.

$$\dot{q}_f = h_a (T_{c1} - T_a) \tag{70}$$

$$\dot{q}_f = \epsilon_{gl} (T_{c2}^4 - T_{c1}^4) + [1 - 0.33 \, (s/90)] \times (T_{c2} - T_{c1})^{4/3} \tag{71}$$

$$\dot{q}_f = \epsilon_{g2}(T_p^4 - T_{c2}^4) + [1 - 0.33(s/90)](T_p - T_{c2})^{4/3} \tag{72}$$

$$\dot{q}_b = U_b(T_p - T_a) \tag{73}$$

$$\dot{q}_l = \dot{q}_f + \dot{q}_b \tag{74}$$

$$\epsilon_{g1} = \frac{1}{2/\epsilon_c - 1} \tag{75}$$

$$\epsilon_{g2} = \frac{1}{\dfrac{1}{\epsilon_p} + \dfrac{1}{\epsilon_c} - 1} \tag{76}$$

$$U_b \simeq \frac{\lambda_b}{e_b} \tag{77}$$

$$U_a = 5.7 + 3.8\, V \tag{78}.$$

In Figures 19 and 20 the heat losses and the overall heat transfer coefficient have been represented versus the temperature
difference $(T_p - T_a)$. It can be seen that they have been considerably reduced compared to the case of the single glass collector.
For the single glass collector the average overall heat transfer
coefficient was of the order of 6.5 W/m^2 °C and for the double
glass collector it is reduced to 4.4 W/m^2 °C.

Unfortunately, the effective transmittance-absorptance product
is also reduced. Using the procedures described previously, a
calculation gives:

- one glass cover: N = 1
    $(\tau\alpha)_e$ = 0.83 at normal incidence,

- two glass covers: N = 2
    $(\tau\alpha)_e$ = 0.71 at normal incidence,

leading to the two following equations for the instantaneous efficiency:

$$N = 1 \quad \eta = 0.83 - 6.5\,\frac{T_p - T_a}{I} \tag{79}$$

$$N = 2 \quad \eta = 0.71 - 4.4\,\frac{T_p - T_a}{I} \tag{80}.$$

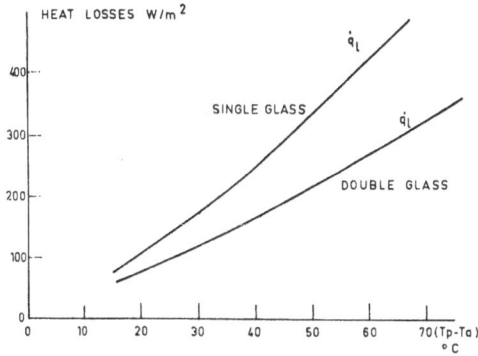

Fig. 19. Total heat losses.

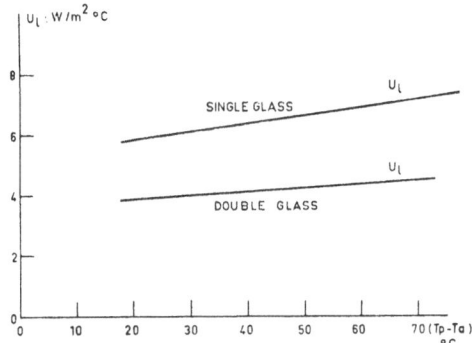

Fig. 20. Overall heat transfer coefficient.

The curves are plotted in Fig. 16. The double glass collector becomes more effective than the single glass collector only for values of

$$\frac{T_p - T_a}{I} > 0.06 .$$

The choice between these two types of solar collectors should be made on the basis of the most likely value of $(T_p - T_a)/I$ under normal operation conditions.

The numerical example has been treated for a low value of the wind velocity. For high wind velocities, the superiority of the double-glass collector would be more evident.

SELECTIVE SURFACES

It was shown previously that for a normal single glass black paint collector the radiative losses represent about 60% of the total losses. So one of the effective ways to increase efficiency, is to reduce the radiative losses, that is the global emissivity coefficient, ϵ_g, between the absorber plate and the transparent cover,

$$\epsilon_g = \frac{1}{1/\epsilon_g + 1/\epsilon_c - 1} \tag{81}$$

This makes sense only if the transmittance of the transparent cover and the absorptance of the absorber plate in the spectrum of solar radiation are not affected.

The emissivity coefficient either of the absorber plate or of the transparent cover must be made wavelength-dependent.

Selective absorber surfaces

From Planck's and Wien's laws, it can be noted (see relevant section) that the spectrum of solar radiation and the spectrum of heat radiation for temperatures up to a few hundred °C do not overlap by any appreciable amount. So if surfaces can be prepared in a manner that their absorption characteristics remain high for wavelengths below 2 μm and their emission characteristics low for wavelengths above 2 μm , the radiative losses can be reduced without impairing the absorption process.

The electromagnetic theory of light shows that metals are good reflectors of long wave radiation, the reflectivity being a function of the wavelength and the electrical conductivity:

$$r_\lambda = 1 - 0.365(\lambda x)^{-1/2} \tag{82}$$

where λ = wavelength in um,
 x = conductivity in ohms per mm^2 per m length,
 r_λ = reflectivity.

The reflectivity of good conductors with polished surfaces such as copper, for instance, is found to be between 0.95 and 0.98 at a wavelength of 2 μm and near 1 for very long wavelengths.

For a given wavelength λ, the relations:

$$r_\lambda + \alpha_\lambda = 1 \qquad (83)$$
$$\alpha_\lambda = \epsilon_\lambda \qquad (84)$$

show that to a reflectivity of 0.95 corresponds an emissivity, ϵ_λ, of 0.05.

Electrolytical, chemical or electrochemical surface treatments can deposit films which are visibly black and at the same time transparent to wavelengths above 2 µm. For instance, black deposits are produced industrially by electroplating in particular of nickel in the presence of zinc and sulphides. These deposits on a bright metal base will show in the visible spectrum an absorption of 80 - 90% and retain the emissivity of the metal base in the infrared spectrum.

An ideal selective surface will have the profile shown in Figure 21.

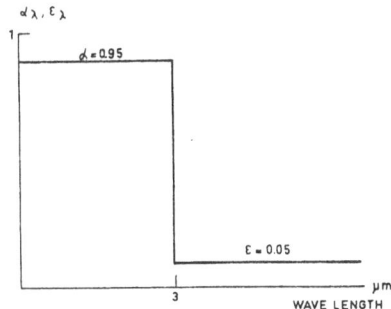

Fig. 21. Ideal selective surface.

Many different combinations of coatings and metal substrates have been investigated, but it seems that the two most promising solutions from an economic and industrial point of view are the so-called "black nickel" and "black chrome" solutions.

In Figure 22 the heat losses and the efficiency curve of a one-glass selective collector have been calculated, assuming:

$$\epsilon_p = 0.15 \text{ and } \alpha = 0.90.$$

For a black paint collector, the radiative losses are predominant, here it is the convective losses which are the most important with the following repartition:

- radiative losses: 20%,
- convective losses: 54%,
- back losses: 26%.

When the convective losses are dominating, it can be of a certain interest to consider evacuated collectors, where by reduction of the pressure, the air is maintained stagnant inside the collector (Gr < 1500). Some attempts have been made to manufacture flat plate evacuated collectors. From a technological point of view it is more realistic to consider cylindrical geometries as in the Philips' collector for instance.

Reflective transparent covers

Another means of reducing the global emissivity coefficient between the absorber plate and the transparent cover, is to increase the reflectivity of the transparent cover in the infrared spectrum. It has been observed for instance that deposits of indium-oxide, In_2O_3, on glass will increase the reflectivity of the glass in the infrared spectrum up to 80% with a slight diminution of the transmittance in the visible spectrum. The economy of such solutions is still to be demonstrated for flat plate collectors.

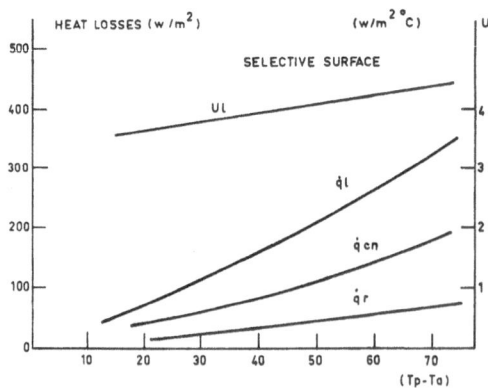

Fig. 22. Heat losses and overall heat transfer coefficient.

HONEYCOMB STRUCTURES

By inserting "honeycomb" structures, also called Francia structures, between the absorber plate and the transparent cover,
both the heat losses due to radiation and natural convection are
reduced in significant proportions. But due to absorption of light
in these transparent structures, specially at inclined incident
angles, the transmittance-absorptance coefficient is also reduced,
leading to an overall effect which is certainly positive but not as
high as could have been expected from the reduction of heat
losses.
Studies have been carried out at the JRC on the heat transfer
phenomena in honeycomb structures specially by A. Baehr (see
report EUR 5760 d) who has investigated the repartition of radiative and convective losses for different configurations. The radiative losses are reduced because of a remission in the infrared
spectrum, from the cellular structure towards the absorber plate.
The governing factor is the solid angle from which the sun is seen
at the bottom of the structure. The higher the ratio ℓ (height of
cell) over d (diameter of the cell), the more pronounced the effect.
But due to absorption of light the optimum ratio ℓ/d is situated
around 4 to 5.

The reduction of convective losses has also been studied extensively by K.G.T. Hollands who used polyethylene cells in order
to eliminate the back radiation effect. Hollands measured the
Nusselt number as a function of the Rayleigh number for different values of the ratio ℓ/d and of the inclination angle of the cells.
In Figure 23 results concerning ratios $\ell/d = 2$ and $\ell/d = 5$ are
presented. It can be noted that for $\ell/d = 2$, convective heat transfer is suppressed up to Rayleigh numbers of the order of 20,000
and for $\ell/d = 5$ the critical Rayleigh numbers are of the order
of 10^6. The convective heat transfer is more important for horizontal positions than for vertical positions. It should also be observed that after the critical Rayleigh number the convective
heat transfer starts very quickly and in some cases becomes
more important than the heat transfer in the absence of honeycombs.

The relative transmission of various honeycomb materials as a
function of incident angle is presented in Figure 24. It can be
seen that for paper and aluminium honeycombs the relative transmission falls too rapidly to permit practical applications. Even

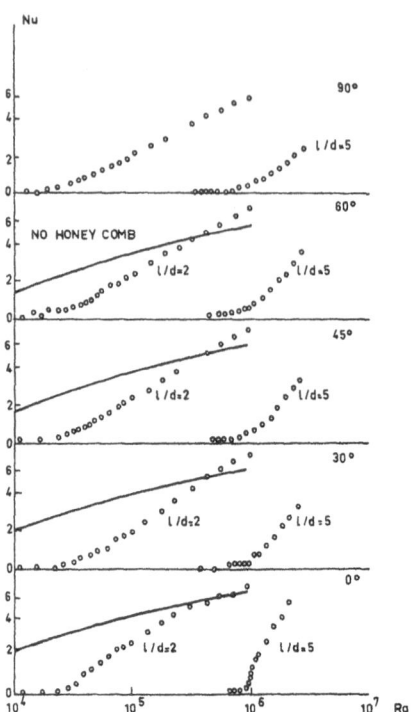

Fig. 23. Measured free convective heat transfer in inclined cells with honeycombs (from K. G. T. Hollands).

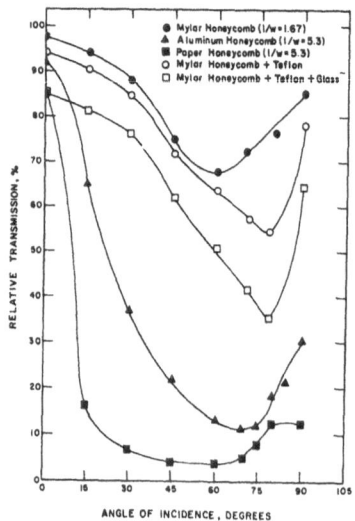

Fig. 24. Relative transmission of selected honeycomb materials as a function of incident angle (from NBS 899).

in the case of Mylar honeycombs, for incident angles greater
than 30°, the reduction in transmission is not negligible.

The transmission-absorptance product for a collector with and
without a honeycomb is shown in Figure 25. By comparisons
with curves of Figure 13, the influence of the honeycombs can
be evaluated.

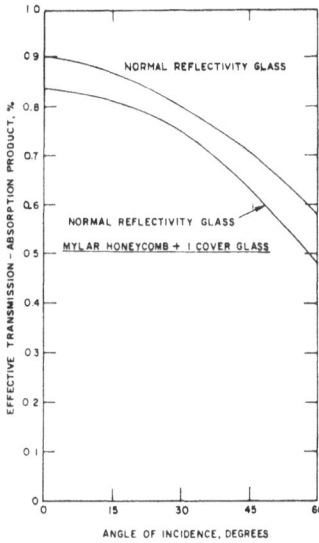

Fig. 25. Transmission-absorptance product for a single-
glazed flat-plate solar collector incorporating a honeycomb convection suppressor (from NBS 899).

CONCLUSIONS

The complex nature of the photothermal conversion solar energy
which involves, on the one hand, the collection and the absorption of sunlight, and on the other hand, its transformation into
useful energy, has been illustrated by the description of the basic
thermal and optical phenomena taking place in flat-plate solar
collectors. Because of the low densities of solar fluxes, only
carefully designed collectors where heat losses are minimized,
will lead to acceptable performances.

The choice of the basic materials for the absorber, the transparent cover and the insulating material will play an important
role in the determination of the quality of the collectors. However, it should not be forgotten that the quality of a collector is
not only based on its thermal performance but also on its cost

and durability. The final product should be the result of a careful balance between these three aspects.

BIBLIOGRAPHY

Arnold, J.N., D.K. Edwards, and I. Catton, Effect of tilt and
    horizontal aspect ratio on natural convection in a rectangular
    honeycomb, Trans. ASME, J. Heat Transfer, 99 (1977) 120.
Buchberg, H., I. Catton, and D.K. Edwards, Natural convec-
    tion in enclosed spaces: A review of application to solar ener-
    gy collection, Trans. ASME, J. Heat Transfer, 98 (2) (1976)
    182.
Charters, W.W.S., and L.J. Peterson, Free convection sup-
    pression using honeycomb cellular materials, Solar Energy,
    13 (1972) 4.
Randall, K.R., J.W. Mitchell, and M.M. El Wakil, Natural
    convection characteristics of flat-plate collectors. In J.R.
    Howell and T. Min (Eds.), Heat Transfer in Solar Energy
    Systems, American Society of Mechanical Engineers, New
    York, 1977.
Beckman, W.A., S.A. Klein, and J.A. Duffie, Solar Heating
    Design, Wiley, New York, 1977.
Hottel, H.C., and B.B. Woertz, Performance of flat-plate
    solar-heat exchangers, Trans. ASME, 14 (1942) 91.
Lin, R.J.H., Optimization of coatings for flat-plate solar col-
    lectors, Report COO/2930-4 to ERDA (Jan, 1977).
McDonald, G.E., Spectral Reflectance properties of black
    chrome for use as a solar selective coating, NASA Tech.
    Memo NASA TMX 0171596 (1974).
Tabor, H., Selective surfaces for solar collectors. In Low Tem-
    perature Engineering Applications of Solar Energy, ASHRAE
    New York, 1967.
Hottel, H.C., and B.B. Woertz, The performance of flat-plate
    solar-heat collectors, Trans. ASME, 64 (1942) 91.
Whillier, A., Solar energy collection and its utilization for
    house heating, ScD. Thesis, MIT, 1953.
Klein, S.A., Calculation of flat-plate loss coefficients, Solar
    Energy, 17 (1975) 79.
Tabor, H., Radiation, convection, and conduction coefficients
    in solar collectors, Bull. Res. Council of Israel, 6C (1958)
    155.
Barley, C.D., and C.B. Winn, Optimal sizing of solar collectors
    by the method of relative areas, Solar Energy, 21 (1978) 279.

PERFORMANCE TEST PROCEDURES FOR THERMAL COLLECTORS
(1) Outdoor Testing

Dr. W. B. Gillett

Solar Energy Unit, University College, Cardiff, U.K.

ABSTRACT

A review of outdoor solar collector test methods is presented, based largely on the CEC Recommendations for European Solar Collector Test Methods. Test facility design and instrumentation are discussed, with reference to their influence on measured collector efficiencies. Steady state outdoor testing, mixed indoor/outdoor testing and transient testing are reviewed, and it is concluded that although the testing of simple flat plate water heaters is fairly well understood, more work is now required to develop test methods for the new high performance collectors which are coming onto the market.

NOMENCLATURE

α	solar absorptance
A	area
a_o, a	constants
C_p	specific heat capacity of fluid
F'	collector efficiency factor
F_R	collector flow factor
G	solar irradiance
j	time
K_n	weighting function
M	fluid capacity of collector
$\dot{m}$	fluid flow rate
q	rate of heat flow
U	collector heat loss coefficient
τ	time

G. Beghi (ed.), Performance of Solar Energy Converters: Thermal Collectors and Photovoltaic Cells, 103–124.

τ	cover transmittance
t	time
T_a	air temperature
T_i	collector inlet temperature
T_m	mean collector temperature
T_o	collector outlet temperature
$T*$	reduced temperature difference $(Tm-Ta)/G$
ΔT	temperature difference between inlet and outlet
η	collector efficiency
η_o	collector efficiency when $T_m = T_a$

ACKNOWLEDGEMENTS

The information contained in this lecture has been gathered through participation in the collaborative collector testing groups of the Commission of the European Communities (CEC) and of the International Energy Agency (IEA), for which funding has been provided through DG12 for Research Science and Education by the CEC, and through the Energy Technology Support Unit by H.M. Department of Energy. Useful experience was also obtained through participation in the drafting of the new British Standard Draft for Development 'Methods of test for the thermal performance of solar collectors'.

I would like to thank my colleagues Professor B. J. Brinkworth, B. M. Cross, A. A. Green, K. Hayward, J.P. Kenna and R. W. Rawcliffe from the Solar Energy Unit, whose work has contributed greatly to my understanding of the problems involved in solar collector testing.

INTRODUCTION

Although the performance of solar collectors has been quite well understood for nearly 40 years (1), no serious attempts to develop standard test methods were made until about 1974 (2).

In response to the growth of interest in solar energy utilisation in the mid-1970s, both the Commission of the European Communities (CEC) and the International Energy Agency (IEA) established working groups to study solar collector test methods. These two groups have each carried out "round robin" collector testing programmes and have collaborated together in the development of test methods and reporting formats for the presentation of test results. The CEC group published its experience in the form of a set of Recommendations for European solar collector test methods in 1980 (3). This contains advice concerning test facility design, the selection of instrumentation, the use of test procedures and the presentation of test results.

In this lecture, the recommendations of the international collaborative groups are reviewed, and some new material is presented in order to illustrate how more accurate testing might be performed, and how collector testing might develop in the future.

THE PURPOSES OF COLLECTOR TESTING

The purpose of the test should be identified before testing begins, in order to ensure an appropriate method is selected. For example, in some cases it might be cheaper to cut a collector into pieces and to test its components, rather than to test the whole collector as a "black box".

Test methods which are used to develop the design of collectors may be rather different from those which are used to rate collectors or to provide data for predicting the annual performance of solar heating systems.

TEST FACILITY DESIGN

All the established test procedures for complete collector modules require a test facility with essentially the same basic features, and in many laboratories the same facility is used for all testing. It may even be taken indoors from time to time to perform heat loss tests or tests in a solar simulator.

An example of a test facility is shown in Figure 1. It contains a collector mounting and instrumentation to measure the solar radiation, the ambient air temperature and the rate of heat collection by a collector. A fluid conditioning loop which provides fluid at a constant temperature is located in the building below.

Collector Mounting

The mounting can affect the test results, so the following recommendations are made by the CEC collector testing group (3).

The collector tilt should be fixed at $45^\circ \pm 5^\circ$. It has been shown by Klein (4) that the tilt angle affects the loss coefficient in typical flat plate collectors. For matt black collectors the loss coefficient varies by $\pm 5\%$ of its value at 45° if its tilt angle is varied in the range 0 to 90°. For selective collectors, the loss coefficient varies by $\pm 10\%$ of its value at 45°. Hence it can be seen that the effect of using tilt angles of other than 45° is likely to be small in comparison with other influences on collector heat losses such as wind speeds.

Fig. 1 The Outdoor Test Facility at University College, Cardiff.

The effect on a selectively coated collector is proportionately
greater and is more significant because these collectors are
less sensitive to variations in environmental parameters such
as wind speed.

There may be other effects caused by a change in tilt angle,
such as a change in the flow distribution in the absorber. The
testing of horizontal collectors should be avoided, because of
the difficulties of removing air locks.

The lowest edge of the collector should be at least 500 mm
above local ground level. This reduces the amount of ground
reflected solar radiation, and the thermal radiation exchange
with the ground. It also allows a free flow of wind over the
collector.

The collector mountings should not significantly affect the
collector back or edge insulation, and should not shade or obscure
the collector aperture. For boxed collectors, most laboratories
recommend an open frame design of test stand. If a back board
or other mounting is used, it may help to insulate the collector
and produce test results with a lower heat loss coefficient.

The collector should have an unobstructed field of view.
Naturally it should not be possible for shadows to be cast onto
the collector during testing. In addition, however, it is
usually recommended that no obstruction in the field of view
should subtend an angle with the horizontal of more than 15°
at the collector aperture. The collector is assumed during
testing to be exchanging thermal radiation with the sky, and to
be receiving diffuse radiation from the sky. If large surfaces
in the field of view reflect solar radiation or emit or absorb
thermal radiation, then errors in the test results could be
produced. Features which are particularly undesirable here
are hot chimneys or cooling towers outdoors, and cold windows
or hot radiators indoors.

The collector should be securely mounted such that it can
safely withstand wind gusts. There are practical advantages in
providing facilities for azimuthal tracking of the test stand
because a fixed orientation very much restricts the time
available for testing. Safety is an important aspect of any test
facility and, if the glass in the collector breaks, personnel
should be protected.

The Fluid Loop

A variety of test loop arrangements is possible, depending
on the layout of the facility, and the space available. A
schematic layout of an open loop facility is shown in Figure 2,
and a slightly different arrangement which employs a more
complex inlet temperature control system is illustrated in
Figure 3.

Most test loops employ water as the heat transfer fluid because
its heat transfer properties are well known. A small quantity
of corrosion inhibitor can usually be added to water without
much effect on its heat transfer properties, but if anti-freeze
or other heat transfer fluids are used then their heat transfer
properties need to be determined. It is important to ensure
that air is efficiently removed from the fluid in a test loop,
and a filter is usually employed to protect the flow meter and
the pump.

The fluid usually flows from the bottom to the top of a
collector under test to minimise the risk of air locks, and a
short section of transparent tube should be incorporated in the
loop to permit the quality of the fluid to be monitored.

Fluid Inlet Temperature Control

One of the most important aspects of a collector test
facility is that it should be able to supply a constant fluid

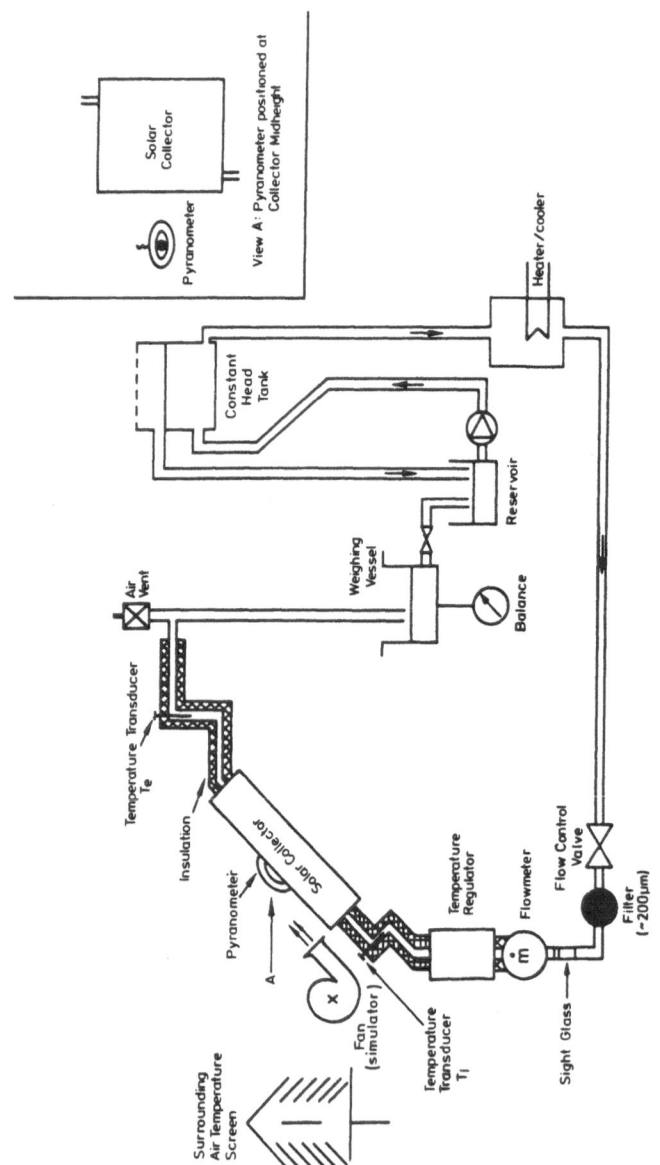

Fig. 2 Schematic Arrangement of a Simple Open Loop Test Facility

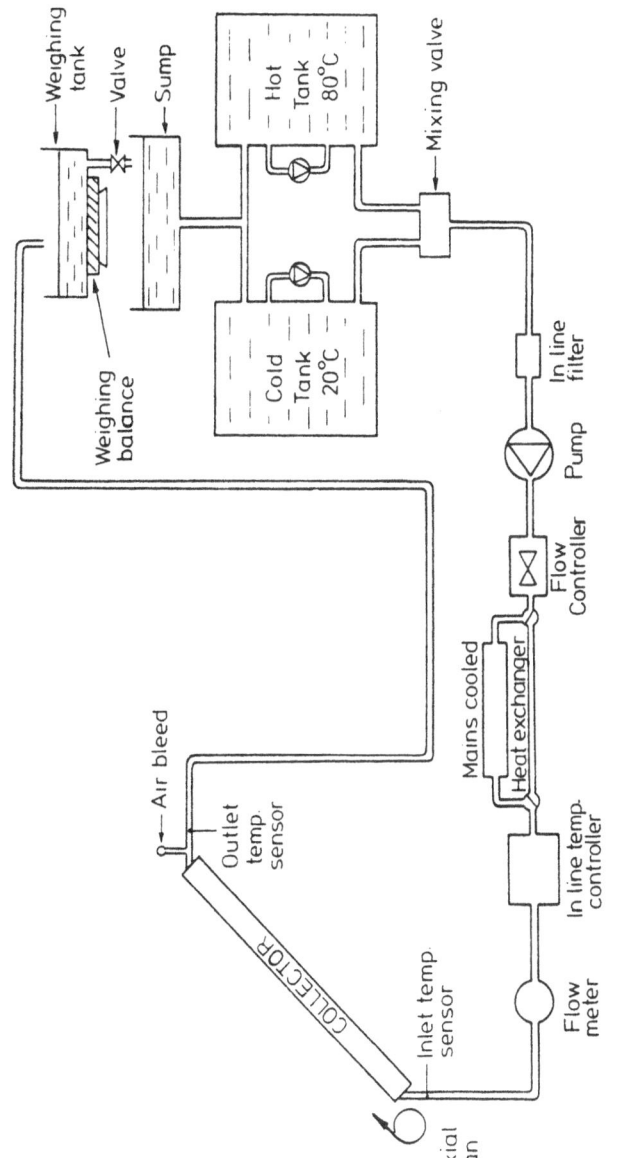

Fig. 3 A Test Loop with Two Stage Inlet Temperature Control

inlet temperature at any temperature chosen in the operating
range. To be able to change rapidly from one inlet temperature
to another is also important in order to minimise the amount
of time wasted between test points

If the fluid inlet temperature is changed, even by a small
amount, then the time taken for the collector to reach a new
steady state will be at least as long as it takes for the fluid
to pass through the collector, and usually much longer, depending on the distribution of thermal mass in the collector.
Furthermore, any change in inlet temperature will not affect
the outlet temperature until a period of time (τ seconds) later,
where τ is given approximately by

$$\tau = \frac{M}{\dot{m}} \tag{1}$$

Since the rate of energy collection in the collector is
deduced by measuring instantaneous values of fluid inlet and outlet temperatures, it follows that small variations in fluid inlet
temperature could lead to errors in the rates of energy collection
which are deduced.

In steady state testing, small variations in inlet temperature are accomodated by the requirement that a mean value be
determined over a 15 minute period. However, attention should
be paid to very small drifts over this period, since a steadily
rising or steadily falling inlet temperature will lead to an
underestimate or an overestimate of collector efficiency
respectively.

For example: Determine the error in collector efficiency
produced by a fluid inlet temperature (T_i) which rises steadily
through 0.2K in 15 minutes. Assume that the collector has a
water capacity of 3 kg m^{-2}, the mass flow rate is 1 kg min^{-1}.m^{-2},
and the temperature rise (ΔT) across the collector is 1 K.

The error in the instantaneous temperature difference is
given by

$$\frac{dT_i}{dt} . \tau = \frac{0.2}{15} \times 3 = 0.04 \text{ K}$$

which as a percentage of the temperature rise is:

$$\frac{\text{error}}{\Delta T} = \frac{0.04}{1} \times 100 = 4\%$$

It can be seen from this example that testing with small temperature rises is likely to lead to significant errors in the results. It also follows that the potential for errors, due to poor inlet temperature control, is greater for collectors with a high fluid capacity than for collectors with a low fluid capacity.

To achieve the required stability, a number of laboratories now employ two stages of temperature control, and use both a cylinder of hot fluid and one of cold fluid. The outlets from the two tanks are first blended using a mixing valve to produce approximately the required temperature, and this is then finely adjusted by an in-line electrical heater with feedback temperature control. The in-line heater is usually placed near to the collector inlet where it can accommodate heating and cooling along the feed pipes between the fluid conditioning equipment and the collector.

INSTRUMENTATION

In most facilities today, the data are recorded using a logger or small computer, although chart recordings are still very useful to give an immediate indication of the progress of the test. One of the most important tasks involved in the use of a collector test facility is the calibration of its instruments. The low flow rates and small temperature rises which have to be measured demand a very high standard of calibration.

Measurement of Solar Radiation

The instruments. The two most widely used instruments are the Kipp and Zonen CM6 pyranometer and the Eppley precision spectral pyranometer (PSP). The Eppley PSP is classified by the World Meteorological Organisation (WMO) as a Class 1 instrument whilst the Kipp CM6 is classified as Class 2 (5). A new pyranometer (CM11) has recently been produced by Kipp, and this appears to meet the WMO Class 1 specifications.

Extensive studies of pyranometers have been undertaken by both the CEC and the IEA since 1978, and work continues in collaboration with the World Radiation Centre at Davos in Switzerland. It is somewhat premature to anticipate the outcome of the Davos work which is not due to be published until early in 1982. However, in order to give an indication of the problems, a few comments will be included here.

The Kipp solarimeter has been available since about 1924, and is widely used by meteorological services in Europe. Two main types have been produced, the CM3 and the CM6, and both have the same rectangular thermopile which gives the instrument an asymmetric azimuthal response. The CM6 has a linearity of about ±1%, a temperature coefficient of about -0.15% per ^{o}C, and its readings can be affected by up to about 4% if the instrument is tilted. The tilt effect depends on the axis about which the tilting takes place and on the irradiance level. The largest source of uncertainty with the CM6 is its cosine response which varies from one instrument to another, and can cause errors of up to about 3 - 4% in the typical operating range.

The Eppley PSP is also a well established and widely used instrument. It has a built in temperature compensation and its linearity is similar to that of the Kipp CM6. It has an azimuthally symmetrical detector which is less influenced by tilting but the PSP also has an imperfect cosine response.

The calibrations of pyranometers used for collector testing should be checked every year, and this should be done by a recognised national calibration service or by the World Radiation Centre. The calibrations should now all be referred to the World Radiation Reference (WRR), and any instruments still calibrated to the old IPS scales should be re-calibrated.

The measurements. For collector testing, both global and diffuse irradiance are usually measured in the plane of the collector. By measuring the global irradiance at approximately the mid-height of the collector, the sky and ground reflected irradiances entering the collector aperture are all included. The diffuse irradiance is measured because it is less well transmitted by most collector covers, and a correction to the zero loss collector efficiency (η_o) based on the measured percentage of diffuse irradiance is usually made when analysing collector test results.

Fig. 4 Solar radiation measuring instruments

The incidence angle of the direct solar beam is important in the analysis of test results, and this can be measured directly by means of a spigot mounted on a disc at the centre of a series of concentric circles. With the disc mounted in the plane of the collector, the incidence angle can be calculated from the length of the shadow cast by the spigot. The shadow is measured using the concentric circles (see Fig.5). Alternatively, the incidence angle can be computed from the sun's position and the azimuth of the test plane for any given time.

Fig. 5   Angle of incidence indicator

Measurement of Fluid Inlet Temperature

The CEC recommendations (3), call for a precision in inlet temperature measurement of 0.1K, which is higher than that of ASHRAE (6), because it is felt to be important to be able to detect small drifts in the collector inlet temperature. Many laboratories have now moved from thermocouples to platinum resistance thermometers for temperature measurements, and can resolve with more confidence to ±0.02K. The transducer is usually placed as close to the collector inlet as possible (say <100 mm away), and the inlet region is very well insulated. If large bore pipes are used for the fluid loop, then it may be necessary to promote turbulence around the temperature probe. This can be done by locally using small diameter pipes, or by placing a pipe elbow before the transducer.

Measurement of the Temperature Difference between the Collector Inlet and Outlet

As high an accuracy as possible is required for the temperature difference measurements, because at high temperatures the collector may only achieve a temperature rise of 1 or 2K. The accuracy of the instrumentation determines the minimum temperature rise which can be meaningfully used. For example, if the accuracy is limited to ±0.1K, then errors of up to ±10% could be measured in a temperature rise of 1K. By using a matched pair of platinum resistance thermometers, and regularly checking the calibration, it is possible today to work with some confidence to ±0.02K, and this gives a reasonable accuracy in collector efficiency measurements, even when using a temperature rise of only 1K.

Thermopiles have also been successfully used for measuring small temperature differences, but they are rather more susceptible to electrical noise and contact potentials. When calibrating transducers for this job it is important to check their performance for measuring temperature differences over the whole range of absolute temperatures used.

The mounting of the transducers for temperature difference measurement also needs to be in a well-stirred flow and as close to the collector as possible. There are some advantages in measuring the temperature difference and inlet temperature with different transducers. This permits the inlet transducers to be compared from time to time, but still allows continuous records of both experimental variables to be made simultaneously.

Measurement of Ambient Air Temperature

Ambient air temperature is usually measured by Meteorologists in a 'Stephenson Screen' (5) which is a louvered white wooden box. However, discrepancies of between 1 and 2K can be produced in such a box at low wind speeds, so a simple radiation shield consisting of a pair of concentric metal tubes, mounted vertically, is recommended for collector testing. The outer tube is painted white, and either a platinum resistance thermometer or a weatherproof thermocouple is placed inside the inner tube. An accuracy of ±0.5K is quoted as acceptable for this measurement, although 0.1K is more often available today. The transducer should be located near to the collector, but away from the ground or any surface which might warm the air.

Measurement of Fluid Flow Rate

For most water heating collectors the collection efficiency is not significantly affected by the flow rate used, but for the purpose of calculating collector performance, the flow rate must be known to at least the same accuracy as that required for the efficiencies deduced. Typically the flow rates selected are in the region of 0.02 kg $s^{-1}m^{-2}$, and the accuracy required is about ±1%.

Experience with a wide range of commercial flow meters, including turbine meters, variable area meters and calorimetric devices suggests that an accuracy of ±1% is difficult to maintain for long periods over the full range of temperatures involved. The most reliable approach is to have a simple flow diverter in the test loop and to measure flow rates directly by weighing while a test is in progress. An alternative way of providing a reliable flow rate is to use a metering pump.

Measurement of Air Speed

The influence of wind speed and direction on collector performance is not well understood, and specifications for its measurement vary from a requirement for an integrated average with an accuracy of ± 0.8 m.s^{-1} (6) to a specification for an accuracy of ± 0.5 m.s^{-1} (3). Some correlations between collector heat losses and air speed over the collector surface were recently reported by Green et al (7) who showed that the effect of increasing the speed was small for speeds of greater than about 2 m.s^{-1}.

This parameter is called "air speed" in accordance with the CEC recommendations (3), because "wind speed" is a defined meteorological parameter which is measured at a height of approximately 10 m in an open site (5). Since air velocities change markedly around buildings it is not yet clear how the local air speed over a collector can be correlated with that measured at a meteorological station.

TEST LOOP CALIBRATION CHECK

In some laboratories the flow rate and temperature measuring instruments are calibrated by a local calibration centre, and in these cases it is common for only the transducers to be calibrated. To give an additional check on the calibration of these instruments when they are on site and connected to their own read-out equipment, a simple facility calibration check may be employed.

A very well insulated electrical water heater (approx 2 kW) is mounted in a small water vessel of approximate volume 4 litres and installed in the test loop in place of a solar collector. The electrical input power is then measured directly using a Watt meter with an accuracy of about 1%, and compared with the rate of energy input determined from the flow rate and temperature rise measurements.

Appropriate experiments can be performed to establish the heat loss coefficient of the electrical heater, and to overcome

any problems of surface boiling in the heater. The results
provide a check on the final data produced from the test facility.

TEST PROCEDURES

The most well established test procedure, which has been
used by both the CEC and the IEA groups, is the steady-state
outdoor test procedure published as ASHRAE 93-77 (6). This is
embodied in the CEC recommendations (3), and has been adopted
as the basis for the development of national test procedures in
many countries. In Italy, the UNI standard (8) is almost
exactly the same as the ASHRAE standard, whilst in France the
AFNOR standard (9) differs in the presentation of the results
and in its specification of the required test conditions.

The ASHRAE test procedure suffers from the limitation that
steady irradiance conditions are required, and in many parts
of the world these are difficult to find. To overcome this
problem, two approaches have been put forward. The German
approach was to develop a mixed indoor/outdoor test, the BSE
test (10), whilst the British approach was to develop a "transient" test method which could be used in unsteady irradiance
conditions outdoors (11).

A third approach which also overcomes the problems of
finding suitable outdoor weather conditions is to employ a
solar simulator, and this option is discussed in detail in the
next lecture.

The Steady State Outdoor Efficiency Test

The test loops and instrumentation described above are
suitable for this type of test, and the detailed requirements
are given in the CEC reommendations (3) or the ASHRAE 93-77
procedure (6). Before a test is carried out, the collector and
pyranometer should be cleaned, and the test loop heated to about
80°C to dry out all the insulation.

To meet the CEC recommendations, a minimum of 16 test
points should be taken, incorporating at least 4 fluid inlet
temperatures, equispaced over the operating temperature range.
Each test point involves a steady state operating period of 30
minutes when the solar irradiance must vary by less than ± 50 Wm^{-2},
the inlet fluid temperature by ± 0.1K, the surrounding air temperature by less than ± 1K, the fluid mass flow rate by less than $\pm 1\%$,
and the temperature difference between outlet and inlet of the
collector by less than ± 0.1K. Also the solar irradiance must
exceed 600 Wm^{-2} and the incidence angle of the direct solar
beam must be less than 40°.

In its simplest form the collector performance can be presented as:

Heat output = solar input - heat losses

or: $\dfrac{\dot{m}\,C_p}{A}$ (To-Ti) $= F'G\tau\alpha - U(T_m - T_a)$ (2)

Hence, dividing by the irradiance:

$\eta = \eta_o - UT*$ (3)

where: $\eta = \dfrac{\dot{m}\,C_p(To - Ti)}{AG}$, $\eta_o = F'\tau\alpha$ and $T* = \left|\dfrac{T_m - T_a}{G}\right|$

This analysis of collector performance is sufficient for most purposes, but as more accurate facilities have been built, it has become possible to employ a more detailed analysis of collector behaviour.

Diffuse and direct solar irradiance. Most collector covers have a value of solar transmittance (τ) which depends on the incidence angle of the beam, and the transmittance for diffuse irradiance is somewhat less than for a normally incident beam. For example, the transmittance for 4 mm glass is about 10% less for diffuse irradiance than it is for a direct beam.

By specifying an irradiance level of greater than 600 Wm^{-2}, the diffuse contribution is usually maintained below about 40%, so the maximum variation in η_o produced by ignoring diffuse irradiance with a 4 mm glass cover is about $\pm 2\%$.

For simple collectors where the cover transmittance characteristics are well understood, the performance can be corrected to equivalent normal irradiance conditions, provided that both direct and diffuse irradiances are measured during testing.

Incidence angle effects. The performance of a collector under direct irradiance at shallow incidence is less than that under normal incidence conditions because of shading from the sides of the collector, and because the cover transmittance varies with angle.

An incidence angle modifier can be determined by testing the collector with its mean fluid temperature equal to ambient at a number of incidence angles. This modifier can then be used to correct the other performance data to equivalent normal irradiance conditions. The amount of testing involved in determining an incidence angle modifier is so great, however, that most laboratories would prefer to make a theoretical correction based on the transmittance characteristics of the cover and the geometry of the collector.

Temperature dependent heat losses. Experiment and theory have shown that a collector heat loss coefficient may be expected to rise with temperature, and a reasonable approximation to the behaviour of typical collector heat losses is given by:

$$U = a_0 + a_1 \, (T_m - T_a) \tag{4}$$

Hence it follows that collector performance characteristics should be of the form:

$$\eta = \eta_o - a_o T^* - a_1 \, G(T^*)^2 \tag{5}$$

Unfortunately, the scatter in outdoor test results is often so large that when fitting the data, the standard error in the estimate for a_1 is as great as the value of a_1 itself, and hence only a linear approximation can be justified.

It is particularly important to note that the value of η_o deduced can be very sensitive to the fit selected, and for this reason it is wise to ensure that at least four data points are measured very close to η_o itself if a second order fit is to be used.

The Mixed Indoor/Outdoor Steady State Test

In this test procedure, proposed by the BSE (10), the measurements of steady state efficiency around η_o are carried out using the methods described in the Steady State Outdoor Efficiency Test, but the amount of outdoor testing required is reduced by deducing the remainder of the collector performance characteristic from indoor heat loss measurements. Details of the procedure are again included in the CEC recommendations (3), and the test facilities involved are similar to those discussed above. For heat loss testing the fluid is often arranged to flow from the top to the bottom of the collector to maintain appropriate temperature gradients, and indoors an artificial wind may be necessary to produce an appropriate heat loss coefficient.

The collector performance characteristic is constructed by determining an average value $(\overline{\eta}_o)$ from about four η_o measurements made outdoors, and then efficiencies are determined from heat loss measurements for a given value of irradiance, usually taken as 800 Wm^{-2} using the equation:

$$\eta = \overline{\eta}_o - \frac{\dot{m}C_p (Ti - To)}{G} \qquad (6)$$

It can be shown analytically (12) that the collector efficiency factor F' decreases as the collector heat loss coefficient increases, and hence the value of F' determined at η_o is higher than that at high values of T*. For "good" collectors this effect is small, but for "poor" collectors a variation of 5 - 10% is possible in F'. This variation unfortunately leads the mixed indoor/outdoor test method to overestimate the efficiency of "poor" collectors at high T* values, and is a serious disadvantage of the procedure.

Testing Under Variable Irradiance Conditions

One way of overcoming the problem of the lack of steady state irradiance conditions outdoors is to assume a time dependent collector performance equation, and to use data recorded at a high frequency to determine the collector performance. This approach has been adopted in the British Standard Draft for Development (11, 13).

A collector response function K(t) is assumed, which is used to provide a weighting to the solar irradiance data. This function ensures that the current irradiance at time t has a high weighting, whilst that at a previous time has a lower weighting, and at some previous time $(t - \tau)$ the irradiance can be neglected altogether.

Testing is performed in much the same way as for the steady state procedure discussed above, except that data are recorded either at a high frequency (about every 2 - 3 seconds) or as one minute integrals. Provided that the variation of transmittance with incidence angle for the collector cover is known, a wide range of diffuse and direct irradiance conditions can be used. Equivalent normal irradiance data are calculated and stored for analysis in the form of one minute averages $\overline{G}(t)$. Similarly minute averages of the collector power output $q_{out}(t)$ are determined from:

$$q_{out}(t) = |\dot{m} C_p (To - Ti)|(t) \qquad (7)$$

Finally, minute averages of the difference between the collector inlet temperature and the surrounding air temperature $(T_i - T_a)(t)$ are determined.

In general it follows that the performance equation for the collector can be written as:

$$q_{out}(t) = \int_{-\infty}^{t} \eta_o \ K(t-\tau) \ G \ (t) - U(T_i - T_a)(t)dt \qquad (8)$$

Now by holding the fluid inlet temperature constant and assuming a period of N minutes for the response time of the collector, equation (8) can be simplified to give:

$$q_{out}(j) = \eta_o \sum_{n=0}^{N} K_n G_n(j) - U(T_i - T_a)(j) \qquad (9)$$

for any point in time j.

The fluid inlet temperature needs to be constant for a period of at least N minutes to permit one set of values to be inserted into equation (8). However, since most collectors have a response time of between 5 and 15 minutes, an hour of testing at each of four fluid inlet temperatures is enough to give at least 180 j values.

A best fit to the N weighting factors K_n, η_o and U can be determined by using a computer to solve equation (8) simultaneously for a large number of j values.

The procedure adopted in reference (11) for choosing the best value of N is to iterate until a minimum is found for the standard error in the estimate U.

The method of analysis and the computer solution techniques in this procedure may appear quite complex in comparison with the simple Hottel-Whillier-Bliss equation, but once a computer programme has been written (and one is included in ref (11)), it has the great advantage of permitting collector testing on many more days during the year. Test results have also been shown to agree well with those from steady state tests.

PRESENTATION OF TEST RESULTS

The most common presentation of collector performance is in the form of an efficiency curve as shown in Fig. 6. The CEC and IEA working groups have, however, developed a comprehensive reporting format which contains rather more details of the collector, the test facility and the tests performed. This format is recommended for use by testing laboratories since it provides for all the relevant data to be presented (3).

Reference Area

One of the most important pieces of information which needs to accompany a set of collector performance results is the area to which the data have been referred. The CEC and IEA have adopted the APERTURE area for reference purposes, but others may use gross or absorber areas. There are arguments for and against any choice of reference area, so some arbitrary reference must be chosen and aperture is as good as any.

Inlet or Mean Temperature

Collector performance characteristics can be presented in terms of T_i or T_m. The CEC and IEA groups have selected T_m, mainly because it implies the use of the collector efficiency factor F' rather than the flow factor F_R. F' is less dependent on the fluid flow rate used than F_R and hence the value of η_o is less likely to vary from one laboratory to another.

Data presented in terms of T_m can be converted to data in terms of T_i using the identity:

$$\frac{F_R}{F'} = \frac{\dot{m}C_p}{F'U} \left| 1 - \exp\left(-F'U/\dot{m}C_p\right) \right| \tag{10}$$

DISCUSSION

Steady state test procedures for flat plate collectors are now quite well established. Their main drawback is that steady state conditions do not occur very frequently in many parts of the world. Most of the errors in today's steady state test results are probably caused by the irradiance measurements, but these may well improve in the near future as pyranometer calibration procedures are improved.

With more advanced collectors, such as evacuated collectors, test facilities using oils may need to be developed. These will present new problems in the determination of the measurement of heat gain, because the specific heat and density of oils vary with temperature. Incidence angle modifiers and corrections to take account of diffuse irradiance are also more complex with some of the new collectors, such as evacuated tubular collectors.

Outdoor test results are being used increasingly as a means of checking the performance of solar simulators, and this may encourage experts to specify reference values for outdoor thermal irradiance, and to try to correct all collector performance measurements to reference conditions. Such corrections are particularly

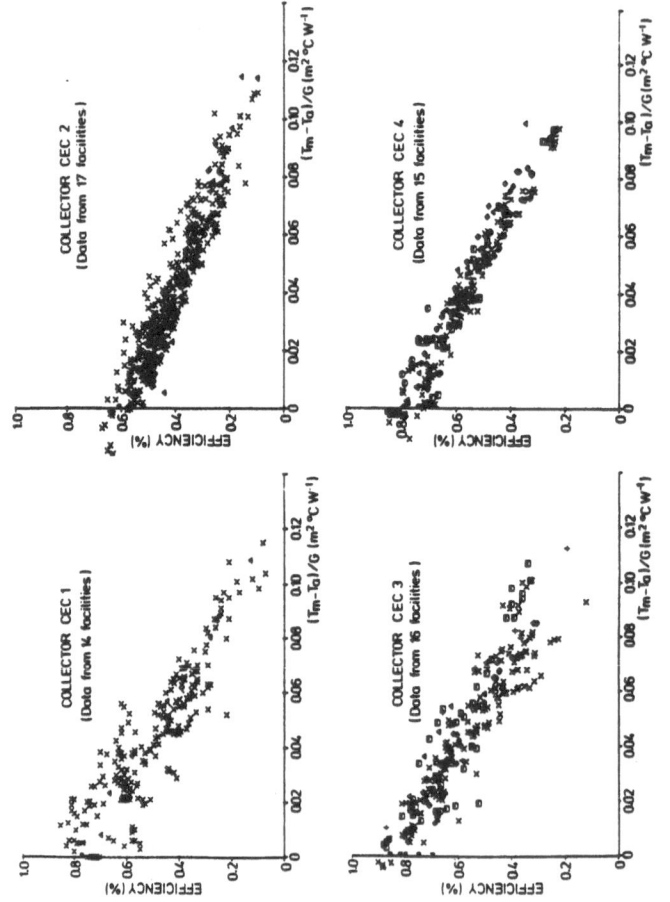

Fig. 6 CEC Round Robin Collector Test Results

important for unglazed collectors, such as are used for swimming pool heating or in combination with heat pumps. Thermal irradiance effects are discussed in detail in the lecture on solar simulators.

REFERENCES

1. Hottel, H.C. and Woertz, B.B.: *Performance of flat plate solar heat collectors.* Trans. ASME 64, 91, 1942.

2. Hill, J.E. and Kasuda, T. *Method of testing for rating solar collectors based on thermal performance.* NBS report NBSIR 74-635 Dec. 1974.

3. Derrick, A. and Gillett, W.B. (Ed): *Recommendations for European solar collector test methods.* Commission of the European Communities, DG12, Brussels, Jan. 1980.

4. Duffie, J.A. and Beckman, W.A.: *Solar Energy Thermal Processes.* Wiley 1974.

5. *Guide to meteorological instrument and observing practices.* World Meteorological Organisation 1971.

6. ASHRAE 93-77.: *Methods of testing to determine the thermal performance of solar collectors.* New York, Jan. 1977.

7. Green, A.A., Kenna, J.P. and Rawcliffe, R.W.: *The influence of wind speed on solar collector performance.* Proc. Conf. IEE. Future Energy Concepts, London 1981.

8. UNI Fascicolo n.4184.: *Collettori solari criteri e raccomandazioni. integrazione edilizia. richieste di prestazioni. metodo di prova.* (1977).

9. P 50-501.: *Mesure des performances thermiques.* AFNOR. Dec. 1977.

10. Bundesverband Solarenergie (B.S.E.): *Guidelines and directions for determining the usability of solar collectors.* May 1978.

11. DD77:1981. *Methods of test for the thermal performance of solar collectors.*: British Standard Draft for Development, London.

12. Gillett, W.B.: *The equivalence of outdoor and mixed indoor/outdoor solar collector testing.* Solar Energy, Vol. 25, pp 543-548. 1980.

13. Rogers, B.A.: *A method of collector testing under transient conditions*. Proc. Solar World Forum. ISES Congress, Brighton 1981.

PERFORMANCE TEST PROCEDURES FOR THERMAL COLLECTORS
(2) Solar Simulators

Dr. W. B. Gillett

Solar Energy Unit, University College, Cardiff, U.K.

ABSTRACT

   The design and use of solar simulators is reviewed in the
light of the experience reported by participants in the collaborative collector testing programmes of Commission of the
European Communities and the International Energy Agency. Experience with the Compact Source Iodide lamp at Cardiff is used to
illustrate the need for correcting both outdoor and solar simulator
test results to reference conditions of solar and thermal irradiance. It is suggested that further work is required on the
development of procedures for predicting typical outdoor performance from solar simulator measurements where collectors contain
new materials or complex geometries.

ACKNOWLEDGEMENTS

   Much of the experience contained in this lecture has been
gathered as a result of designing, commissioning and operating
the SRC Solar Simulator at University College, Cardiff, for
which funding was provided by the Science Research Council,
University College, Cardiff, and through the Energy Technology
Support Unit by H.M. Department of Energy.

   Experience has also been contributed through the collector
testing groups of both the Commission of the European Communities
and the International Energy Agency, and by individuals, laboratories and companies who have come to Cardiff to discuss the
problems.

G. Beghi (ed.), Performance of Solar Energy Converters: Thermal Collectors and Photovoltaic Cells, 125–146.

I would like to thank my colleagues, Professor B. J. Brinkworth, B.M. Cross, A. A. Green, K. Hayward, J.P. Kenna, J.E. Moon and R. W. Rawcliffe from the Solar Energy Unit, whose work has contributed to the contents of this lecture.

NOMENCLATURE

α	solar absorptance
$(\alpha\tau)_\theta$	angular dependent absorptance/transmittance product
$(\alpha\tau)_f$	absorptance/transmittance product for diffuse solar radiation
a_0, a_1	constants
ϵ	thermal emittance
F'	collector efficiency factor
G	solar irradiance
η	collector efficiency
η_0	collector efficiency when $T_m = T_a$
μ	collector parameter defined in Eqn.5
q	rate of heat flow
R	atmospheric thermal radiation flux
τ	cover transmittance ($\overline{\tau}$ is the mean value over the solar spectrum)
τ_0	cover transmittance at normal incidence
T_a	ambient air temperature
T_m	mean collector temperature
T^*	reduced temperature difference $(T_m - T_a)/G$
U	collector overall loss coefficient
U_b	collector back loss coefficient
U_{ca}	loss coefficient between collector cover and the environment
x	percentage of diffuse radiation in the global irradiance
σ	Stefan Boltzmann Constant

1. INTRODUCTION

In many parts of the world, and in Northern Europe in particular, steady conditions of fine weather are an infrequent occurrence, and the idea of using an artificial sun to study solar collectors is therefore attractive. For research and development purposes there are also many advantages in being able to reproduce test conditions exactly, and this is not easy to do outdoors.

One of the earliest solar simulators to be used for solar collector testing was the tungsten halogen simulator at the NASA Lewis Research Center in the USA (1). Experience with this simulator was used to provide a basis for the solar simulator test procedure included in ASHRAE 93-77 (2). The simulator appeared to work well despite its very large number of lamps

(143) and their rather short life (35-80 hrs). An upgraded version
of the NASA Lewis facility was later developed at the Marshall
Space Flight Center (3). It was also copied by the Japanese
who installed a similar simulator at Showa Aluminium in Oyama.

Early tungsten simulators were also installed at CEA and
EDF in France (4) but these employed larger power linear tungsten
lamps with a lower colour temperature and longer life. The main
problem with these simulators was their large output of thermal
(infra-red) radiation.

Perhaps the most advanced solar simulator to be used for
solar collector testing is the Xenon simulator at DFVLR in
Cologne (5). This simulator was originally used as a space
simulator; it has very sophisticated optics and a fully automatic
control system. Its beam uniformity is better than ±4%, its
spectrum can be adjusted to suit the test specifications and it
has a small climatic chamber in which the collector can be placed.
Its main disadvantage is that the test area is limited to about
1.3 m square.

New simulators are still being built in many countries around
the world. Some are constructed by commercial concerns, and are
therefore not known to other researchers, but the majority of
simulator designers are pleased to share their difficulties in
the hope of improving their facilities. A list of those facilities
that are known to participants in the collaborative collector
testing groups of the Commission of the European Communities (CEC)
(6) and of the International Energy Agency (IEA) (7) is given in
Table 1. It can be seen from this table that a high proportion
of the well known simulators employ the British Compact Source
Iodide (CSI) lamp made by Thorn Lighting Ltd., which was first
selected for the solar simulator at Cardiff in 1975. Details
of the CSI lamp are summarised by Beeson (8) and are discussed
in the light of experience with the SRC solar simulator at Cardiff
by Gillett et al (9, 10).

Radiation simulators are built for several purposes other
than for testing the thermal performance of solar collectors,
such as to test the resistance of materials to ultra-violet light,
and to determine the electrical output characteristics of photovoltaic devices. However, each application demands a rather
different design of facility. It will be seen that in the design
of a solar simulator for testing the thermal performance of
collectors, a wide range of options is available, and a fairly
versatile facility is therefore worthy of consideration.

Test methods for use in solar simulators are still to a
certain extent under development. There are three documented
procedures in existence, including ASHRAE 93-77 (2), AFNOR 50-P (11)

Country	Laboratory	Lamp	No of Lamps	Nominal test area
AUSTRALIA	CSIRO, Victoria	CSI	14	2.0 m x 2.0 m
	Univ. of Sydney	CSI	18	1.8 m x 1.2 m
BELGIUM	Polytechnique de Mons	CSI	20	2.0 m x 1.0 m
	Katholieke Univ. Leuven	CSI	36	2.0 m x 1.2 m
CANADA	Nat. Research Council, Ontario	Vortek	1	2.4 m x 2.4 m
DENMARK	Tech. Univ. Lyngby	CSI	36	2.4 m x 1.2 m
FRANCE	CETIAT, Lyon	CSI	22	2.0 m x 1.8 m
	CEA, Saclay	Tungsten	12	10.0 m x 1.5 m
	EDF, Moret-s-Loing	Tungsten		2.0 m x 1.0 m
GERMANY	DFVLR, Cologne	Xenon	10	1.1 m x 1.1 m
	Univ. of Stuttgart	Osram (I-R)	44	1.3 m x 1.3 m
	TÜV, Bayern e.V.	?CSI		(under construction)
HOLLAND	TNO-TH, Delft	CSI	27	1.8 m x 1.5 m
	Philips, Eindhoven	Xenon	16	2.0 m x 1.0 m
ITALY	JRC, Ispra (ESTI)	HQI	296	4.0 m x 3.0 m
	Zanussi, Pordenone	Tungsten		(ceased operation)
JAPAN	Showa Aluminium, Oyama	Tungsten	187	1.8 m x 1.2 m
	Tokyo Sanyo, Ohota	Tungsten	288	2.6 m x 1.2 m
	Matsushita, Kohriyama	Tungsten	360	2.0 m x 2.0 m
SWEDEN	Statens Provningsanstalt, Boras	CSI	36	2.4 m x 1.2 m
SWITZERLAND	EIR, Zurich	Xenon		
UK	Univ. College, Cardiff	CSI	19	2.0 m x 2.0 m
USA	NASA, Marshall Space Flight Center	Tungsten	405	2.4 m x 1.2 m
	Boeing, Seattle	Xenon	7	7 m^2
	Honeywell, Minnesota	Tungsten	200	1.6 m^2
	DSET, Arizona	CSI	40	(under construction)

Table 1. Some of the world's solar simulators which are used for thermal collector testing

and the British Standard Draft for Development DD77 (12), but so far neither the CEC nor the IEA group has drawn up agreed recommendations for testing in solar simulators.

2. SOLAR SIMULATOR DESIGN

The most important decision in the design of a solar simulator involves the choice of a suitable lamp, and for this a compromise has to be reached by taking account of many variables

The main constraints which are likely to vary from one laboratory to another are (a) the size and shape of the laboratory space available, (b) the electrical power supplies, (c) the range of collector designs for which testing is planned and (d) the finance available for constructing and for later development of the facility.

Before finalising the choice of lamp a number of performance factors also need to be considered, and these are discussed below.

2.1 Spectrum

The spectrum of the direct solar radiation at the Earth's surface varies with the air mass through which the radiation has passed, and with the humidity and turbidity of the atmosphere. Diffuse solar radiation has a different spectral distribution from that of the direct, so the spectrum of global solar radiation could be said to vary with the percentage of diffuse irradiance present. Typical variations in the spectrum of direct solar radiation are given by Robinson (13).

Once it has been recognised that the spectrum of solar radiation varies widely during the day, it follows that there is little point in attempting to simulate any given spectral distribution very accurately for the purposes of obtaining typical collector performance data. Indeed, for thermal collector testing, any spectrum which is a rough approximation to the solar AM 1.5 or AM 2 spectrum will give reasonable results, provided that the proportion of the total energy in the wavelength range below the cut-off of typical selective surfaces is the same in both the simulated and the solar spectrum. It is also important that the lamp output at wavelengths greater than about 2.6 μm should be very small, to prevent unrepresentative heating of collector covers.

This requires a lamp with a high colour temperature so that the "light" emitted does not extend beyond 2.5 μm, but also there should be a low heat output from the lamp cover glasses and housing. The heat output can be minimised by having a small lamp housing, or by using cooled sheets of glass which absorb thermal radiation between the lamps and the collector(14). The thermal radiation flux can also be reduced by placing the lamps a large distance away from the test plane. A typical distance to give reasonable results might be 4 m or greater, and methods for calculating the appropriate view factor are given in most text books on radiation heat transfer.

As shown in Table 1, one of the more common simulator lamps is the Thorn CSI, and its spectrum is compared with that for AM1 and AM2 solar radiation in Figure 1. Another well established

lamp is the Xenon arc, for which a typical spectrum is shown in
Fig. 2. The CSI lamp has lines in its spectrum which might give
problems if it were used to study photovoltaic or organic
materials, but these are relatively unimportant for thermal
collector testing. The peaks at about 0.9 μm in the Xenon
spectrum can be removed by filtering (5).

By convoluting radiation spectra with measured absorptance
and transmittance spectra it is possible to determine the total
absorptances and transmittances which would be measured if
materials were tested under different forms of radiation (10).
The results of such calculations for some typical solar collector materials are given in Table 2 where it can be seen that
they behave similarly under both solar radiation and light
from a CSI lamp.

		AM1	AM2	CSI
Acrylic sheet	(τ)	0.84	0.85	0.83
Glass sheet	(τ)	0.82	0.82	0.82
G.R.P. sheet	$(\bar{\tau})$	0.80	0.82	0.82
Black nickel foil	(α)	0.97	0.97	0.96
Oxidised stainless steel	$(\bar{\alpha})$	0.85	0.85	0.83
Black chrome	$(\bar{\alpha})$	0.96	0.96	0.96
Black paint	$(\bar{\alpha})$	0.95	0.95	0.95

Table 2. Mean absorptance and transmittance data for solar
          and CSI spectra

2.2 Irradiance Levels

Most simulators aim for a standard irradiance level in the
region of 800-1200 W m^{-2}, and some have the capability of varying
the irradiance level.

The minimum collector efficiency (or maximum T* value) which
can be measured in a given simulator is limited by the maximum
temperature available in the fluid loop and the minimum irradiance level available. For example, with a typical ambient
temperature of (say) 20°C, and a peak water loop operating
temperature of (say) 90°C, the maximum value of T* which can be
obtained when G = 1000 W m^{-2} is 0.07 m^2 KW^{-1}. High irradiance
levels are attractive, however, because they minimise the
errors in the many other variables measured during testing.

Some types of lamp will not operate properly at part load
and this may restrict their usefulness in solar simulators. The
CSI lamp can be reduced in power to about 50% of its radiation

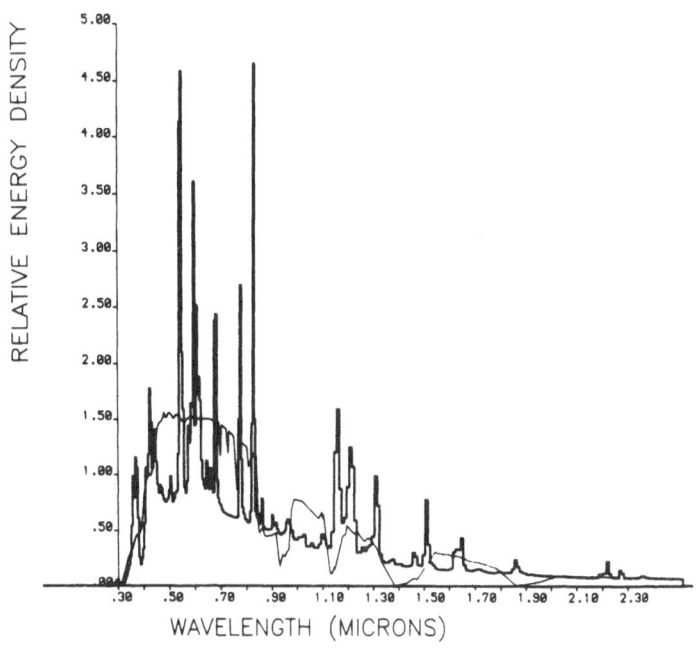

Fig. 1   Comparison between the CSI Lamp Spectrum
and Solar Spectrum

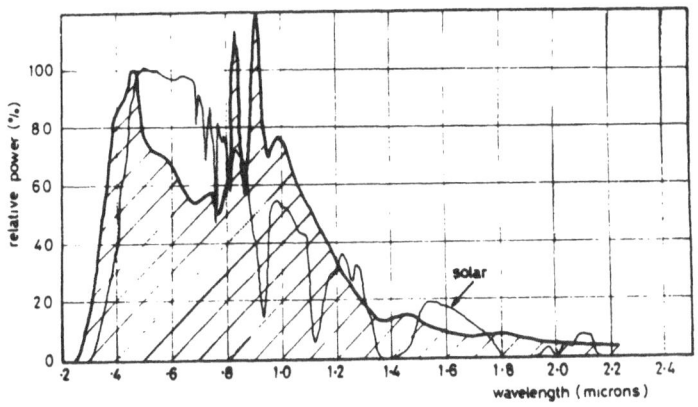

Fig. 2   Comparison between the Xenon Lamp Spectrum
and the AM2 Solar Spectrum

(∿70% of its electrical input power), without significant problems
other than a very slight shift in the output spectrum. Unfortunately this shift is detectable with a silicon solar cell and
so, if silicon cells are used for irradiance measurement, new
cell calibrations are required for each irradiance level.

In some simulators the mean irradiance level can be
altered by switching lamps off, but this may destroy the uniformity of the irradiance.

2.3 Beam Uniformity

For most flat plate collector testing, the uniformity
of the beam appears to be relatively unimportant because the
collectors operate as simple integrators. However with tubular
collectors or those with any reflecting or concentrating elements,
the beam uniformity is more important.

By using sophisticated optics, the Xenon simulator at
DFVLR achieves a uniformity of ±4%, and a similar result (<±5%)
has been deomonstrated with a CSI simulator at TNO-TH in Delft.
Such results can only be achieved with CSI lamps by electrically
trimming the power to each lamp, however, since the lamps vary
in their output by up to ±15% as a result of manufacturing
tolerances in the ballast resistors and capacitors and in the
lamps themselves. The beam at Cardiff where no individual lamp
trimming has yet been installed has a standard deviation of
about 110 W m^{-2} when the mean irradiance is 790 W m^{-2} over a
typical 2.1 m x 1.3 m test area (see Fig. 3). Very good uniformity has been reported with the Vortek simulator where a single
lamp has its beam spread by reflectors over a 2.4 m x 2.4 m test
area with a uniformity of ±5% (15).

In most simulators, however, there has to be a compromise
between uniformity and parallelism in the beam. With an array
of lamps, a good uniformity can be obtained by spreading the
beam from each lamp, but this results in a wider range of
incidence angles than may be desirable. For the CSI lamp a
range of lamp cover glasses is available to give different
amounts of beam spreading.

2.4 Beam Parallelism

The ideal beam would, of course, be almost parallel and
perfectly uniform, but this is very difficult to achieve in
practice. Analysis of collector test results becomes more
complicated as the beam divergence increases, because the transmittance of collector covers is a function of incidence angle.
For flat plate collectors, however, the effect on performance of

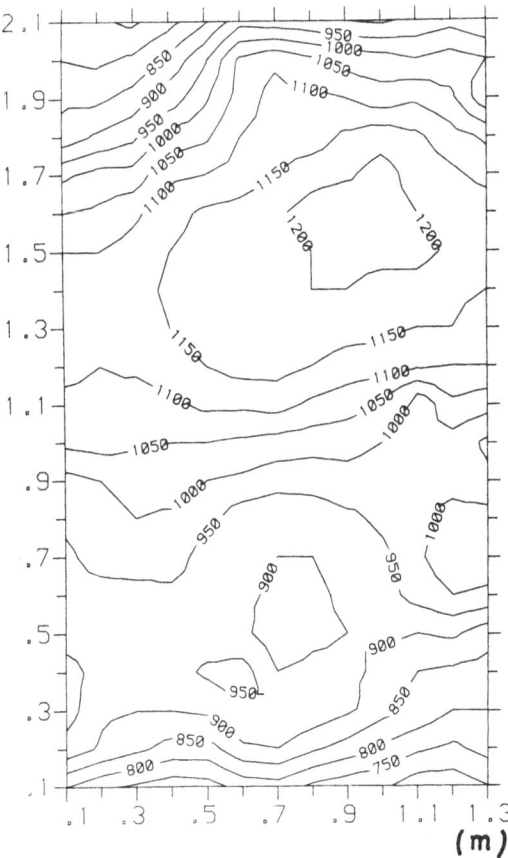

Fig. 3   Irradiance distribution in the SRC simulator
at Cardiff (Normalised to 1000)

incidence angles in the range 0 to 40° is small (see Fig. 4), so
most simulators will give reasonable results. Problems occur
when tubular or concentrating collectors are to be tested because
their reflecting surfaces are more sensitive to the incidence
angles of the irradiance.

A number of laboratories are now beginning to look seriously
for methods of measuring collector incidence angle modifiers in
solar simulators. For these measurements a diverging beam presents
two problems. Firstly there is uncertainty concerning the mean
incidence angle being measured, and secondly the non-uniformity
of irradiance over the collector increases as the incidence angle
is increased if the beam itself is divergent. This is because the
distance between the collector and the lamps is not everywhere

the same when the collector is inclined to the axis of the simulator beam. Procedures for making reliable incidence angle
modifier measurements are still under development.

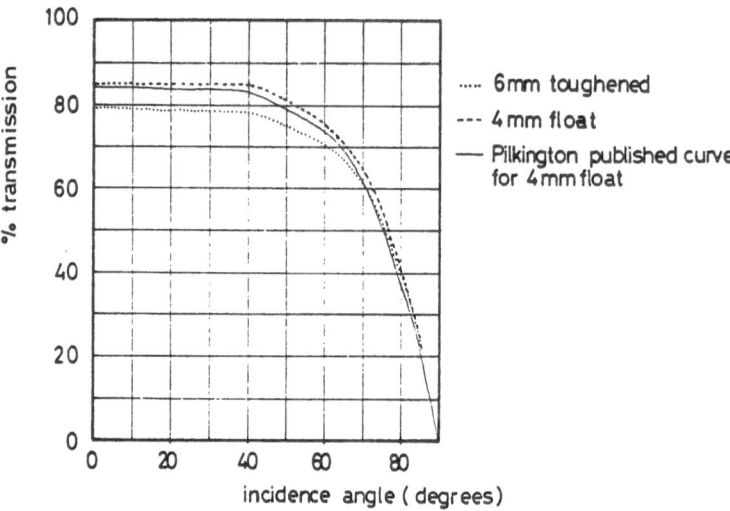

Fig. 4   Typical variation of transmittance with
incidence angle for glass

2.5   Lamp Life and Stability

The power output from most lamps decreases with age, and a
long life lamp is therefore attractive from a performance view-
point as well as for economic reasons.

The early tungsten halogen lamps used in the NASA Lewis
simulator were cheap, but their life was only about 35 hours.
The life of the Vortek lamp is also rather short (50 hrs), but
this takes only a few minutes to change, in contrast to the NASA
Lewis facility for which relamping was a fairly major task.   The
CSI lamp has a nominal life of 1000 hrs, but the first array at
Cardiff operated satisfactorily for 1400 hrs.

Stability is important in the short term because steady
state conditions for an hour or two at a time are required for
testing.   In this regard it is unfortunate that the CSI lamp

tends to amplify fluctuations in the mains voltage by up to about
40% and some measure of mains stabilisation is therefore beneficial
for CSI lamp simulators. It is also important to cool the CSI
lamp to maintain stable operation indoors, and a flow of ambient
air with a mean velocity of greater than about 1 m s^{-1} appears
to be required.

With a multiple lamp array it is important to monitor the
entire beam to accommodate variations in uniformity with time.
This is usually accomplished by regular mapping of the irradiance
over the surface of a collector under test.

2.6 Simulator Geometry

Solar simulators may be used to study building components
with vertical surfaces, such as Trombe Walls, and they may also
be used to study horizontal surfaces such as solar ponds. However,
for most of the time, simulators are used to test collectors, and
both the CEC and the IEA collector testing groups have agreed
that these should be tested at a standard inclination of 45°.

Some types of lamp are designed to operate within only a
limited range of inclination angles and to select one of these
would limit the versatility of a solar simulator. The CSI lamp
has the advantage that it will operate at any inclination
angle and in Cardiff, for example, the geometry is arranged to
provide a full 90° of variation in solar altitude.

The ability to move the lamp array to ground level, or to
be able to climb easily to the array is important for maintenance
purposes, for lamp cleaning and for characterisation of the
simulator.

3. MEASUREMENT OF SIMULATED SOLAR IRRADIANCE

Pyranometers have recently been shown by Grüter (16) and by
Riches (17) to have a number of problems relating to their calibrations, and work continues on the subject of pyranometer
calibration for outdoor measurements (18). In a solar simulator
there are additional problems with calibration since the
pyranometers must operate with more thermal irradiance than
outdoors, in a tilted position, and located over a hot collector.

Black detectors such as the conventional pyranometer are
attractive for use in simulators because they are little affected
by small variations in the colour of the simulated solar irradiance and indeed Eppley PSP pyranometers are used in several
simulators. However, where there is a non-uniform beam, the
irradiance must be mapped over the whole test area to deduce a

mean irradiance level, and this task demands a lightweight detector with a very rapid response. With the commonly used 100 mm grid, about 200 measurements are required to determine the irradiance over a typical 2 m^2 collector. The physical shape of conventional pyranometers is not ideal for use in simulators with a diverging beam, because there is a significant distance between the plane of the detector and the aperture of the collector when the pyranometer is mounted in front of the collector. This distance could introduce an error in the irradiance measurements. To minimise these problems a number of irradiance measurement techniques have been developed.

In the Xenon simulator at DFVLR in Cologne, where the beam distribution is almost constant, the irradiance needs to be mapped only rarely and can for most of the time be measured by means of a fixed cavity radiometer (5). In Denmark, the irradiance is mapped regularly by a tracking Eppley PSP pyranometer, and in Sweden and Belgium arrays of calibrated solar cells are moved over the test plane. In the early work at Cardiff a single solar cell which had been calibrated under CSI lamps against a Kipp CM6 pyranometer was used. To minimise the errors in the readings of the Kipp, the pyranometer was mounted horizontally in a polished aluminium box with forced ventilation over its front and back surfaces. Comparisons were made with the solar cell under a number of lamps, and an average value used because the cell "calibration" varied from lamp to lamp, probably because of small differences in the output spectrum from one lamp to another.

An average "calorimetric" calibration of a solar cell is now made at Cardiff in the simulator beam itself. The irradiance is mapped, using the cell, over the surface of an unglazed absorber which has a known absorptance and a high absorber efficiency. The absorber is operated with its mean temperature equal to ambient and the rate of heat collection is measured. Corrections have to be applied to take account of the thermal radiation input from the lamps, but the results show less scatter than with the previous method.

4. MEASUREMENT OF AIR TEMPERATURE AND WIND SPEED IN SIMULATORS

It is well known that collector results obtained in still air conditions show a significantly reduced heat loss coefficient because a boundary layer of warm air becomes established over the collector surface. Simulators are therefore usually fitted with an artificial wind generator. The rate of energy input to most simulator laboratories is so large that some form of air conditioning is required to maintain steady temperatures, and this is also used to mix the air in the testing zone.

Even when equipment is installed to mix the air in a solar simulator it is not unusual to find significant temperature differences between one part of the laboratory and another, and it is therefore important to select an appropriate location for measuring the temperature which will be used for analyses of collector performance. One commonly used location is in the exit from the wind generator, before the air is heated by passing over the collector surface. An alternative is to measure the air temperature beside or behind the collector by using a shaded detector. In this case it is advisable to arrange for air to be drawn over the detector by means of a small fan.

It has been suggested by Green et al (19) that an average wind speed in the range 2 to 8 m s^{-1} is required to obtain appropriate collector test results in a solar simulator. Because the wind is usually produced at one end of the collector, its velocity decreases as it passes along the collector, and one must expect a significant variation in wind speed over the collector surface. A spacial average can be determined by a series of measurements with a moving anemometer.

5. THERMAL RADIATION IN SIMULATORS

The thermal radiation flux in a solar simulator can be estimated by summing the flux from a black enclosure at ambient temperature, and the thermal radiation output from the lamps. The lamp output will depend on the lamps themselves, their mountings and the distance between the lamps and the collector.

The flux from the lamps can be calculated using radiation view factors (9) if the temperatures of the lamps and their housings are known. It can also be measured, but this is not easy because it is usually small in comparison with the simulated solar radiation. Measurements can be made using a pyrheliometer with and without glass filters, or with a pyrradiometer and a pyranometer.

The thermal radiation flux outdoors on clear days is approximately equal to 100 W m^{-2} less than that of a black enclosure at ambient temperature (20), and on overcast days the flux approaches that from a black enclosure at the ambient air temperature (see Figure 5). The low thermal radiation flux on clear days outdoors poses a design problem for solar simulators because, as can be seen from Figure 5, the temperature of the walls would need to be reduced to about 20K below ambient to simulate cold clear sky conditions, and a further reduction might be required to offset thermal radiation from the simulator lamps.

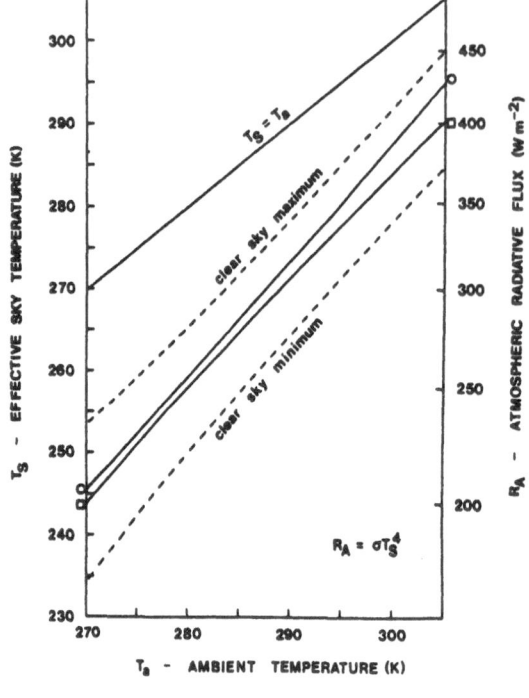

Fig. 5 Variation of atmospheric thermal radiation
flux with ambient temperature

    In practice, there are very few simulators with facilities
for cooling the surfaces in the field of view of the collector.
The CSI simulator at CETIAT in France has cooled walls; the simulator at the University of Stuttgart, which employs Osram-Siccatherm
Infra-red emitters, uses three sheets of glass with a special
cooling circuit to minimise the thermal radiation output (14);
and the simulator at CSIRO in Victoria has the real atmosphere
as its field of view, because it is used outdoors to supplement
the natural solar radiation (21). An alternative approach is
to determine the thermal radiation flux in the simulator and
to apply a mathematical correction to the measurements of collector
performance.

6. ANALYSIS OF COLLECTOR PERFORMANCE

It was shown in the previous lecture on outdoor collector testing that the steady state performance of collectors can be approximated by the Hottel-Whillier-Bliss equation as:

$$\eta = \eta_0 - a_0 T* - a_1 GT*^2 \tag{1}$$

This form of the equation accommodates a temperature dependent overall heat loss coefficient, but implies an average absorptance/transmittance produce $(\tau\alpha)$ which is independent of the percentage of diffuse irradiance and of variations in the incidence angle of the beam radiation. Equation (1) also ignores any thermal radiation exchange between the collector and its environment.

With outdoor collector testing there are quite wide variations in all the weather parameters including wind, thermal radiation and diffuse radiation, and consequently a rather simplified approach is usually taken to the presentation of outdoor test results. In solar simulators, on the other hand, the conditions are rather more reproducible, but the environment is in some respects not typical of outdoor operation. In order for solar simulation test results be of real use, they must, therefore be corrected in such a way as to indicate typical outdoor performance, and this implies a need for a set of reference outdoor conditions.

At the present time the problem of specifying appropriate reference conditions and correction procedures has not been widely studied. The ASHRAE 93-77 (2) and AFNOR P50(11) procedures assume that the problem should be entirely solved by the simulator designer. Only Draft British Standard DD77 (12) suggests analytical procedures for correcting simulator results to reference conditions.

6.1 Diffuse Irradiance Corrections

The effect of diffuse irradiance on collector performance is to change the measured value of η_0. The magnitude of the change depends on the collector design. For typical flat glass covers, the change in transmittance is approximately 1% for each 10% of diffuse irradiance present.

This problem is accommodated in DD77 (12) by presenting all data in terms of collector performance under direct irradiance at normal incidence. This procedure requires the measured direct and diffuse irradiances to be converted to a value of "equivalent normal irradiance". The correction is collector dependent, and is based on the incidence angle modifier of the collector, which can either be determined experimentally, or estimated from the design of the collector and the optical properties of its materials.

The equivalent normal irradiance G' can be determined from the measured global irradiance G, the percentage of diffuse irradiance x and the collector transmittance absorptance products using the relation:

$$G' = G \left[\frac{(\alpha\tau)_\theta - x \left[(\alpha\tau)_\theta - (\alpha\tau)_f \right]}{(\alpha\tau)_0} \right] \tag{2}$$

For example:
When a single glazed flat plate collector is tested outdoors in typical clear day conditions with 20% diffuse irradiance, the efficiency measurements will be increased by about 2% in the correction process.

If laboratories continue to present outdoor test results without any corrections for diffuse irradiance, then it must be expected that collector tests carried out in a solar simulator which employs a beam of near normal irradiance will produce collector efficiencies which are typically between 1% and 3% higher than the outdoor test results. On the other hand, solar simulators which employ a diffuse beam of radiation must be expected to produce low values of efficiency unless appropriate corrections are applied.

Whilst it is relatively easy to determine corrections for flat glass covers, the calculation of corrections for collectors employing reflecting surfaces is very much more complex, and further work is required here.

6.2 Thermal Irradiance Corrections

The effect of thermal radiation on collector performance depends on the collector design. It is most important on unglazed collectors, because in this case the exchange takes place directly with the absorber.

For glazed collectors, the effect of the thermal irradiance exchange is reduced because most collector covers have a high absorptance for thermal radiation. It can be shown (22), that the thermal radiation exchange with a glazed collector is reduced in the ratio of the overall front heat loss coefficient $(U-F'U_b)$ to the loss coefficient between the front of the cover and the environment $(F'U_{ca})$ such that:

$$q_{out} = \eta_o G - U(T_m - T_a) - \mu\varepsilon(\sigma T_a^4 - R) \qquad (3)$$

where:

$$\mu = \frac{U - F'U_b}{F'U_{ca}} \qquad (4)$$

In typical test conditions, the following approximations may be used to estimate the magnitude of the corrections for simple flat plate collectors:

(1) The cover emittance $\varepsilon \approx 0.9$
(2) The collector back loss $U_b \approx 1$ to 1.5 $Wm^{-2}K$
(3) The absorber efficiency $F' \approx 0.95$
(4) The top loss coefficient from the cover $U_{ca} \approx 25$ $Wm^{-2}K$

Hence the following values of μ may be estimated for typical collectors

single glazed selective collector $U = 4$ $Wm^{-2}K$; $\mu \approx 0.13$
single glazed matt black collector $U = 8$ $Wm^{-2}K$; $\mu \approx 0.29$
unglazed matt black collector $U = 25$ $Wm^{-2}K$; $\mu \approx 1.0$

Once the value of μ is known for a collector, then its performance can be determined for any given value of incident thermal radiation flux R. However, the main problem here is that the value of R outdoors is not usually known. It can be seen from Figure 5, that even on clear days the atmospheric thermal irradiance can vary from about 60 Wm^{-2} to about 120 Wm^{-2} below that of a black enclosure, and on overcast days this difference drops to zero as the atmosphere itself approximates well to a black enclosure.

If an arbitrary reference for outdoor conditions of 100 Wm^{-2} below that of a black enclosure is assumed, as suggested in DD 77(12), then the corrections for typical collectors in the SRC simulator at Cardiff, for example, where there is a 30 Wm^{-2} contribution from the lamps, can be computed from equation (3) for any given values of G as

$$\eta_{simulator} - \eta_{outdoor} = 0.9\mu \left| \frac{30}{G_{simulator}} + \frac{100}{G_{outdoor}} \right| \qquad (5)$$

or assuming an average irradiance of 800 Wm^{-2} as

$$\eta_{simulator} - \eta_{outdoor} = 0.146\mu \qquad (6)$$

Hence for the collectors considered above, the thermal irradiance corrections to efficiency in the Cardiff simulator are:

$$\eta_{simulator} - \eta_{outdoor} \begin{cases} = 0.019 \text{ for the single glazed selective} \\ \qquad\qquad\qquad\qquad\qquad \text{collector} \\ = 0.042 \text{ for the single glazed matt black} \\ \qquad\qquad\qquad\qquad\qquad \text{collector} \\ = 0.146 \text{ for the unglazed matt black collector} \end{cases}$$

7. COLLECTOR TEST RESULTS

It has been shown by Streed et al (23), and by Derrick (24), that efficiencies obtained in solar simulators are always somewhat higher than those obtained outdoors. Typical uncorrected results from the Solar Simulator and the Outdoor Test Facility at Cardiff confirm this trend (see Fig. 6).

However, as discussed above, solar simulator results at Cardiff are expected to be higher for two reasons. Firstly, there is less diffuse irradiance in the solar simulator, than in typical UK outdoor test conditions, and secondly there is more thermal irradiance in the Cardiff simulator than outdoors. If the two sets of results are compared on the basis of equivalent normal irradiance, then the outdoor efficiencies should all be increased by about 2 to 3%. Also, if the difference in thermal irradiance between indoors and outdoors is assumed to be 130 Wm^{-2}, then the simulator results should be reduced by between 2 and 5% depending on the collector under test. Hence the difference in uncorrected efficiency measurements could be as much as 4 to 8% for typical flat plate collectors, if outdoor test results are compared with those from a normal incidence solar simulator.

It is perhaps worthy of note here, that a simulator which produces completely diffuse irradiance may give good agreement between indoor and outdoor test results for single glazed collectors without any corrections being applied. This is simply because in this case the diffuse solar irradiance corrections and the thermal irradiance corrections are of approximately equal magnitude but of opposite sign. With other collector designs, of course, a residual correction would need to be applied.

An important advantage of solar simulator testing, in addition to that of being able to test at any time is the very high reproducibility which can be achieved (see Fig. 7). This is important for collector development when small changes in design are being studied. It is also important when investigating degradation of performance due to weathering or following some form of collector durability test.

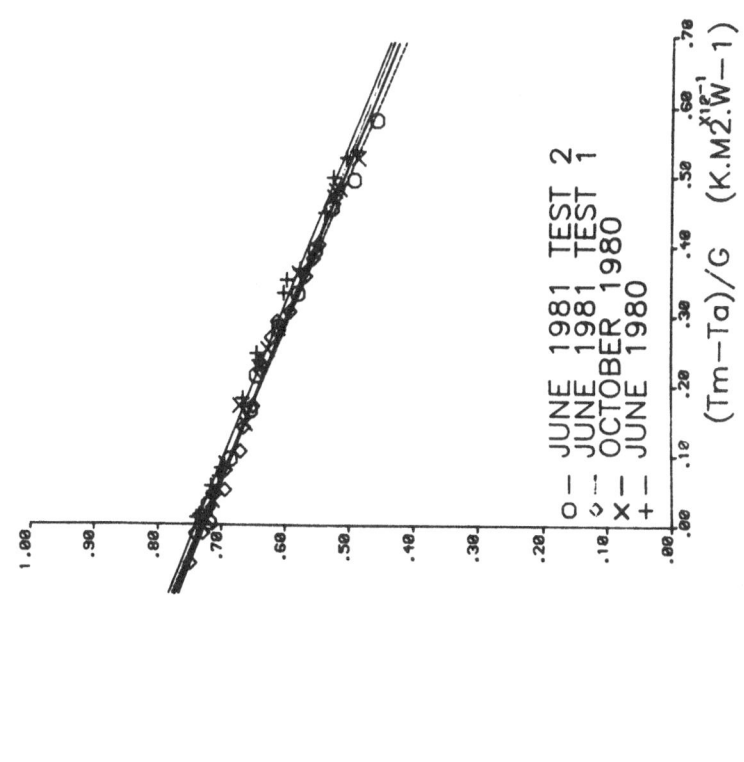

Fig. 7   Reproducibility in the SRC Solar
Simulator at Cardiff

Fig. 6   Uncorrected outdoor and simulator test
results from Cardiff

8. DISCUSSION

The understanding of solar simulators is now quite well advanced, and their development owes much to the collaborative collector testing programmes of both the CEC and the IEA. More than 20 simulators are now in use for solar collector testing, and in some countries the advantage of being able to test at any time of year has resulted in a preference for simulator testing over outdoor testing.

Despite the obvious attractions of solar simulators, however, it seems likely that outdoor test facilites will also need to be retained for some years to come. New materials such as thin film plastic covers, and new designs such as evacuated tubular collectors still pose a problem to the solar simulator user, and indoor/outdoor comparisons are still required to help with the development of appropriate procedures for predicting typical outdoor performance from measurements made in solar simulators.

The next major improvement in collector testing accuracy will come when better calibration procedures have been established for pyranometers. This will need to be followed by the development of simple equipment for mapping the irradiance in solar simulators. However, when the accuracy of outdoor test results has been improved it will become even more apparent than it is today, that collector performance should be corrected to reference environmental conditions.

Most system models which employ a collector performance characteristic, assume that the characteristic applies to normal incidence irradiance, and they adjust the collector efficiency during the day using an incidence angle modifier. It therefore seems appropriate that collector efficiency measurements should be presented in terms of equivalent normal irradiance conditions.

A reference condition for thermal irradiance is more difficult to choose. Furthermore, system models do not usually include variations in atmospheric thermal irradiance, and thermal irradiance conditions outdoors are seldom measured. A somewhat arbitrary choice must therefore be made in order that typical outdoor collector performance can be correctly predicted from efficiencies measured in solar simulators. The British Standard Draft for Development suggests a reference condition of 100 Wm^{-2} below the flux emitted by a black enclosure, but as an average condition this is perhaps a little low. Further work is required here, preferably by an international collaborative group, to identify suitable reference conditions for general use.

9. REFERENCES

1. Yass, K. and Curtis, H.B. *Low cost air mass 2 solar simulator*, NASA TM X-3059 (June 1974).

2. ASHRAE 93-77 *Methods of testing to determine the thermal performance of solar collectors* (1977) ASHRAE, 345 East 47th Street, New York, NY 10017.

3. Humphries, W.R. *Use of the Marshall space flight centre solar simulator in collector performance evaluation*. DOE/NASA TM-78165 (April 1978)

4. Derrick, A. *European solar collector testing activities and test methods*. Final report. EUR 6185 Vol.1 (1980) Commission of the European Communities.

5. Ley, W. and Rehmann, K.V. *Investigation of selective surfaces and honeycomb structures with the use of a solar simulator*. DFVLR Report No. 1B 353 77-15 (Dec. 1977) DFVLR, Postfach 906058, D-5000 Koln 90, W. Germany.

6. Moon, J.E. and Gillett, W.B. *Collector Testing in the European Communities*. Proc. Solar World Forum ISES Congress, Brighton UK (Sept. 1981).

7. Ley, W. *Survey of solar simulator test facilities and initial results of IEA round robin tests using solar simulators*. IEA Task III report (Dec. 1979). DFVLR D-5000 Koln-90, W. Germany.

8. Beeson, E.J.G. *The CSI lamp as a source of radiation for solar simulators*. Lighting Research and Technology, Vol. 10 No. 3 (1978)

9. Gillett, W.B., Rawcliffe, R.W. and Green, A.A. *Collector testing using solar simulators*. Proc. Conf. UK-ISES (C22) London (1980)

10. Gillett, W.B. and Kenna, J.P. *Recent developments in solar simulators for collector testing*. Proc. Solar World Forum, ISES Congress, Brighton, UK (Sept. 1981).

11. AFNOR pp 50-51. *Capteurs solaires. Mesure des performances thermiques*. Association Francaise de Normalisation (1977).

12. DD77:1981 *Methods of test for the thermal performance of solar collectors*. British Standard Draft for Development.

13. Robinson, N. *Solar Radiation*. Elsevier (1966).

14. Kraus, K., Hahne, E and Sahns, J. *Laboratory tests for flat plate solar collectors*. SUN II, Proc. ISES Congress, Atlanta, Vol. 1, pp 385-389, (May 1979).

15. Camm, D.M., Richards, S.L.F. and Albach, G.G. *A Large Area Solar Simulator Construction and Performance*. Proc. Solar World Forum, ISES Congress, Brighton, UK (Sept. 1981).

16. Grüter, J.W.(Ed). *Proceedings of workshop on the accuracy of pyranometers*. Commission of the European Communities, DG12, Brussels (Sept. 1980).

17. Riches, M.R. and Stoffel, T.L. *International Energy Agency Conference on Pyranometer Measurements*. Boulder, Colorado (March 1981)

18. Talarek, H.D. *IEA Round Robin Testing of Solar Collectors*. Proc. Solar World Forum, ISES Congress, Brighton 1981.

19. Green, A.A., Kenna, J.P. and Rawcliffe, R.W. *The influence of wind speed on solar collector performance*. Proc. Conf. IEE Future Energy Concepts, London (1981).

20. Green, A.A. and Gillett, W.B. *The significance of longwave radiation in flat plate solar collector testing*. Proc. IEE Conf. Future Energy Concepts, London (Jan. 1979).

21. Gillett, W.B. *The equivalence of outdoor and mixed indoor/ outdoor solar collector testing*. Solar Energy <u>25</u>, pp 543-548 (1980).

22. Streed, E., Waksman, D., Dawson, A. and Lunde, A. *Comparison of solar simulator and outdoor ASHRAE 93 thermal performance tests*. Proc. AS.ISES meeting, Phoenix, Arizona, June 2 - 6 (1980)

23. Derrick, A. *Results and analysis of the fourth round robin solar collector in the CEC programme*. Commission of the European Communities, DG12, Brussels (1980).

PHOTOVOLTAIC ENERGY CONVERTERS

W. H. Bloss

Institut fuer Physikalische Elektronik
Universitaet Stuttgart
D-7000 Stuttgart, F. R. Germany

INTRODUCTION

Utilization of solar energy and conversion into electrical
energy is based on two different methods:
    a) Thermodynamic conversion. Hereby solar radiation is ab-
sorbed and converted into heat at elevated temperatures. In a
second step electrical energy is generated by means of heat
engines or thermoelectric devices.
    b) Photovoltaic conversion. This procedure represents a
direct conversion process of solar radiation into electricity
and is realized by means of solar cells.

The thermodynamic conversion process, limited by the temperature
difference between absorber and environment, utilizes the total
intensity of the incoming solar radiation independent of its
spectral distribution. Photovoltaic conversion process due to
the quantum electronical transitions involved is sensitively
affected by the spectral distribution of solar radiation.

The energy of a photon ($E = h\nu$) by interaction with matter is
transferred in a single step process to one electron. In semiconductors hereby electron-hole-pairs are created and the
potential energy of pair production can be converted into
electrical energy. The direct interaction of photons and
electrons leads to a thermal equilibrium between radiation
and electron gas in solids and provides in principle high
efficiencies based on the radiation temperature of the sun.

G. Beghi (ed.), Performance of Solar Energy Converters: Thermal Collectors and Photovoltaic Cells, 147–162.
Copyright © 1983 ECSC, EEC, EAEC, Brussels and Luxembourg

PHOTOVOLTAIC CONVERSION

The spectral distribution of solar radiation at Air Mass 1
(AM 1) is shown in figure 1. The intensity related to a small
spectral interval expressed in terms of photon energy is plotted
as a function of photon energy. This diagram exhibits the
potentials of photovoltaic conversion if the photon energy is
related to the characteristic energy terms of a semiconductor.

If the photon energy exceeds the energy band gap E_g of a semiconductor, band-band-transitions of an electron lead to an
absorption process. Only the high energetic part of the spectrum
above the limit E_g indicated by hatching in the diagram for
different semiconductors is available for the conversion process.

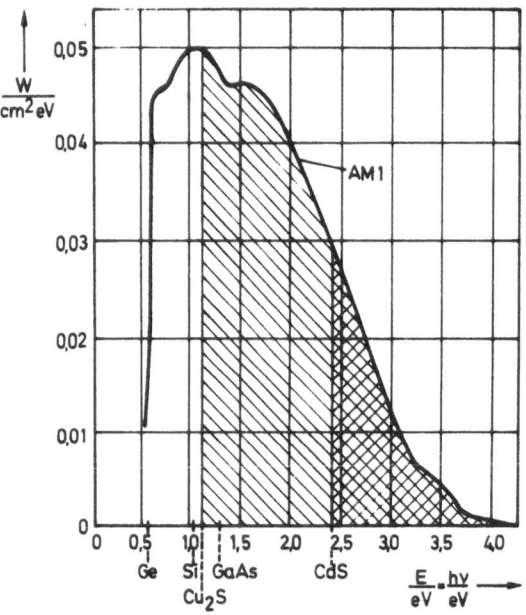

Figure 1. Spectral distribution of solar radiation
at Air Mass 1 (AM 1) in terms of photon
energy $h\nu$.

Figure 2 shows the energy band diagram of an intrinsic semiconductor with a band gap $E_g = E_c - E_v$. The interaction photon-
electron leads to an energetic transition of the electron to E_1.
The resulting intraband transition to E_c is dissipated by
phonon interaction as heat which exhibits an inherent loss
mechanism in the conversion process. The mobile charge carriers,
electron and hole, separated by the energy band gap E_g and

existing for a limited life time, represent by their energetic
separation the potential energy which can be converted into
electrical energy. E_g defines the minimum photon energy of the
spectral radiation distribution (fig. 1) which allows conversion
into electrical energy.

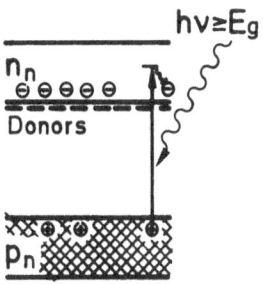

Figure 2. Energy band diagram of an intrinsic
semiconductor; $(E_1 - E_c)$ defines the
inherent thermal loss mechanism.

RADIATION ABSORPTION

The interaction of radiation with different - semiconducting -
materials is characterized by the absorption coefficient α.
The variation of α with photon energy is shown in figure 3 for
different materials which have been used in photovoltaic
conversion processes. The absorption edge is defined by the
energy band gap. The slope of the absorption curves (in a
logarithmic scale) is extremely high for semiconductors with
direct transitions, "direct semiconductors", whereas the so-
called indirect semiconductors show absorption curves with
much smaller slopes.

Due to the exponential relation

$$I(Z) = I_o (1 - r) e^{-\alpha z} \qquad (1)$$

where I_o is the incoming radiation, r the reflection coefficient
at the surface, and z the direction of penetration, the absorption coefficient implies the condition for the thickness of
photovoltaic generators. Let αz_s be in the order of 5, which
expresses that within the thickness

$$z_s = L_\alpha = \frac{5}{\alpha}$$

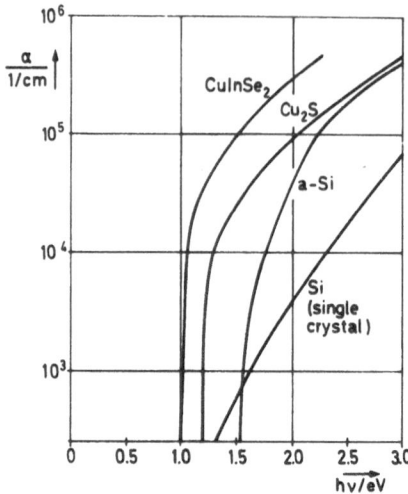

Figure 3. Absorption coefficient α of different
semiconducting materials vs. photon energy.

more than 99 % of the radiation is absorbed, then the absorption depth describes the minimum thickness of the base
material for photovoltaic generators which can be reduced by
the use of reflecting back surfaces. Whereas crystalline Si
cells require values of thickness ranging between 50 and 100 μm,
cells consisting of direct semiconductors or amorphous Si can
be made with thicknesses below 1 μm. The use of direct semiconductors implies a substantial reduction of material and
energy consumption for the production process.

JUNCTIONS

The photovoltaic conversion process implies in a first step
the generation of electron-hole-pairs as mobile charge carriers
and in a second step the separation of these pairs by electric
fields, which originate from junctions between semiconductors
or between metals and semiconductors as demonstrated in
figures 4 and 5.

The energy conversion process is mainly performed by the
minority carriers in the semiconducting materials. The migration
of the charge carriers is described by diffusion processes if
junctions of the type shown in figure 4 are used. During the
life time τ, which is sensitively affected by impurities which
provide recombination transitions, the diffusion length is

$$L = \sqrt{D\,\tau} \;, \tag{2}$$

where D is the diffusion coefficient of the charge carrier type under consideration.

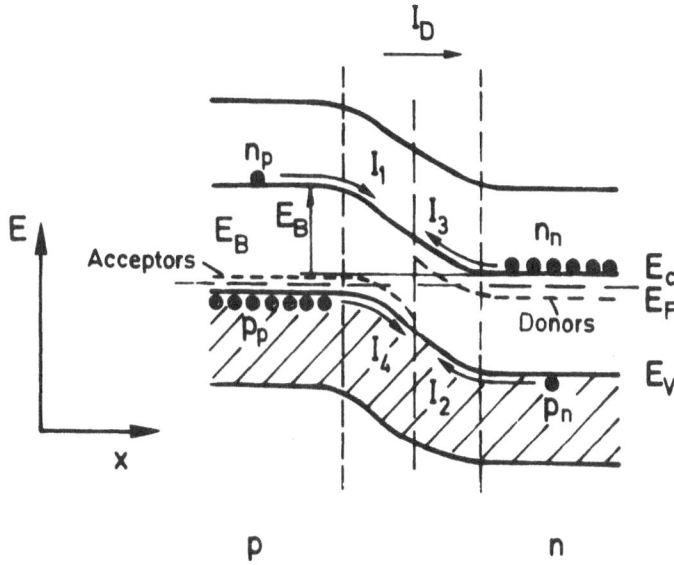

Figure 4. Energy band diagram of a p-n-junction
(homojunction) at equilibrium (V = 0)
indicating the charge carrier types
and the resulting currents.

By comparison with the absorption mechanism - the absorption length L_α defines the minimum thickness - an efficient energy conversion process requires that the diffusion lengths of the electrons and holes

$$L_e + L_h \geq L_\alpha \;, \tag{3}$$

that all charge carriers created by absorption can be collected at the junction and fully contribute to the output current.

Figure 5 shows the junction between a metal exhibiting a high work function $q\Phi$ and n-type semiconductor. Band bending in the semiconductor describes the space charge region which results in an extended electrical field by which minority charge carriers (holes) are collected. With this Schottky-type diode configuration the absorption processes are limited to the

semiconductor side. The motion of the holes i.e. the resulting
photoelectric current is described by combined diffusion and
field effects.

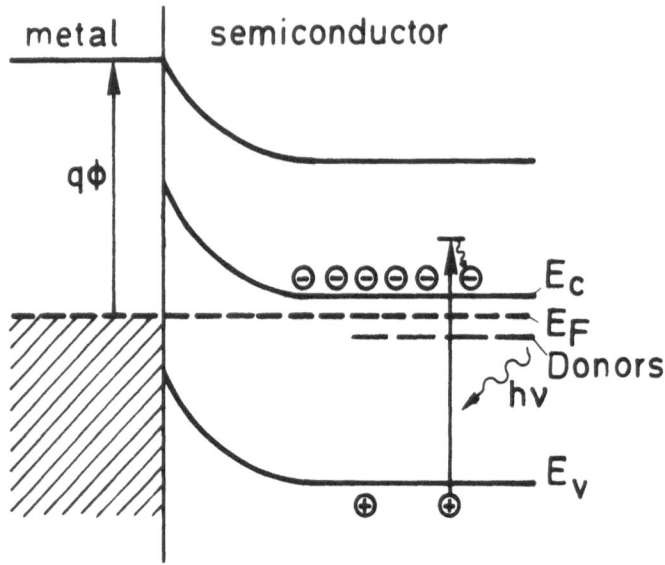

Figure 5. Schottky-type junction between a metal
and a n-type semiconductor.

The performance of Schottky diodes is considerably improved by
introducing thin (40 Å) insulating films between metal and
semiconductor. Hereby detrimental effects of interfacial states
on the energy band configuration and on the transport mechanism
and lifetime of charge carriers can be reduced.

The resulting current composed of minority and majority carriers
through a homojunction equals zero in the equilibrium condition
i.e. without illumination. The I-V-characteristic of the
junction is expressed by

$$I = I_{sp} (e^{\frac{qV_D}{AkT}} - 1) - I_{ph} \qquad (4)$$

and is shown in figure 6.

By illumination the additive term I_{ph} is introduced which expresses the increase of the current of minority carriers

generated by photon interaction.

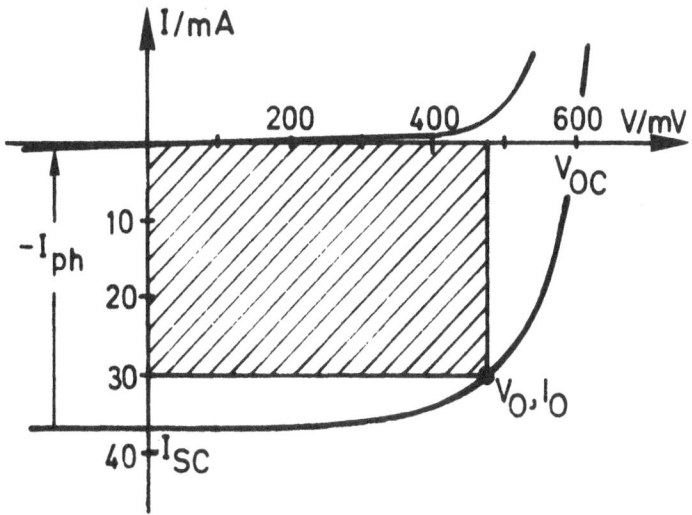

Figure 6. I-V-characteristic of a homojunction (Si).

The intersections with the axes yield the short circuit current I_{sc}, which is directly proportional to the illumination intensity, and the open circuit voltage V_{oc}, which shows logarithmic dependance of the illumination intensity. The point of maximum power output (V_o, I_o) is defined by

$$P_{el\ max} = I_o V_o = (FF) \cdot V_{oc} \cdot I_{sc} , \qquad (5)$$

where FF is the filling factor under optimum operating conditions. The analog circuit diagram is demonstrated in figure 7a.

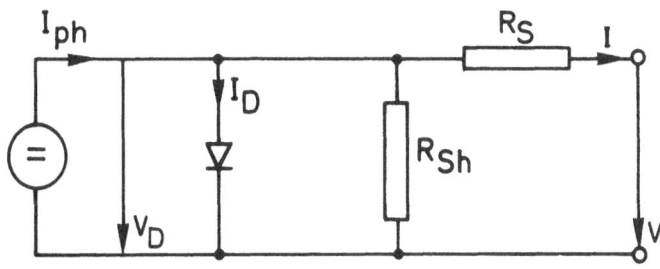

Figure 7a. Analog circuit diagram of a
photovoltaic generator.

The analog circuit also describes the effects of temperature
increase and limitations of the series resistance on the performance characteristics of photovoltaic generators. The diode
current

$$I_D = I_{sp} \; (e^{\frac{qV_D}{AkT}} - 1) \tag{6}$$

is sensitively affected by variations of temperature, both by
the dependance of I_{sp} from temperature and the exponential
diode term. Figure 7b demonstrates the effect of temperature
on the I-V-characteristics. This general behaviour imposes
limitations on the operating temperature which is defined by
the semiconductor properties.

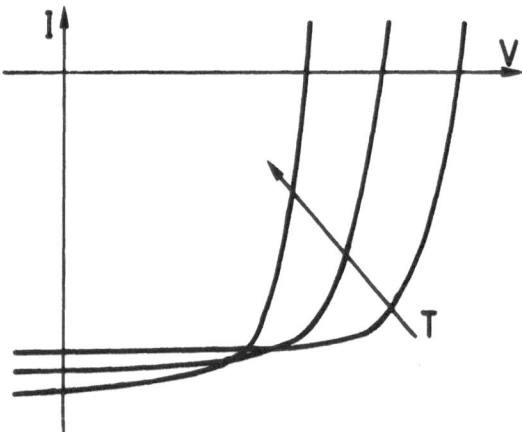

Figure 7b. The influence of operating temperature T
on the I-V-characteristic of a photovoltaic generator.

The series resistance R_S which is influenced by the geometry
and electrical properties of the contacts as well as by the
conductivity of the semiconducting materials exhibits an important influence on the I-V-characteristic, which is schematically shown in figure 7c.

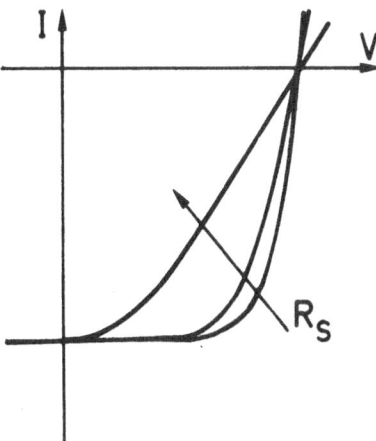

Figure 7c. The influence of a series resistance R_S
on the I-V-characteristic of a photovoltaic generator.

EFFICIENCY

The efficiency of photovoltaic energy conversion is defined as
the ratio of maximum electrical output power to the incoming
radiation. It essentially defines the economy of converter
systems. Research efforts are directed towards the increase of
the efficiency.

Maximum electrical power output is achieved under optimum
operating conditions as indicated in figure 6:

$$P_{el} = (FF) \cdot I_{sc} \cdot V_{oc} \, . \tag{7}$$

The current I_{sc} is directly proportional to the effective
radiation and can be expressed by

$$I_{sc} = q \int_0^\infty P_\nu \cdot Q_\nu \, d\nu \tag{8}$$

$$\nu = E_g/h$$

where q is the electronic charge, P_ν the spectral distribution
function of incoming radiation and Q_ν the collection efficiency
of minority charge carriers.

The open circuit voltage V_{oc} is related to I_{sc} by

$$V_{oc} = \frac{AkT}{q} \ln \left(\frac{I_{sc}}{I_{sp}} + 1 \right) \qquad (9)$$

and shows, by comparison with the above definition if I_{sc}, logarithmic dependance of the radiation.

The efficiency therefore can be expressed by

$$\eta = \frac{(FF) \cdot V_{oc} \cdot q \displaystyle\int_{\nu=E_g/h}^{\infty} P_\nu \cdot Q_\nu \cdot d\nu}{\displaystyle\int_{\nu=0}^{\infty} P_\nu \cdot d\nu} \qquad (10)$$

Calculated values of the photovoltaic conversion efficiency for different semiconducting base materials are plotted in figure 8. By use of a single material efficiencies above 20 % can be obtained.

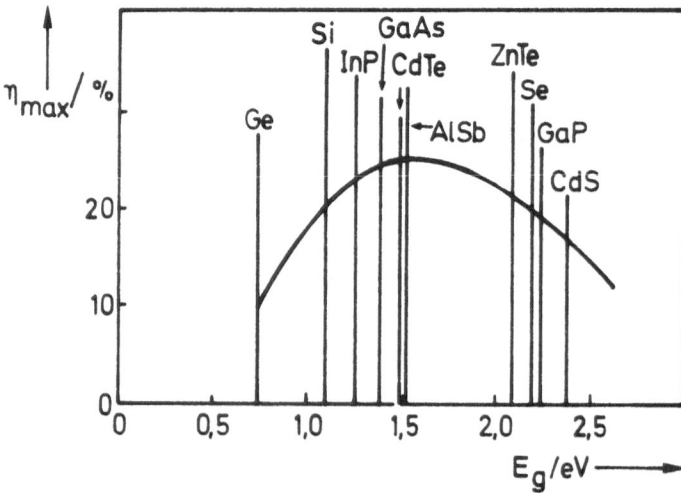

Figure 8. Calculated photovoltaic conversion
efficiency vs. energy band gap of
the semiconductor.

A further increase of the conversion efficiency can be achieved
by a) concentrating devices and b) tandem systems. By the use
of appropriate mirror or lens systems direct solar radiation
can be focused and the radiation intensity can be increased by
factors up to several thousand. Ideal photovoltaic generators
exhibit, due to the linear dependence of I_{SC} and logarithmic
dependence of V_{OC} as a function of radiation intensity, increased efficiency at highly concentrated illumination. Limitations are given by the increase of temperature, which reduces
the conversion efficiency, and by the series resistance R_s. The
conversion efficiency as a function of the concentration factor
K is given in figure 9.

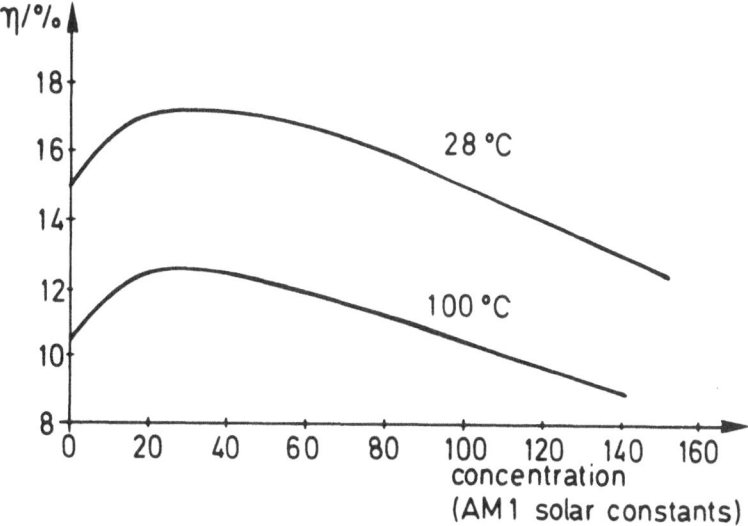

Figure 9. Conversion efficiency of a photovoltaic
generator as a function of concentration
factor K with operating temperature as
a parameter.

High concentration factors demand well adjusted optical systems
and tracking devices. The applications are limited to regions
with high values of direct solar radiation.

Tandem systems represent combinations of different semiconductor
materials to provide better matching to the solar spectral
distribution. By means of spectral filters or mirrors the
radiation, mainly in concentrating systems, is splitted into
two or more spectral ranges and focused on different photovoltaic generators as indicated in figure 10. By this procedure
efficiencies up to 30 % have been realized. Tandem systems
generally provide much higher conversion efficiencies than

conventional systems based on one semiconducting material.

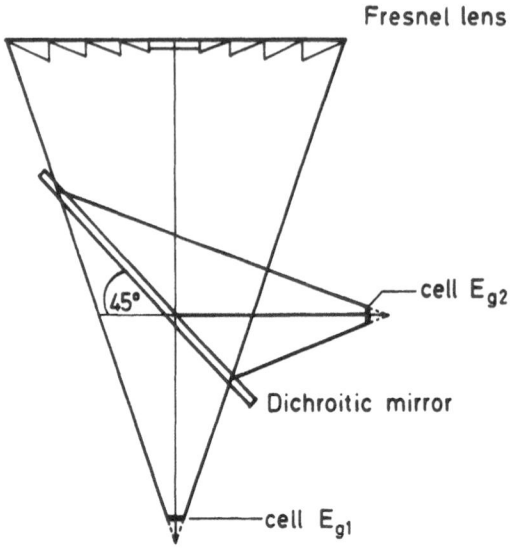

Figure 10. Concentrating system with a Fresnel lens
and a dichroitic mirror as a beam splitter.
Optimum conversion is obtained in 2 generators with different band gaps E_{g1}, E_{g2}
matched to the radiation spectrum.

STRUCTURES OF SOLAR CELLS

Figure 11a shows a typical structure of an encapsulated solar
cell. The conversion process takes place in the semiconductor,
where the absorbed photon creates an electron and a hole. These
two types of carriers are separated by the internal field between two glass sheets.

Crystalline Si-solar cells, whose structure is shown schematically in figure 11b are usually made by diffusing a thin (some
1000 Å) n^+-layer into a p-type Si wafer. Both back and front
contacts are made by evaporation, although there are some
other methods as silk-screen printing, electroplating etc. under
consideration. In the production process of thin film solar
cells the glass on the backside of the encapsulated solar cell
is usually used as a substrate, on which the different films –
back contact and semiconducting films – are evaporated. The
front contact can be fabricated either on the semiconducting
film on the top or on the coverglass. In the second case,

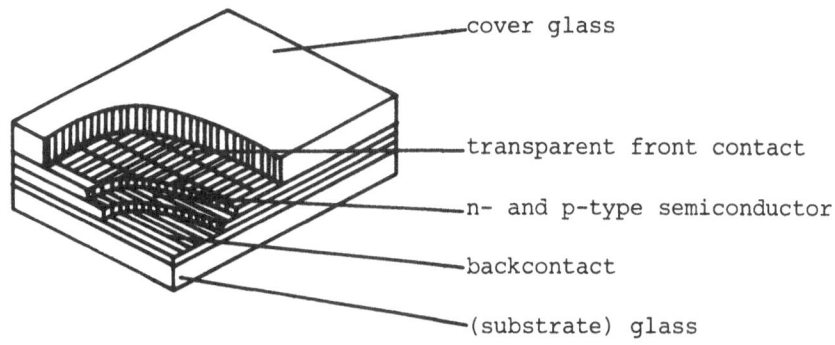

Figure 11a. Schematic structure of an
encapsulated solar cell.

encapsulation and contacting is performed in the same process.
The schematic structure of a Cu_2S-CdS thin film solar cell,
which has been prepared in the way described above, is shown
in figure 11c.

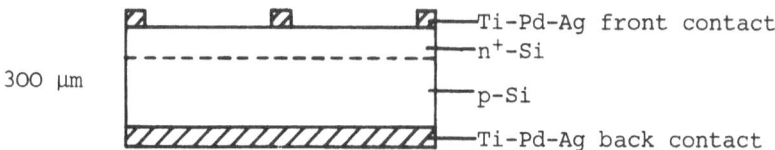

Figure 11b. Schematic structure of a
crystalline Si solar cell.

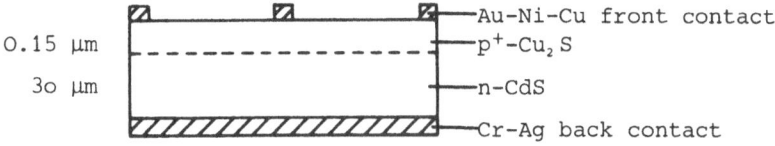

Figure 11c. Schematic structure of a
Cu_2S-CdS thin film solar cell.

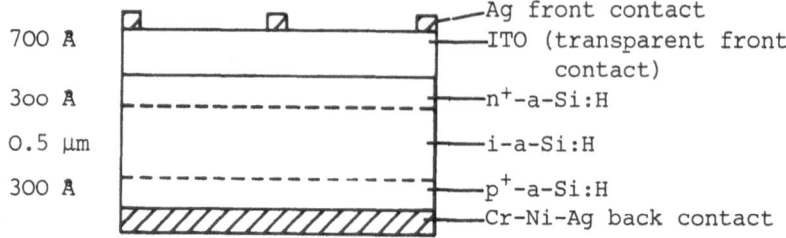

700 Å		Ag front contact
		ITO (transparent front contact)
3oo Å		n^+-a-Si:H
0.5 μm		i-a-Si:H
300 Å		p^+-a-Si:H
		Cr-Ni-Ag back contact

Figure 11d. Schematic structure of an amorphous
p^+-i-n^+-Si:H thin film solar cell.

Another type of thin film solar cells can be realized on the
basis of hydrogenated amorphous Silicon a-Si:H. The structure
of that solar cell is shown in figure 11d. The glass on the
backside is used as a substrate. The back contact is fabricated
by evaporation and the a-Si:H films are produced by an rf- or
dc-glow discharge or by sputtering. The n^+- and p^+-films are
deposited using PH_3 or B_2H_6 respectively as doping additives
to the sputter gas. The ITO transparent film on top of the
various a-Si films is used as an antireflective coating and
transparent front contact. To lower the series resistance of
the solar cell, some bus bars are put on top of the ITO-film.

MATERIALS FOR SOLAR CELLS

Single and polycrystalline Si solar cells are by far the most
outstanding photovoltaic generators which also provide large
scale production and application. Generators with peak power
output of 300 kW_{el} are under construction. Efficiencies range
from 12 to 15 %.

Alternative approaches are directed towards the use of polycrystalline semiconductors which would require considerably
less material and energy for cell production. Thin film solar
cells based on CdS-Cu_2S are considered as a promising candidate
for photovoltaic generators. Efficiencies of 8 - 10 % appear
realistic.

More recent investigations on amorphous Si solar cells have
demonstrated that this type would meet the requirements of low
cost and large scale production. The maximum efficiencies
reported range between 6 and 7 %.

The following table summarizes the data of some cell types
which are considered for economic photovoltaic power generation.
Prospectives for further developments based on minimum prices

per peak Watt electricity are sketched in figure 12 according
to evaluations of the Department of Energy, USA. The diagram
indicates that polycrystalline Si material could meet the
economic requirements, however, new technologies, for instance
thin film solar cells, could provide economic large scale
solutions as well.

Economic evaluations consider the price goal of 0.5 $/$W_p$ as a
break even point for competitive power generation.

Table 1.

	η
Single crystalline Si	12-15 %
Poly-crystalline Si	10-12 %
Single crystalline GaAs	16-20 %
Thin film CdS-Cu$_2$S	8-10 %
Thin film CdS-CuInSe$_2$	8-10 %
Amorphous Si	6- 8 %

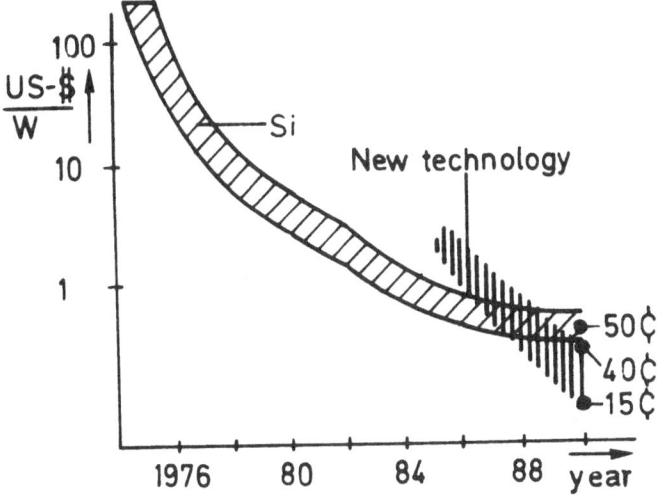

Figure 12. Estimated future price development
for photovoltaic generators
(Department of Energy, USA).

Apart from this economic analysis photovoltaic generators,
even at much higher prices, present a highly useful and desirable tool for a variety of applications for remote stations
mainly in developing countries.

PERFORMANCE TEST PROCEDURES FOR SOLAR CELLS AND MODULES

F.C. Treble

Consultant Engineer, Farnborough, Hants., UK

ABSTRACT

    After some preliminary background information on the
performance characteristics of photovoltaic solar cells,
recommended standard procedures for the outdoor measurement of
the rated or 'peak' performance of cells and modules are
described. These procedures include the selection and outdoor
calibration of the reference cells or modules which are used to
monitor the irradiance in terms of standard sunlight of a
specified spectral irradiance distribution.

    Details are also given of methods of measuring the relative
spectral response of solar cells, their temperature coefficients
of current and voltage and their internal series resistance.

    Definitions of some relevant meteorological terms are given
in an Appendix.

1.   INTRODUCTION

    The main problem in photovoltaic performance measurement
arises from the fact that solar cells have a highly selective
spectral response and are therefore very sensitive to the
spectral irradiance distribution of the incident radiation.
Outdoors, this varies considerably with location, weather, time
of year and time of day. Indoors, it depends on the type of
simulator used and can change as the equipment ages. Unless
measurement procedures take account of these variations and
other difficulties, such as the marked temperature dependence of

G. Beghi (ed.), Performance of Solar Energy Converters: Thermal Collectors and Photovoltaic Cells, 163–187.
Copyright © 1983 ECSC, EEC, EAEC, Brussels and Luxembourg

solar cells, the results can be grossly erroneous. Indeed, performance ratings of the same module by different laboratories have been known to differ by as much as 20%.

Uncertainties of this magnitude cannot be tolerated at any stage in the development, design and manufacture of photovoltaic equipment. They can mislead R&D workers, distort cost-per-watt comparisons between competing manufacturers and result in over- or under-designed arrays. Discrepancies between manufacturer's and customer's measurements can lead to costly and time-wasting arguments and investigations. Standard measurement procedures are essential if these problems are to be avoided.

The procedures described in this paper are based on recommendations by the author (1), which are now embodied in a European Standard Specification (2). Properly carried out, they should reduce measurement discrepancies to no more than ± 2%. They cover outdoor testing, the measurement of spectral response, temperature coefficients and series resistance and the outdoor calibration of reference cells and modules. Although the procedures are intended for flat-plate modules, similar principles apply to concentrator modules.

We start with some background information on the performance characteristics of solar cells.

2.   CELL CHARACTERISTICS

2.1 Voltage-current characteristic

Fig. 1 shows the voltage-current characteristic of a typical modern silicon solar cell at 25°C in sunlight at an irradiance of 1000 Wm^{-2}. The short-circuit current is proportional to the irradiance and the active area of the cell. As the load resistance or voltage is increased, the current at first stays fairly constant and then falls to zero at an open-circuit voltage of about 0.6 V. The maximum power delivered is represented by the area of the largest rectangle that can be fitted under the curve - in this case 12.7 $mWcm^{-2}$ at a voltage of 0.48 V. The squareness of the characteristic depends on the quality of the junction and the internal series resistance of the device. A measure of the squareness is the 'fill factor', which is defined as :-

$$\text{Fill factor} = \frac{\text{Maximum output power}}{\text{Short-circuit current x open-circuit voltage}}$$

In this case, it is about 0.75.

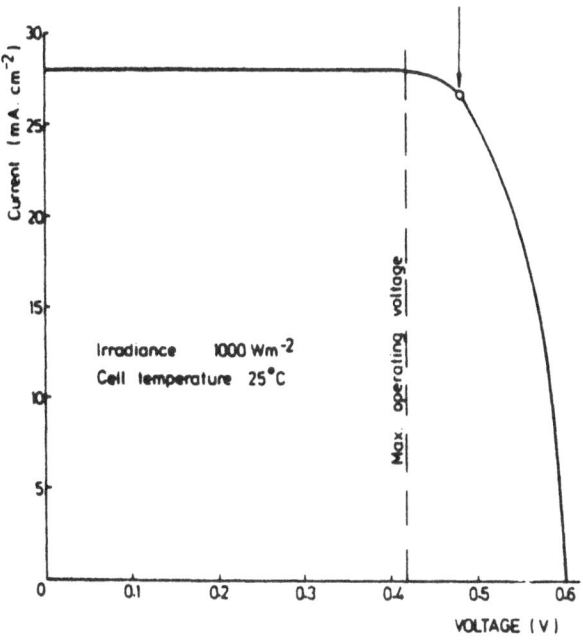

Max power = 12.7 mW/cm^2 at 0.48 V
Conv. Effy = 12.7 %

Irradiance   1000 Wm^{-2}
Cell temperature   25°C

Fig. 1   Voltage-current characteristic

The conversion efficiency is the maximum output power
expressed as a percentage of the input power, thus :-

$$\text{Conversion efficiency} = \frac{\text{Max. output power}}{\text{Irradiance x Area}} \times 100\%$$

The area used in this calculation is the entire frontal
area of the cell or module. In the case of cells, it includes
the area of the upper contact grid and, in the case of modules,
the frame and any protruding mounting lugs.

Fig. 2 illustrates how the characteristic is affected by
temperature. An increase from 25°C to 60°C causes a slight rise
in short-circuit current but a sharp fall in open-circuit
voltage and maximum power. The voltage at which maximum power
is delivered also falls.

Typical temperature coefficients of short-circuit current
(I_{sc}), open-circuit voltage (V_{oc}) and maximum power (P_{max}) for
a 10 ohm.cm silicon cell are :-

$$\frac{dI_{sc}}{dT} = + 0.01 \ \text{mAcm}^{-2}\text{degC}^{-1}$$

$$\frac{dV_{oc}}{dT} = - 2.20 \ \text{mVdeg C}^{-1}$$

$$\frac{dP_{max}}{dT} = - 0.5\% \ \text{deg C}^{-1}$$

Fig. 3 shows the effect of a change of irradiance from 1000 Wm^{-2} (bright tropical sunlight) to 200 Wm^{-2} (overcast conditions). The short-circuit current falls linearly but the open-circuit voltage decreases as the logarithm of the irradiance and so is little affected. A similar effect may be observed as the angle of incidence of the direct solar beam is increased. The short-circuit current, following the irradiance, varies very nearly as the cosine of the angle of incidence, so the lower curve corresponds to an angle of incidence of $\cos^{-1}$ 0.2, that is about 78.5°.

2.2 Spectral response

The short-circuit current is composed of increments produced by photons of varying wavelengths within the effective range. If the incremental current generated by unit irradiance is plotted as a function of wavelength, the resulting curve is called the 'absolute spectral response'. Fig. 4 shows a typical example for a silicon cell. Note that it covers the visible and near infra-red parts of the spectrum, peaking sharply at about 0.85 μm.

If the ordinates of the absolute spectral response s_λ are multiplied by the corresponding ordinates E_λ of the spectral irradiance distribution of the incident radiation and integrated, the result is the total generated (short-circuit) current I_{sc}, thus :-

$$I_{sc} = \int s_\lambda . E_\lambda . d\lambda$$

The measurement of absolute spectral response is fraught with difficulties but, for most performance measurement purposes, it is sufficient to measure this parameter in relative terms. (See Section 7).

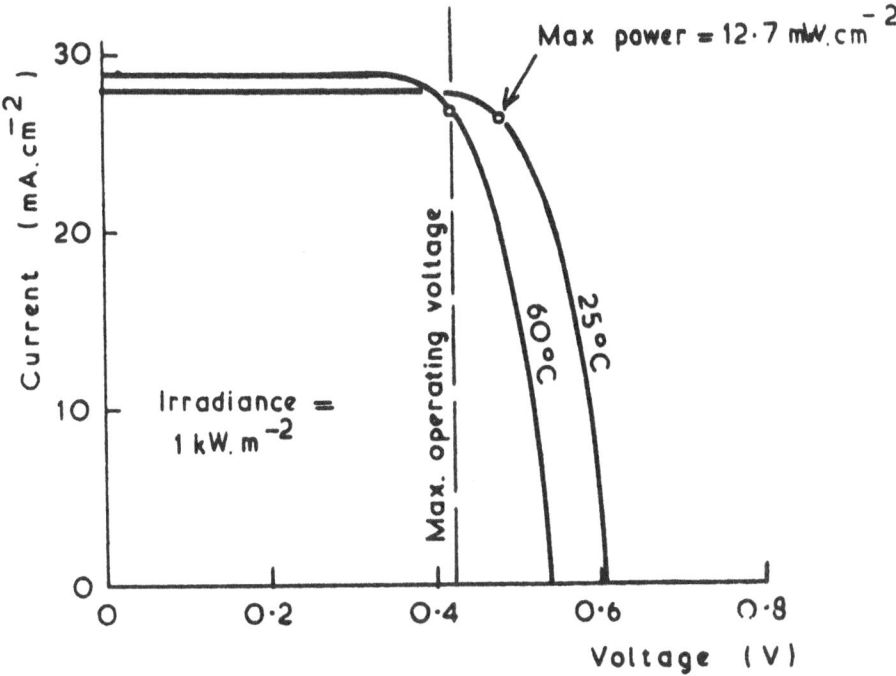

Fig. 2 Effect of temperature on V-I characteristic

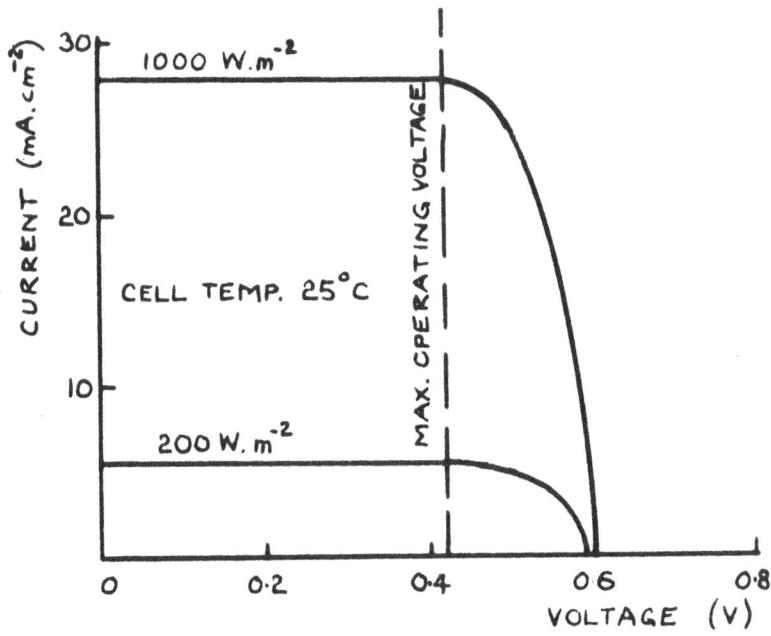

Fig. 3 Effect of change in irradiance

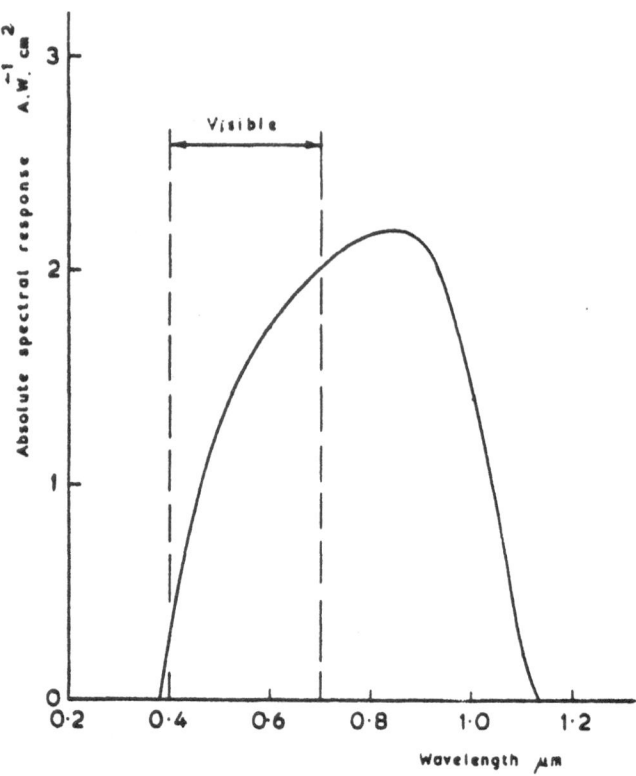

Fig. 4   Spectral response

3.   BASIC METHOD OF PERFORMANCE MEASUREMENT

In current practice , the photovoltaic performance of a solar cell is determined by exposing it at a known temperature to non-fluctuating sunlight, natural or simulated, and tracing its voltage-current characteristic while at the same time measuring the incident irradiance. For performance rating, the measured performance is then corrected to standard conditions of irradiance and temperature. The rated power, sometimes referred to as the 'peak' power, is the power output at the nominal working voltage under those standard conditions.

In production testing, it is sometimes the practice to measure only the short-circuit current, open-circuit voltage and current at the nominal working voltage instead of tracing the complete characteristic.

Discrepancies due to spectral variations in the irradiation are minimised by relating the measured performance to a standard solar spectral irradiance distribution, which we shall refer to

as 'standard sunlight'. This is done by measuring the irradiance
with a reference solar cell or module which has been specially
calibrated in terms of short-circuit current per unit of standard
sunlight irradiance $(AW^{-1}m^2)$. The reference cell or module is
selected to have essentially the same relative spectral response
as that of the equipment to be tested. So, unlike a thermopile-
type pyranometer, which has a flat response, it automatically
corrects for variations in spectral distribution. Because of this,
location and weather conditions are not critical when the reference
cell method is used outdoors, nor is the type of simulator so
critical in indoor measurements. Moreover, as the irradiance
monitor and the test cells have the same time constant, the
radiation does not have to be so stable as when using a pyranometer, with its much slower response.

If the performance of a photovoltaic generator is related in
this way to standard sunlight and its relative spectral response
is known, its performance in light of any other known spectral
composition can be computed within a reasonable tolerance.

4.   STANDARD SUNLIGHT

Fig. 5 shows three possible candidates for adoption as the
standard sunlight distribution. The first is Parry Moon's curve
for direct sunlight at air mass 1 (AM1), which has been adopted
by the International Electrotechnical Commission for their solar
radiation tests on electrical equipment. The second is ERDA/NASA's
AM1.5 direct sunlight distribution (3), derived from the Labs and
Neckel AM0 data by applying revised Thekaekara attenuation factors
to allow for atmospheric scattering and absorption. This has been
adopted as a provisional photovoltaic standard by the European
Commission. The third curve, for global sunlight at AM1, is based
on actual field measurements made by the Royal Aircraft Establishment, UK in Malta. Opinion in the US as well as in Europe is
turning towards a global distribution, as it is more representative of flat-plate conditions. Note that the effect of the sky
radiation is to make the light considerably stronger in the
violet/blue wavelengths.

5.   STANDARD TEST CONDITIONS

The present universal standard of irradiance for photovoltaic
performance rating is 1000 Wm^{-2}, although there is a body of
opinion in the US which considers that this should be changed to
800 Wm^{-2}, as being nearer the maximum value which most arrays
experience in operation.

The standard temperature for the performance rating of solar
cells, in common with most other electronic devices, is the IEC

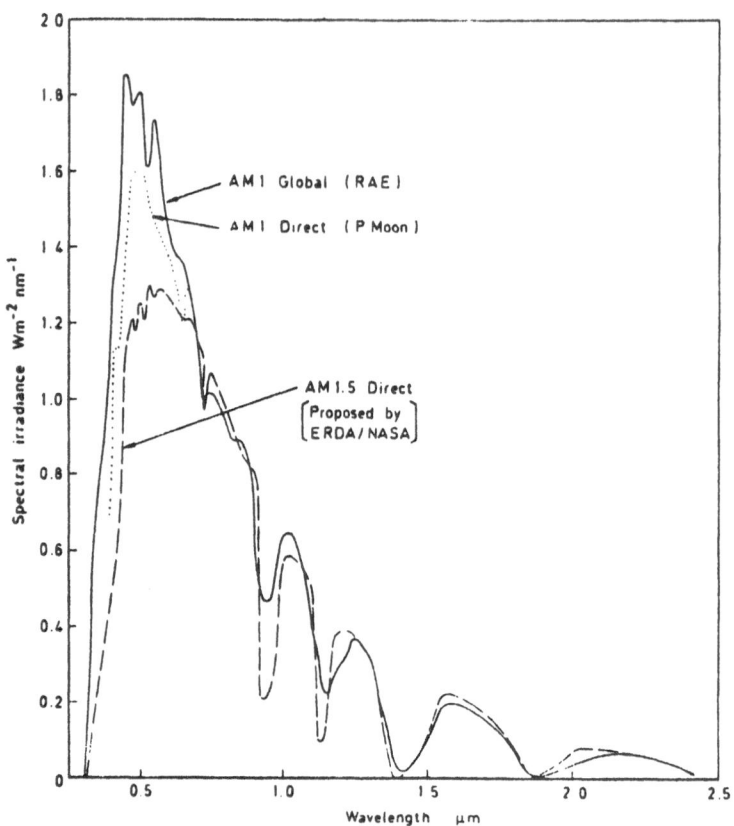

Fig. 5 Solar spectral irradiance distributions

standard of 25°C. Some American companies used to use 28°C but
are now conforming to the lower figure. As the standard temperature is lower than the temperature at which most arrays stabilise in sunlight, there is a move towards establishing a 'nominal
operating cell temperature' (NOCT) under certain specified
conditions and quoting performance at this more realistic temperature as well as the standard level for the guidance of system
designers.

6.   MEASUREMENT IN NATURAL SUNLIGHT

Ideally, performance measurements should be carried out in
natural sunlight only when the total irradiance is at least 800 Wm^{-2}
and is stable enough not to fluctuate by more than ± 1% in the
course of a single measurement. Measurements may be taken at
lower irradiances but in this case the errors in extrapolating the
readings to 1000 Wm^{-2} are likely to be higher.

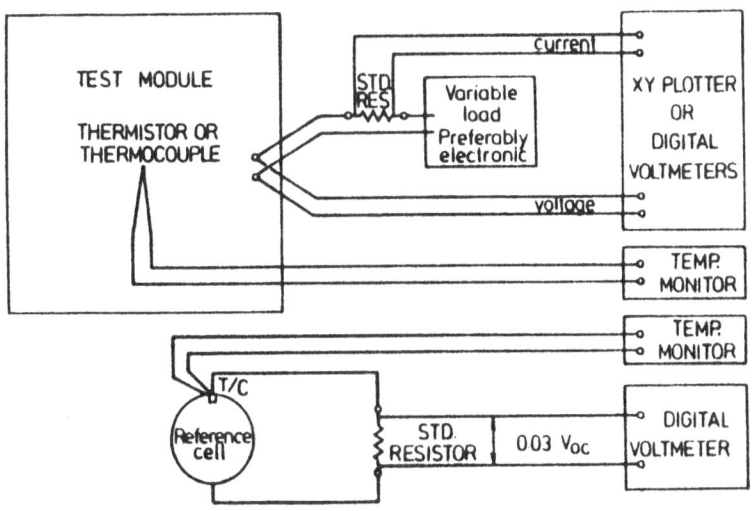

Fig. 6   Performance test connections

The test procedure is as follows :-

6.1 Mount the test cell or module near to the reference solar cell or module so that their active surfaces are co-planar and normal to the direct solar beam, within $\pm$ 10°. Connect to the instrumentation as shown in Fig. 6. Detailed mounting and instrumentation recommendations are given in Sections 13 and 14. The entire test sample and reference cell or module should be fully irradiated by the direct sunlight and no collimators should be used. Reflected sunlight falling on the active surfaces should be kept to a minimum.

6.2 Ensure that the temperature of the reference cell is 25 $\pm$ 2°C. When testing a single cell, maintain its temperature at the same level.

6.3 Record the voltage-current characteristic and temperature of the test sample at the same time as the short-circuit current and temperature of the reference cell. In the case of a module, where it is not practicable to control the temperature, shade it from the sun and wind before the measurement until it is at ambient air temperature and then carry out the measurements immediately after removing the shade. In most cases, the thermal inertia of the assembly will limit the temperature rise during the first few seconds to less than 2°C and its temperature will remain reasonably uniform.

6.4 Correct the measured voltage-current characteristic to Standard Test Conditions by applying the following equations :-

$$I_2 = I_1 + I_{sc}\left[\dfrac{I_{sr}}{I_{mr}} - 1\right] + \alpha(T_2 - T_1)$$

$$V_2 = V_1 - R_s(I_2 - I_1) - KI_2(T_2 - T_1) + \beta(T_2 - T_1)$$

Where :

I_1, V_1	are co-ordinates of the measured characteristic,
I_2, V_2	are the corresponding co-ordinates of the corrected characteristic,
I_{sc}	is the measured short-circuit current of the test sample,
I_{mr}	is the measured short-circuit current of the reference cell or module,
I_{sr}	is the short-circuit current of the reference cell or module in standard sunlight at 1000 Wm^{-2},
T_1	is the temperature of the test sample,
T_2	is the standard temperature ($25^\circ \pm 2^\circ C$),
α and β	are the current and voltage temperature coefficients of the test sample for standard sunlight at 1000 Wm^{-2} (β is negative),
R_s	is the internal series resistance of the test sample and
K	is a curve correction factor.

Procedures for the measurement of α and β are given in Section 8 and a method of determining R_s is described in Section 9. The curve correction factor K may be obtained from voltage-current characteristics measured on the test sample at different temperatures. A typical value for silicon cells is 1.25×10^{-3} ohm.deg C^{-1}. With a cell giving, for example, a current of 950 mA at maximum power, this gives an additional voltage transformation of about 1 mV.deg C^{-1} per cell at the maximum power point, i.e. a change of about 2% for $(T_2 - T_1) = 10^\circ C$. When $(T_2 - T_1)$ is under $2^\circ C$, as it should be if the specified procedure is followed, the curve correction term in the voltage equation can generally be ignored.

A procedure similar to that described in 6.4 may be used for transposing the measured voltage-current characteristic to other irradiances and temperatures, e.g. NOCT. It is usual nowadays for manufacturers' data sheets to show module characteristics at three irradiances and three temperatures.

7.   MEASUREMENT OF RELATIVE SPECTRAL RESPONSE

The relative spectral response of a solar cell is measured by placing it on a temperature-controlled mount, irradiating it uniformally from a monochromatic source and measuring the short-circuit current and the irradiance at fixed wavelength intervals over the response rage. The currents are then divided by the irradiances or a proportional parameter and plotted as a function of wavelength. Alternatively, the irradiance may be kept constant (for instance, by varying the width of a monochromator exit slit) in which case the relative spectral response is obtained directly from the current readings.

The irradiance monitor may be a vacuum thermocouple, a pyro-electric radiometer or other suitable detector. Another alter-native is a reference solar cell whose spectral response, covering the required range, has been pre-calibrated by a recognised standards laboratory. In this case, the spectral response of the test cell is computed as follows :-

$$s_{t\lambda} = s_{r\lambda} \cdot \frac{I_{mt\lambda}}{I_{mr\lambda}}$$

where : $s_{t\lambda}$ is the spectral response of the test cell at wavelength λ ,

  $s_{r\lambda}$ is the spectral response of the reference cell at the same wavelength,

  $I_{mt\lambda}$ is the measured short-circuit current of the test cell at wavelength λ ,

  $I_{mr\lambda}$ is the measured short-circuit current of the reference cell at the same wavelength.

Fig. 7 and 8 show two possible test arrangements, the first embodying a quartz prism monochromator and the second a filter wheel as the monochromatic source. In both cases, the light source is a 1000 W tungsten halogen lamp operated from a stable supply at a colour temperature of 3200 K. The test cell and irradiance monitor are mounted on opposite sides of a rotatable

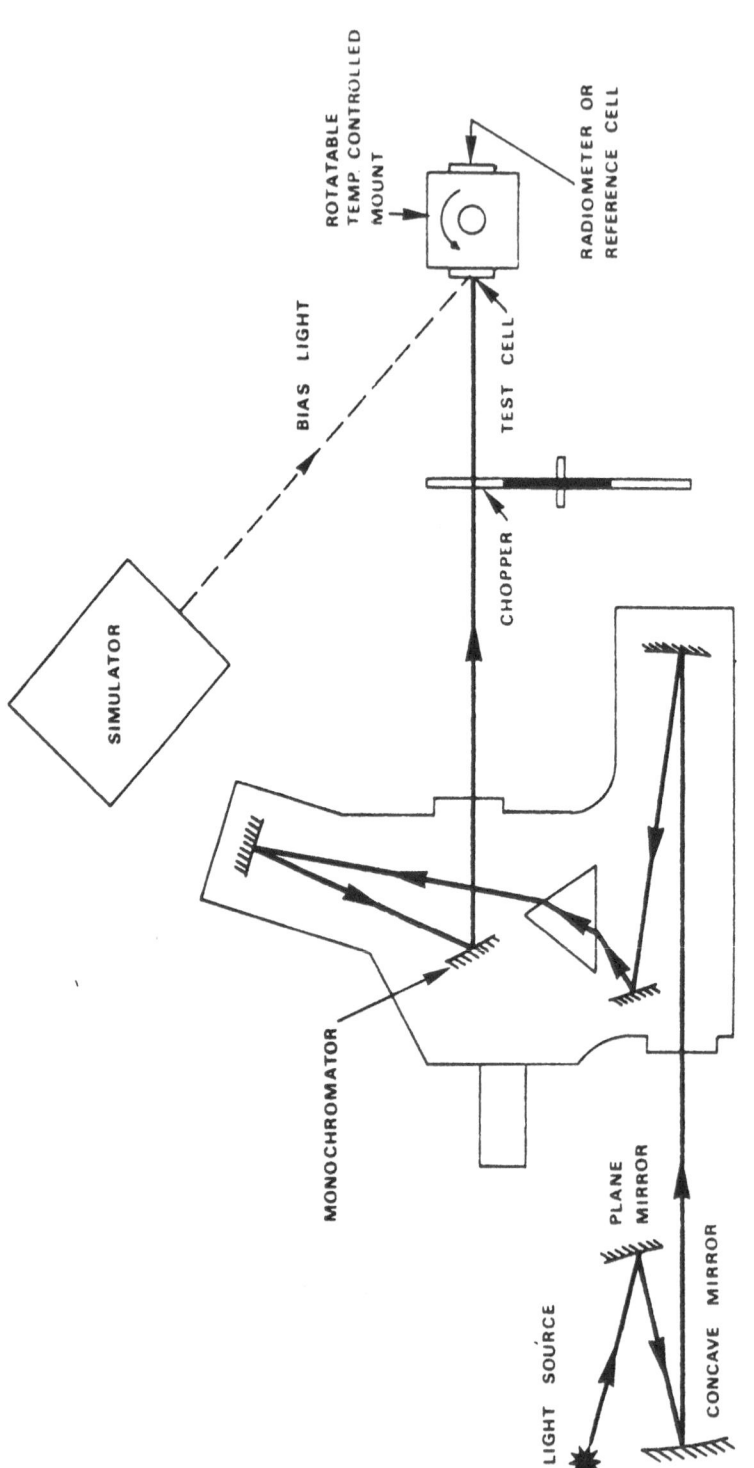

Fig. 7 Spectral response measurement using a monochromator

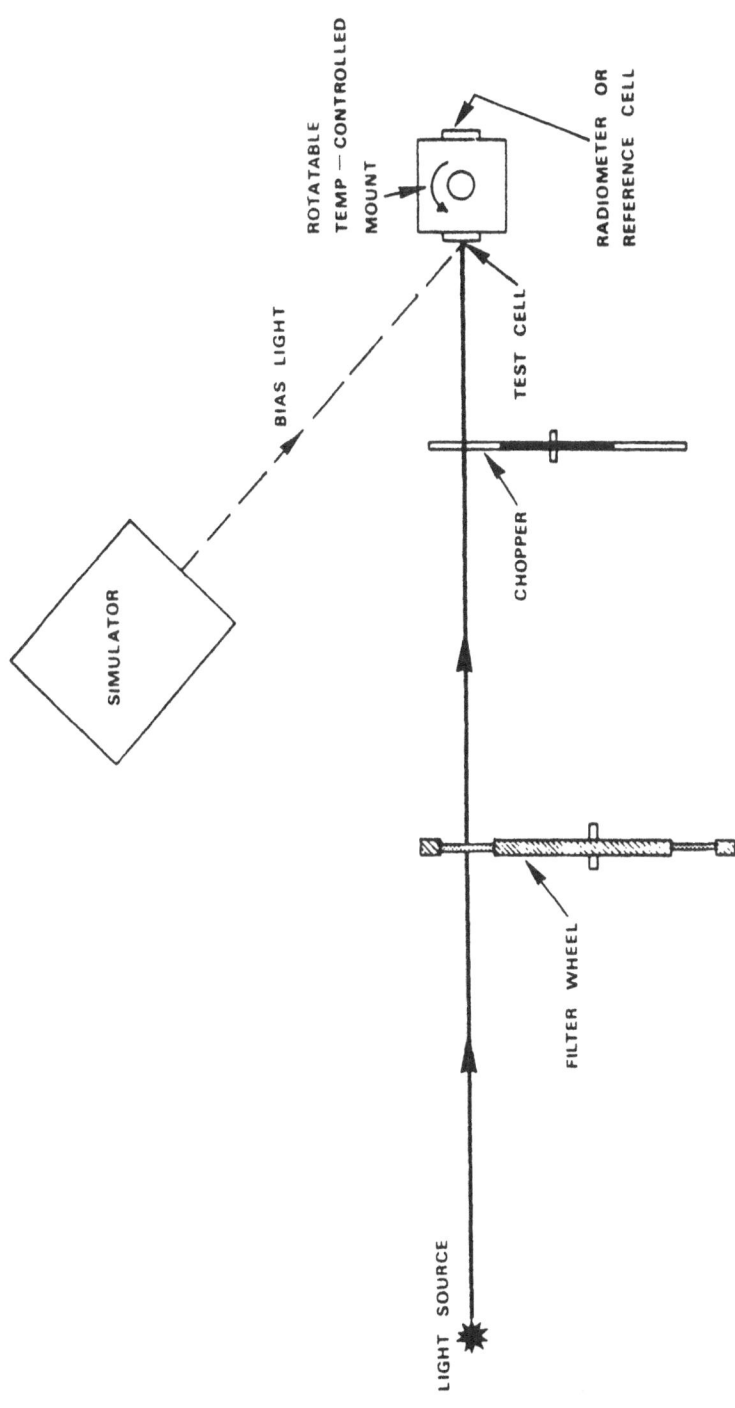

Fig. 8 Spectral response measurement using a filter wheel

temperature-controlled block, so that either may be presented to the monochromatic beam in precisely the same place. Alternatively, they may be mounted on a slide with suitable positioning stops.

The filter wheel carries seventeen narrow-band filters centred at wavelengths ranging from 350 nm to 1150 nm in 50 nm steps and arranged so that each can be indexed in turn between the light source and the test cell or irradiance monitor. It is important to ensure that the filters have negligible (under 0.2%) side-bands. The monochromator is normally used with fixed slits and manually set to the same wavelength steps.

With crystalline silicon and other cells, where the response may be assumed to change linearly with irradiance, the short-circuit current of the cells (voltage drop across a standard 4-terminal fixed resistor) and the open-circuit voltage of the vacuum thermocouple or radiometer may be measured directly with a dc digital voltmeter or potentiometer. The requirements for instrumentation accuracy and the measurement of short-circuit current laid down in Section 14 apply here. If the dc method is used, the exit beam, test cell and irradiance monitor must be completely enclosed in an antireflective light-tight box and meticulous precautions taken to avoid thermal and other random emfs which would cause errors. Alternatively, the exit beam may be chopped at a low frequency and the output voltage amplified and rectified. In this case, it is important to ensure that the amplifiers are linear and drift-free.

With non-linear devices such as cuprous sulphide-cadmium sulphide solar cells it is necessary to use a chopped monochromatic beam and increase the irradiance to the desired operational level (e.g. 1000 Wm^{-2}) by unmodulated bias light from a suitable steady-state simulator.

8. MEASUREMENT OF TEMPERATURE COEFFICIENTS

The temperature coefficients of current (α) and voltage (β) vary with irradiance and, to a lesser extent, with temperature. However, for the purpose of correcting voltage-current character-istics to Standard Test Conditions, the values corresponding to an irradiance of 1000 Wm^{-2} and temperatures of up to $20^{\circ}C$ above ambient are appropriate. The coefficients are best measured in simulated sunlight, using a single representative solar cell of the same area and configuration as those in the relevant module. The procedure with steady-state simulators is as follows :-

8.1 Attach a suitable lightweight temperature sensor to the front contact of the test cell.

8.2 Mount the test cell on a temperature-controlled block with vacuum hold-down device, using the attached sensor to provide the control signal.

8.3 Position the test cell and a suitable reference solar cell side-by-side, with their active surfaces in the test plane of the simulator and at normal incidence (within $\pm$ 5°) to the centre-line of the beam. Connect to the instrumentation as shown in Fig. 6.

8.4 Set the irradiance to produce from the reference cell its calibrated short-circuit current under Standard Test Conditions.

8.5 With the test cell stabilised at or near ambient air temperature, measure its short-circuit current (I_{sc}) and open-circuit voltage (V_{oc}). Note the cell temperaure T_1.

8.6 Stabilise the test cell at a temperature T_2 about 10°C above T_1 and then at a temperature T_3 about 20°C above T_1. Repeat I_{sc} and V_{oc} measurements at each temperature.

8.7 Calculate α_c and β_c, the temperature coefficients for single cells, by a least square fit to the measured temperature dependence of I_{sc} and V_{oc}, preferably on more than one sample.

8.8 For a module or other assembly of cells, calculate the temperature coefficients as follows :-

$$\alpha = n_p \alpha_c \qquad\qquad \beta = n_s \beta_c$$

- where n_p is the number of cells in parallel and n_s the number in series.

The procedure with a pulsed simulator is similar except that, in this case, the short-circuit current signal from the reference cell is used to correct the data for small variations from the normal irradiance of 1000 Wm^{-2}. A pulsed simulator is preferred because there is no additional heat input to the cell during the measurement.

A similar procedure may be used to determine α and β at other irradiances and temperatures. At sub-ambient temperatures, however, precautions must be taken to prevent moisture condensing on the active surface of the test cell. This can be done by directing a jet of dry nitrogen gas onto the surface or by enclosing the cell in a vacuum chamber.

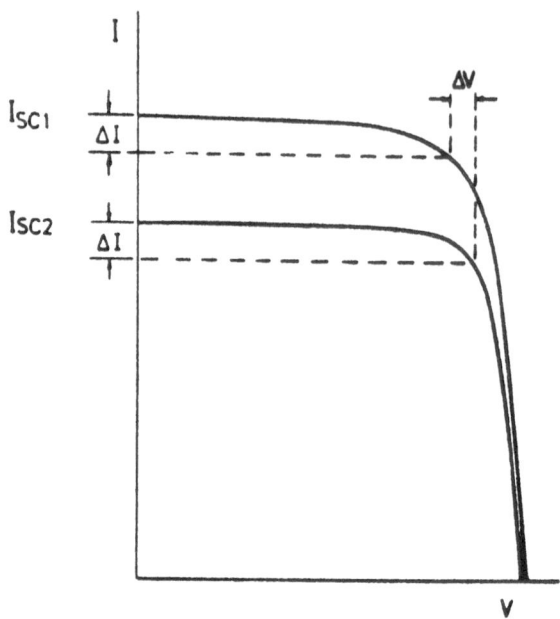

Fig. 9   Determination of R_s

9.   DETERMINATION OF INTERNAL SERIES RESISTANCE

R_s may be simply determined in simulated sunlight by the method described by Wolf and Rauschenbach (4).

The procedure is as follows   (see Fig. 9) :-

9.1   Trace the voltage-current characteristic of the cell or module at two different irradiances (the magnitudes of which do not have to be known) and at the same temperature (within 0.5°C) in the range 23 to 27°C.

9.2   Choose a point on each characteristic, preferably near the knee of the curve, where the current is the same increment ΔI below the short-circuit current.

9.3   Measure the voltage displacement ΔV between these two points.

9.4   Calculate R_s as follows :-

$$R_s = \frac{\Delta V}{I_{sc1} - I_{sc2}}$$

-   where I_{sc1} and I_{sc2} are the two short-circuit currents.

9.5 Repeat steps 9.2 to 9.4, using a characteristic taken at a third irradiance and the same temperature in combination with each of the first two curves. Take the mean of the three values of R_s thus calculated.

10. REFERENCE SOLAR CELLS AND MODULES

The key to good photovoltaic measurement lies in the selection and calibration of reference solar cells and modules.

Reference cells should be stable devices having essentially the same relative spectral response as the cells or modules to be tested. The matching should take account of the spectral response spread in production cells and the effects, if any, of the module window and encapsulant. At least two master reference cells should be selected to minimise mismatch errors. For production testing, the relative spectral response of a typical batch of production cells should be measured and the 'reddest' and 'bluest', representing the response spread, selected for calibration. Master reference cells need not be specially packaged.

To save wear and tear on the master reference cells, it is good practice to use a working reference cell or module for the day-to-day setting and monitoring of irradiance levels. Working reference cells should be protectively mounted in accordance with the recommendations of Section 13. The working cell or module should be calibrated against the appropriate master reference cells by simultaneously measuring their short-circuit currents in natural or simulated sunlight and calculating the calibration value as follows :-

$$I_{sw} = \frac{1}{2} \left[\frac{I_{mw1} \cdot I_{sr1}}{I_{mr1}} + \frac{I_{mw2} \cdot I_{sr2}}{I_{mr2}} \right]$$

- where : I_{sw} is the calibration value of the working reference cell or module,

I_{mw1} and I_{mr1} are the simultaneously measured short-circuit currents of the working reference cell or module and master reference cell 1,

I_{mw2} and I_{mr2} are the simultaneously measured short-circuit currents of the same working reference cell or module and master reference cell 2,

I_{sr1} and I_{sr2} are the short-circuit currents of master reference cells 1 and 2 resp. per unit of standard sunlight irradiance.

If all three devices are compared in a single test,
$I_{mw1} = I_{mw2} = I_{mw}$ and the above equation simplifies to :-

$$I_{sw} = \frac{I_{mw}}{2}\left[\frac{I_{sr1}}{I_{mr1}} + \frac{I_{sr2}}{I_{mr2}}\right]$$

Master reference cells should be kept in protective packing under lock and key to preserve them against damage and degradation. They should be brought out periodically (at least weekly during continuous testing) and used to check the calibration of the working reference cell or module. The cause of any change in the current ratios should be investigated. In natural sunlight, such a change would indicate that one or both of the reference cells have degraded but, with a simulator, an alternative explanation could be a change in the spectral content of the light. The windows of working reference cells and modules must be kept clean and scratch-free.

The calibration of master reference cells is best left to recognised agencies with the necessary facilities and expertise. In Europe, there are at present four solar cell calibration agencies :-

      o Space Department
         Royal Aircraft Establishment (RAE)
         FARNBOROUGH, HANTS. GU14 6TD
         UNITED KINGDOM

      o Centre Nationale d'Etudes Spatiales (CNES)
         18, Avenue Edouard Belin
         31055 TOULOUSE CEDEX
         FRANCE

      o Power Supplies Division TE
         European Space Research And Technology Centre (ESTEC)
         2200 AG NOORDWIJK
         NETHERLANDS

      o Deutsche Forschungs- und Versuchsanstalt fur
         Luft- und Raumfahrt eV (DFVLR)
         STUTTGART-VAIHINGEN
         F.R. GERMANY

Each agency has its own method of calibration, CNES and RAE relying on outdoor techniques, which are outlined in the following sections, while ESTEC and DFVLR use indoor methods (6). Periodically, calibrations are compared in 'round robin' exercises in which a batch of various cells is sent to each agency in turn for AM1.5 calibration and the results are analysed by the CEC Photovoltaic Test Group under the aegis of the Joint Research Centre, Ispra.

11. THE NIP METHOD OF CALIBRATION

The procedure followed by CNES, which is derived from American practice (3), is as follows :-

11.1 Use five matched normal incidence pyrheliometers (NIPs), one unfiltered and the others fitted with bandpass filters Nos. 1 to 4, covering different parts of the solar spectrum.

11.2 Mount the cell to be calibrated on a temperature-controlled block and fit it with a collimator matching those of the pyrheliometers to mask it from the diffuse sky radiation.

11.3 When conditions are suitable (see below), place a filter with the same spectral transmittance as Filter 1 over the cell collimator, point the cell and the pyrheliometer with Filter 1 at the sun and take simultaneous measurements of short-circuit current and irradiance. At the same time, measure the direct irradiance with the unfiltered pyrheliometer. Repeat with Filters 2, 3 and 4, noting any change in the unfiltered irradiance and correcting the corresponding filtered irradiance reading as necessary.

11.4 Compute the calibration value as follows :-

$$\text{Calibration Value} = \frac{\displaystyle\sum_{i=1}^{i=4} \frac{I_i}{E_i} \cdot E_{STi}}{E_{ST}}$$

- where : I_i = short-circuit current of cell with Filter i,

E_i = measured irradiance with Filter i,

E_{STi} = irradiance of standard AM1.5 sunlight with Filter i,

E_{ST} = total irradiance of standard AM1.5 sunlight, unfiltered (834.6 Wm^{-2})

11.5 Take the average of at least three calibration values on at least two different days as the definitive calibration.

Care must be taken to ensure that the NIP and reference cell collimators are accurately aligned to the sun (within $\pm 0.5°$) and have the same field of view. The five pyrheliometers and the pairs of filters must also be accurately matched.

The following weather conditions are specified for the NIP method in Ref. 3 :-

o Clear blue sky, with no observable cloud formations within a 15° half-angle cone surrounding the sun.

o Irradiance between 750 and 900 Wm^{-2}, as measured by the NIP.

o Atmospheric conditions sufficiently stable so that the variation in cell current is less than 0.5% during any 30s measurement period.

o The product of the optical air mass and the atmospheric turbidity during measurements must be less than 0.25, the turbidity being determined from measurements at 500 nm. Alternatively, the ratio of uncollimated to collimated short-circuit current must be less than 1.2.

o The optical air mass must be between 1 and 2.

12. THE GLOBAL METHOD OF CALIBRATION

This method was developed by the RAE and has been successfully used by them for many years in Malta and more recently in Cyprus, showing a year-to-year consistency of ± 1%. The procedure is as follows :-

12.1 Measure the relative spectral response of the cell to be calibrated.

12.2 Mount the cell on a temperature-controlled block co-planar with a horizontal pyranometer.

12.3 Measure the short-circuit current of the cell in global sunlight under suitable conditions (see below). At the same time, measure the global irradiance.

12.4 Measure the relative spectral irradiance distribution of the global sunlight at or about the same time as the other measurements (5).

12.5 Compute the calibration value by the method set out below.

12.6 Take the average of at least three calibrations on three different days as the definitive calibration.

The method of computation is as follows :-

Let :

I_{sr} = short-circuit current of the reference cell in standard sunlight at 1000 Wm^{-2},

I_{sg} = short-circuit current of the reference cell in global sunlight, as measured,

s_λ = absolute spectral response of the reference cell at wavelength λ ,

$k_1 s_\lambda$ = relative spectral response of the reference cell at wavelength λ ,

$E_{s\lambda}$ = absolute spectral irradiance at wavelength λ of standard sunlight,

$E_{g\lambda}$ = absolute spectral irradiance at wavelength λ of the global sunlight in which the short-circuit current was measured,

$k_2 E_{g\lambda}$ = relative spectral irradiance at wavelength λ of the global sunlight, as measured,

E_{glob} = measured irradiance of the global sunlight.

Then :
$$I_{sr} = \int s_\lambda E_{s\lambda} \, d\lambda$$

$$I_{sg} = \int s_\lambda E_{g\lambda} \, d\lambda$$

and
$$E_{glob} = \int E_{g\lambda} \, d\lambda = \frac{1}{k_2} \int (k_2 E_{g\lambda}) \, d\lambda$$

Therefore
$$I_{sg} = \frac{1}{k_2} \int s_\lambda (k_2 E_{g\lambda}) \, d\lambda$$

$$= \frac{E_{glob}}{\int (k_2 E_{g\lambda}) \, d\lambda} \cdot \int s_\lambda (k_2 E_{g\lambda}) \, d\lambda$$

Dividing the first of the above equations by the last :

$$I_{sr} = I_{sg} \frac{\int (k_2 E_{g\lambda}) \, d\lambda}{E_{glob}} \cdot \frac{\int s_\lambda E_{s\lambda} \, d\lambda}{\int s_\lambda (k_2 E_{g\lambda}) \, d\lambda}$$

Since s_λ appears in both numerator and denominator, relative values of spectral response embodying the same constant may validly be used.

Thus :

$$I_{sr} = I_{sg} \frac{\int (k_2 E_{g\lambda}) d\lambda}{E_{glob}} \cdot \frac{\int (k_1 s_\lambda) E_{s\lambda} d\lambda}{\int (k_1 s_\lambda)(k_2 E_{g\lambda}) d\lambda}$$

Thus, the calibration value $I_{sr}/1000$ in units of AWm^{-2} may be computed from the tabulated values of $E_{s\lambda}$ and measured values of I_{sg}, $k_1 s_\lambda$, $k_2 E_{g\lambda}$ and E_{glob} . Similarly, the calibration value may be computed for any other known spectral irradiance distribution.

This method is simple and free from collimator alignment and field-of-view errors but it requires a good test site and the following weather conditions for consistent results :-

o Clear, sunny weather, with the diffuse irradiance no greater than 0.25 of the global irradiance.

o Global irradiance no less than 800 Wm^{-2}.

o Solar elevation no less than 54°.

o Radiation sufficiently stable to allow the spectral irradiance distribution to be measured.

o No atmospheric pollution.

o Prevailing good weather, so that measurements can be taken on three suitable days without undue delay.

Both the NIP and global methods depend upon the accurate calibration of thermopile-type radiometers. RAE's practice is to calibrate their pyranometer on site against a Kendall black-body pyrheliometer before every series of measurements. The Kendall instrument is calibrated to the international PACRAD scale.

13. TEST MOUNTS

Single test cells and unpackaged master reference cells should be mounted on temperature-controlled blocks embodying a vacuum hold-down device, a suitable thermocouple and four contact probes (current + and - , voltage + and -). The mount for working reference cells should consist of a robust temperature-controlled block, equipped with a suitable thermocouple, prefer-ably attached to the upper contact bar, and an easily-cleaned distortion-free protective window, allowing the cell an

unobstructed view over a solid angle of 2π steradians. Test
mounts and working reference cells should be installationally
interchangeable.

14. INSTRUMENTATION

The temperature of cells and modules should be measured to an
accuracy of $\pm$ 2°C. Large modules shoud be fitted with not less
than three temperature sensors per m^2 of area. One sensor is
sufficient for measurements with pulse simulators.

In natural sunlight, sensors applied to the front or back of
the module may not register the cell temperature with sufficient
accuracy. In this case, the necessary correction should be
established using an adequately instrumented dummy module of the
same design.

Voltages and currents should be measured to an accuracy of
$\pm$ 0.5%, using separate voltage and current leads (see Fig. 6).
Short-circuit currents should be measured at zero voltage, using
a variable bias, preferably electronic, to offset the voltage drop
across the series resistor. In the case of reference cells, the
short-circuit current may alternatively be determined by measuring the voltage drop across a standard 4-terminal resistor,
provided that the measurement is made at a voltage less than $0.03V_{oc}$
within the range where there is a linear relationship between
current and voltage and, if necessary, the reading is corrected to
zero voltage. Voltmeters should have an internal resistance of at
least 20 kΩ/V. The calibration of all instruments should be
checked at frequent intervals.

APPENDIX -   DEFINITION OF METEOROLOGICAL TERMS

AIR MASS   The length of path through the Earth's
                            atmosphere traversed by the direct solar beam,
                            expressed as a multiple of the path traversed
                            to a point at sea level with the sun at zenith.

SOLAR ELEVATION   The angle between the direct solar beam and
                            the horizontal plane (degrees).

                            Air Mass = cosecant of the solar elevation.

DIRECT IRRADIANCE   The radiant power from the sun (and a small
(E_{dir}) area of sky surrounding it, defined by the
                            acceptance angle of the pyrheliometer) incident
                            upon unit surface area (Wm^{-2}).

DIFFUSE IRRADIANCE (E_{diff})	The radiant power from the sky incident upon unit surface area (Wm^{-2}).
GLOBAL IRRADIANCE (E_{glob})	The total solar radiant power incident upon unit area of a horizontal surface (Wm^{-2}).

Global irradiance = Direct irradiance (horiz.) + Diffuse
irradiance (horiz.)

TOTAL IRRADIANCE (E_{tot})	The total solar radiant power incident upon unit area of an inclined surface (Wm^{-2}).
SPECTRAL IRRADIANCE (E_λ)	The irradiance, global, direct or diffuse, per unit bandwidth at a particular wavelength λ $(Wm^{-2}\mu m^{-1})$.
SPECTRAL IRRADIANCE DISTRIBUTION	Spectral irradiance, plotted as a function of wavelength.
PYRANOMETER	A radiometer normally used to measure global irradiance (or, with a shade ring or disc, diffuse irradiance) on a horizontal plane. Can also be used at an angle to measure the total irradiance on an inclined plane, which in this case includes an element due to radiation reflected from the ground.
PYRHELIOMETER	A radiometer, complete with collimator, used to measure direct irradiance. (Sometimes called a 'Normal Incidence Pyrheliometer' or NIP).
TURBIDITY	The reduced transparency of the atmosphere, caused by absorption and scattering of radiation by solid and liquid particles, other than clouds, held in suspension. As defined by Ångström, the turbidity of the atmosphere is related to t, the extinction coefficient at a wavelength of 1000 nm (normally called the 'turbidity coefficient') and $\mathcal{E}$, the wavelength exponent in the expression for the aerosol extinction function :-

$$a_{D,\lambda} = t\lambda^{-\mathcal{E}}$$

Values of t less than 0.10 denote very clear conditions, whereas values greater than 0.20 are a distinctly hazy condition. The average value of $\mathcal{E}$, which is dependent on the particle size distribution, was assumed by Ångström to be about 1.3.

REFERENCES

1. Treble, F.C.   'Recommendations for the Performance Rating
                    of Flat-Plate Terrestrial Photovoltaic Solar
                    Panels'. Second ERDA/NASA Terrestrial
                    Photovoltaic Measurements Workshop, Baton
                    Rouge, 10-12 November 1976.

2. Anon   'Standard Procedures for Terrestrial Photo-
                    voltaic Performance Measurements'
                    JRC Specification No. 101 Issue 2
                    (EUR 7078 EN), 1980.

3. Anon   'Revised Terrestrial Photovoltaic Measure-
                    ment Procedures'. ERDA/NASA 1022/77/16,
                    June 1977.

4. Wolf, M. and   'Series Resistance Effects on Solar Cell
    Rauschenbach, H. Measurements'. Advanced Energy Conversion,
                    Vol. 3, pp 455-479, April-June 1963.

5. Walkden, M.W. 'The Spectral Energy Distribution of Sunlight
                    in Malta'. RAE Technical Report 67248,
                    September 1967.

6. Bogus, K.   'Solar Cell and Module Performance Assessment
                    Based on Indoor Calibration Methods'
                    Ispra Course 'Performance of Solar Energy
                    Converters', 16-18 November 1981, this volume.

SOLAR CELL AND MODULE PERFORMANCE ASSESSMENT BASED ON
INDOOR CALIBRATION METHODS

K. Bogus

European Space Agency, ESTEC,
Noordwijk, The Netherlands.

ABSTRACT

For accurate electrical performance measurements of solar cells
and solar cell modules under indoor test conditions, various
calibration steps are required. This paper reviews a combined
space/terrestrial indoor calibration method which relies on
five calibration steps. It also covers the following topics:

-    Past experiences and future plans at ESTEC with primary
     standards obtained from essentially all calibration centres
     in the western world (ground, balloon flight, aircraft and
     space calibration);

-    Correlation between air-mass-zero (space) conditions and
     AM 1.5 (or other terrestrial conditions);

-    Error budgets in indoor calibration;

-    Requirements on and capabilities of steady-state
     simulators and pulsed simulators (flashers);

-    Application of indoor measurement methods to SWS calibration,
     solar cell development and type approval, solar cell module
     development and qualification, final acceptance tests.

The advantages of the indoor sun simulator method (independence
from weather conditions, season, location, continuous availability, low costs also at small volume of work in a single calibration campaign) will be assessed and the need for complementary
outdoor methods will be verified.

G. Beghi (ed.), Performance of Solar Energy Converters: Thermal Collectors and Photovoltaic Cells, 189–219.
Copyright © 1983 ECSC, EEC, EAEC, Brussels and Luxembourg

INTRODUCTION

Solar cells and solar cell modules have to be tested for their performance, i.e. their ability to produce electrical power under well-defined conditions of illumination and temperature. The need for performance measurements arises at various stages of development, design and production from different requirements: at the development level, the main objective is to assess the performance of a new concept in comparison with state-of-the-art competitors; at the design level, performance data are required as input for the sizing and dimensions of photovoltaic systems operating under certain environmental conditions; at the production level, the cells and modules must be acceptance tested against a set of specifications in order to separate deliverables from rejects.

For accurate performance measurements and calibration, several methods exist which can be generally categorised into two classes: outdoor methods in natural sunlight, and indoor methods, using simulated sunlight or indirect analytical methods. Depending on the purpose of a performance test, each of these methods can offer specific advantages. Therefore, the various methods are not to be looked at as competing, but rather as complementary tools for the achievement of accurate electrical performance measurements. This paper concentrates on indoor measurement and calibration methods as they have developed over the last twenty years in parallel with the development and application of photovoltaic devices. Outdoor methods are excluded from detailed discussion and are discussed in another lecture of this course. Reference to outdoor methods must be made, however, in a few cases where it is required for the understanding of indoor calibration problems.

Originally, solar cells and solar cell modules and panels for spacecraft power supply were tested in the laboratory, under incandescent lamp illumination, using tungsten lamps. Water filters were applied in order to reduce the excessive infrared content of the tungsten lamp spectrum. Nevertheless, the spectral irradiance of these sources remained drastically different from the natural sunlight irradiance, be it outside the earth's atmosphere in space (at AMO, air-mass-zero), or on ground. Therefore, for accurate measurements, tungsten sources could only be used in combination with calibrated reference solar cells or standards selected, such that the performance ratio between the test cell and the standard cell was independent from the spectrum of the light source. Obviously, the conditions for this independence is an identical spectral response for test cell and standard cell and therefore, a spectral response measurement was always required as a second test if the light source used for performance testing was spectrally different from the actual sunlight in the application.

This historical development in the early phases explains already the need for the major elements required in solar cell calibration and performance testing:

- Sun simulators are needed as stable and reproducible light sources, in order to perform tests under development or production environment conditions.

- Solar Cell Standards are needed to adjust the intensity level of sun simulators to the equivalent of the natural sunlight intensity, i.e. the level at which the same electrical output is obtained from the test cell.

- Spectral response (together with spectral irradiance of sun simulators and natural sunlight) data are needed in order to verify the validity of simulators and standards and in order to make corrections for spectral mismatch errors.

In the sixties, considerable efforts took place to improve the accuracy of indoor and outdoor test equipment and methods, since experience had shown that newly developed solar cells with spectral response characteristics different from previously produced and calibrated standards introduced errors up to 20%. The newer sun simulators were equipped with high pressure Xenon arc lamps which operated continuously and illuminated areas up to 20cm Ø reasonably uniform with AMO simulated sunlight. For the illumination of large areas, the so-called pulsed Xenon solar simulators became available in the late sixties.

Solar cell standard calibration was performed as a regular task under reproducible test conditions by several agencies using complementary methods. Moreover, more accurate satellite flight data became available in the sixties and seventies, allowing a verification of ground calibration data.

In parallel, improved spectro-radiometric equipment was used for the accurate measurement of solar cell spectral response and the spectral irradiance of sun simulators.

Looking back onto the historical development of indoor performance testing with the knowledge of today's measurement capabilities with analog and digital equipment, it appears that all problems should have been solved a long time ago. This is correct in principle. In practice, however, unexplained discrepancies in performance testing still occur as new solar cell types with exotic performance properties are being developed and as accuracy requirements are becoming more and more demanding.

In the late 1970's, a new branch was added to the established calibration and performance test methods for space applications, with the increasing interest in terrestrial photovoltaic applications. The major differences in this new branch result from the following factors:

- terrestrial devices operate under a variety of illumination conditions and not under the extremely stable and reproducible AMO illumination as space solar cells;

- terrestrial devices show a larger variability in configuration (e.g. cell sizes and shapes, module types, planar and concentrator concepts);

- the requirements on terrestrial photovoltaic performance measurements are not yet well-defined and globally accepted specifications do not exist. The accuracy requirements should be less demanding than for space applications.

1. SOLAR CELL SHORT CIRCUIT CURRENT CHARACTERISTICS

The purpose of performance testing is to establish the essential solar cell (or module) output data under well-defined illumination and temperature conditions.

The most critical parameter to be determined in accurate solar cell performance measurement is the short circuit current, Isc.

Other important parameters are open-circuit voltage, Voc, and maximum power Pm (and efficiency η). These can generally be extracted from I-V-curve measurements.

A basic difference between Isc measurement and measurement of the other I-V-curve parameter becomes apparent if measurement accuracy and error sources are considered:

Errors in Isc measurement are primarily related to measurement principles, whereas errors in the other parameters are mainly connected with the practical test set-up and the way the test is executed.

The main difficulty in accurate Isc measurements stems from the fact that two separate physical dimensions of the light as an input determine Isc as the solar cell output parameter: one is the intensity (or better, total irradiance) at the solar cell surface; the other is the spectral composition of the incident light. In most practical cases, the short-circuit current density of a solar cell can be described as:

(i) $\quad I_{sc} \quad = \quad \int I(\lambda)\, d\lambda \qquad$, with

(ii) $\quad I \quad = \quad S(\lambda)\cdot E(\lambda) \quad = \quad e\, Q(\lambda)\cdot N(\lambda)$

$S(\lambda)$ and $Q(\lambda)$ are the absolute spectral response, respectively quantum yield of the cell, $E(\lambda)$ and $N(\lambda)$ are the spectral irradiance resp. the photon flux density incident on the cell per unit area. In order to separate intensity and spectral properties as independent variables, it is advisable to use normalised parameters:

(iii) $\quad \varepsilon(\lambda) \quad = \quad \dfrac{E(\lambda)}{\int E(\lambda)\, d\lambda} \qquad ; \quad \sigma(\lambda) = \dfrac{S(\lambda)}{\int S(\lambda)\, d\lambda}$

This implies that:

(iv) $\quad \int \varepsilon(\lambda)\, d\lambda \quad \equiv 1 \qquad ; \qquad \int \sigma(\lambda)\, d\lambda \equiv 1$

$\varepsilon(\lambda)$ is the relative spectral irradiance (independent from intensity) and $\sigma(\lambda)$ is the relative spectral sensitivity. With this, the short-circuit current becomes:

(v) $\quad I_{sc} \quad = \quad E_o\cdot S_o\cdot \int \sigma(\lambda)\cdot \varepsilon(\lambda)\, d\lambda$

where

$\quad E_o \quad = \quad \int E(\lambda)\, d\lambda$

$\quad S_o \quad = \quad \int S(\lambda)\, d\lambda$

Obviously, the short-circuit current is proportional to the total irradiance E_0 and to an integral determined by the relative spectral distribution of solar cell sensitivity and irradiance.

In order to measure the short-circuit current of a solar cell accurately, it is necessary, therefore, to set the total irradiance of the light source to the correct level and to ensure by spectral measurements that the spectral distribution of the illumination used is identical to the spectral distribution of sunlight in those wavelength regions where $\sigma(\lambda)$ differs from zero.

Usually, indoor short-circuit current calibrations are performed with sun simulators, having a slightly different relative spectral energy distribution $\varepsilon(\lambda)_{sim}$ than the actual sunlight, $\varepsilon(\lambda)_{sun}$. The intensity of the simulator is set to a sun-equivalent level by means of a reference standard cell, with a spectral response distribution $\sigma(\lambda)_s$ which is slightly different from the one of the test cell, $\xi(\lambda)_T$.

Instead of comparing the spectral response curves of the two cells, it is more illustrative to compare the relative spectral current distribution in sunlight.

(vi) $$F(\lambda)_{T,s} = \frac{I(\lambda)_{T,s}}{\int I(\lambda)_{T,s}\, d\lambda}$$

If $B(\lambda)$ is the ratio of spectral irradiance of the adjusted simulator compared to the sunlight, then the short-circuit current measurement error becomes (see Reference 1)).

(vii) $$\Delta = \frac{I_{T,sim} - I_{T,sun}}{I_{T,sun}} = \int F(\lambda)_T \cdot B(\lambda) d\lambda \quad - 1 \quad, \text{ or}$$

(viii) $$\Delta = \int [\, F(\lambda)_T - F(\lambda)_s\,] \cdot [\, B(\lambda) - 1\,]\, d\lambda$$

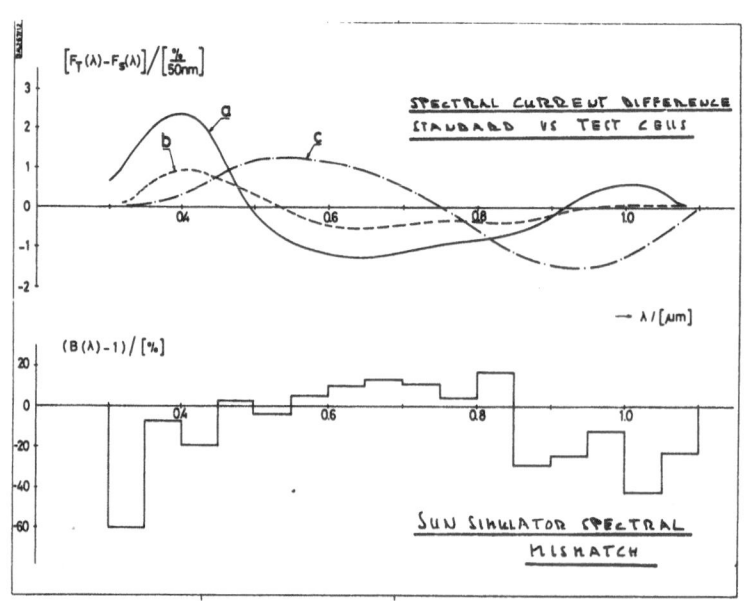

Figure 1: Spectral Current Difference and Sun Simulator
Mismatch Curves

- 194 -

This formula clearly shows the two components which produce the total spectral mismatch error:

a) the difference in the spectral current distribution between test and reference cell, folded with,

b) the difference in the spectral energy distribution between the sun simulator spectrum and the sunlight spectrum.

Examples for these spectral components are shown in Figure 1.

2. GENERAL OBJECTIVES AND REQUIREMENTS FOR INDOOR TESTS

It must be pointed out that performance measurement procedures are used for measurements of many different types, depending upon the objectives of the measurement. This implies a certain number of requirements which have to be considered in the preparation of the procedures.

It is the purpose of this paragraph to summarise the essential photovoltaic performance test types and their requirements as a preparatory step for the introduction of a standard measurement and calibration procedure in chapter 5.

- CLASSIFICATION OF PHOTOVOLTAIC MEASUREMENTS

As explained in Reference 2, photovoltaic measurements are being performed by:

- users of solar cell or panels;

- manufacturers of solar cells or panels;

- institutions involved in solar energy programmes of solar calibration.

The essential performance test types and their objectives are listed in Table 1. Moreover, this table also indicates which measurement principle (i.e. outdoor, air mass zero, indoor) is preferable for a certain performance test. The preferences, as indicated, result from the test requirements, as discussed in the following sections.

TEST TYPE	PURPOSE OF THE TEST	OUTDOOR MEASUREMENT		MEASUREM. AT AM0	INDOOR MEASUREMENT		
		IRRADIANCE	MONITOR		SUN SIMULATOR		"FLASHER"
		PYRANOM OR NIP	REF. SOLAR CELL		PRECISION TYPE	LOW COST TYPE	
1 CALIBRATION OF REFERENCE CELLS	TO PROVIDE REF. STANDARDS FOR PERFORMANCE TEST 2-6	X		X	X		
2 SOLAR CELL DEVT. AND QUAL. TESTS	TO ASSESS THE PERFORMANCE OF NEW SOLAR CELL TYPES				X		
3 MODULE DEVT. & QUALIF'N TESTS	TO ASSESS THE PERFORMANCE OF NEW PANEL TYPES						X
4 SOLAR CELL FINAL PRODUCTION TESTS	MANUFACTURER'S PRODUCTION CONTROL, ACCEPTANCE TEST					X	X
5 MODULE FINAL PRODUCTION TESTS	- DITTO -						X
6 IN-SITU FIELD TESTS (PANELS MAINLY)	TO ASSESS THE OUTPUT UNDER OPERATIONAL CONDITIONS		X				

Table 1:   Summary of Photovoltaic Measurement Types

- PHOTOVOLTAIC MEASUREMENT REQUIREMENTS

a) Standard Sunlight Conditions

Calibration data for reference standards refer to standard sunlight of well-defined spectral energy distribution. The standard sunlight spectrum should be close to the average sunlight
spectra under which performance and operational tests are performed in order to reduce the necessary corrections to second
order correction, whereby the influence of measurement errors
in the correction factors is minimised.

For space applications, the AMO spectrum is used, as it is
defined in Reference 3. This, so-called Makarova spectrum is
presently considered to be more realistic than the slightly
deviating data used in the USA (see Figure 2).

In the United States, an air mass 1.5 standard sunlight spectrum had been proposed for terrestrial solar arrays by NASA/
ERDA (Reference 4), which is calculated from the well-
established air mass zero spectrum considering various atmospheric scattering and absorption processes.

In Europe, the average sunlight conditions are shifted into
higher air mass values, due to the higher average latitude. A
common reference at AM 1.5 has been proposed recently in
Reference 5.

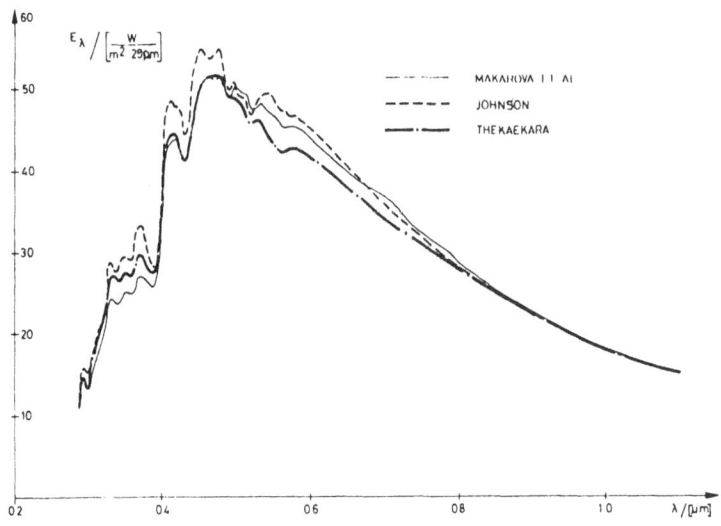

Figure 2: AMO Solar Spectra

b) Availability of Reference Standard Cells

By definition, a reference standard cell is a calibrated solar
cell which is used for the performance rating of solar cells
or solar cell panels with approximately the same spectral
response characteristics under various natural or simulated
sunlight conditions.

For the assessment of novel solar cell types (test types 2 and
3 of Table 1), it is desirable to provide calibrated reference
standard cells with minimum delay time and at any time of the
year.

For final production tests (test types 4 and 5), the situation
is less time constrained, since, in general, a qualification
exercise has been performed in advance, from which reference
standards are available.

c) Calibration of Reference Standards

Since (at least for terrestrial applications) the standard sunlight is a theoretically fixed reference, the actual calibration
of reference standard cells is always performed under sunlight
which more or less deviates in its spectral energy distribution
$E_{Test\ Sun}$ (λ) from the standard sunlight spectrum $E_{St.\ sun}$ (λ).
Therefore it is always necessary to monitor $E_{Test\ sun}$ (λ)

during the calibration exercise and to measure the spectral response of the cell to be calibrated. These measurements yield the correction factor for the accurate short circuit current value under standard conditions:

$$I_{St.sun} = I_{Test\ sun} + I_{Corr} = I_{Test\ sun} + \int S(\lambda)\left[E_{St.sun}(\lambda) - E_{Test\ sun}(\lambda)\right]d\lambda$$

It is obvious that the sunlight spectrum used for calibration should not deviate too much from the standard sunlight spectrum in order to limit the influence of measurement errors in $S(\lambda)$ on the accuracy of the calibrated cell current $I_{St.sun}$ to the same level as it results from the errors in $E_{Test\ sun}(\lambda)$ and $I_{Test\ sun}$.

d) Assessment of Development-type Solar Cells and Panels

Performance tests on novel photovoltaic devices require extreme care, since the relevant solar cell properties may deviate from the properties of conventional silicon solar cells, resulting in serious measurement errors. The most important potential devia-tions are:

- non-linear intensity dependence of the short-circuit current;

- spectral quenching or enhancement effects with the consequence that the integral solar cell current cannot be expressed as the sum of the contributions from the various spectral regions;

- excessive response times.

The first two deviations can lead to erroneous results in spectral response measurements and in short-circuit current measurements under sunlight conditions which strongly deviate from the standard conditions.

The third type of deviation causes trouble if measurements are performed with pulsed light sources ("flasher").

e) Final Production Tests

This type of test which serves as a production control test requires the availability of two different types of calibration standards:

I)   Reference standards which have been calibrated by a recog-nised institution under standard conditions. These standards should have approximately the same spectral response distribution as the production cells to be tested. Moreover, their long-term stability has to be verified.

II) Working standards which are used for the continuous intensity control of the sun simulator under which the production tests are being performed. The working standards have to be checked against the reference standards under the sun simulator in regular intervals, depending on the long-term stability of the sun simulator and of the working standards.

Since the final production test standards of the same spectral response type as the test cells are always available, there is no stringent requirement on the spectral match of the simulator used. Emphasis lies rather on short and long-term stability and continuous operation.

3. INDOOR MEASUREMENT METHODS

3.1 Steady-state Sun Simulator with Reference Cells

This method, which is the most obvious one, requires an artificial light source simulating the natural sunlight in space or on ground and a calibrated reference solar cell or standard to set the light intensity of the simulator. Moreover, the appropriate equipment for mounting the test and reference cells, for temperature control and for measurement of the electrical output, are required.

The sun simulator has to meet certain requirements which can be more or less demanding, depending on the type of test to be performed (see paragraph 2)). A typical example for the requirements on a precision sun simulator is given in Table 2.

LIGHT SOURCE:	XENON SHORT ARC
ILLUMINATED AREA:	35 CM Ø WITH ± 5% UNIFORMITY
	15 CM Ø WITH ± 2% UNIFORMITY
IRRADIANCE IN TEST PLANE:	ADJUSTABLE BETWEEN 0.6 SOLAR CONSTANTS (AMO)
	AND 1.5 SOLAR CONSTANTS (AMO)
SPECTRAL MATCH:	± 5 % FOR 0.35 — 0.5 MICROMETER
	± 3 % FOR 0.5 — 0.8 MICROMETER
	± 7 % FOR 0.8 — 1.1 MICROMETER
STABILITY - SHORT TERM:	± 0.3% OVER PERIODS FROM 50 MSEC TO 1 HOUR
- LONG TERM:	MAINTAIN SPECTRAL MATCH (S.A.) FOR AT LEAST
	300 HOURS OF OPERATION.
SECONDARY REFLECTION ONTO	
TEST PLANE:	LESS THAN 5% WITH REFLECTOR IN TEST PLANE.

Table 2: **Requirements** on a **Precision** Sun Simulator

Among several solar simulation techniques, the most common method
is the use of a xenon short arc lamp with filters to remove
undesired line spectra in the near infrared (in the pulsed mode,
which does not generate the undesired line spectra, unfiltered
xenon lamps can be used). Unfiltered carbon arcs have also been
used to simulate solar illumination with a reasonable spectral
match. The closest spectral match to the solar spectrum is
obtained by the use of a filtered xenon source. These sources
match the solar spectrum well enough that cell measurements made
under these sources can be considered representative for the natural environment.(Figure 3).

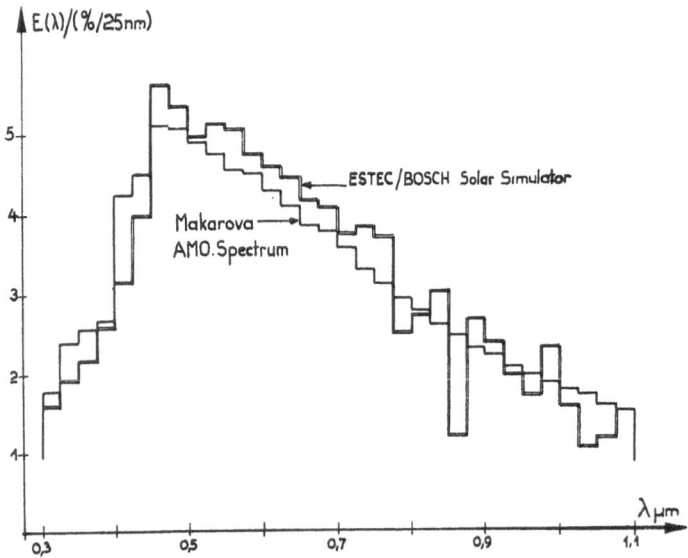

Figure 3: Sun Simulator Spectrum

Filtered xenon arc sun simulators have been manufactured by sev-
eral European and American companies (e.g. AEG, Bosch, BBT in
Europe, and Spectrolab in the US) for simulation of AMO-sunlight
(Figure 4).

For the simulation of terrestrial sunlight at AM1-AM2, additional
filters can be conveniently added into the AMO simulator beam.

Figure 4: Photograph of Bosch Sun Simulator

The transmission characteristics of these filters simulate the
atmospheric attenuation of AMO-sunlight and therefore, the system
is a close simulation of the actual conditions. It has the advantages of using the stable and well-defined AMO spectrum as basic
reference and, moreover, different atmospheric attenuation conditions can be simulated by simply replacing attenuation filters.

The most critical quality of a sun simulator for precision measurements is its spectral match to the actual sunlight spectrum.
Measurement and control of spectral match requires complex spectro-
radiometric equipment in order to achieve reliable and accurate
data over the full wavelength band of interest, i.e. 0.3 micrometers to 1.2 micrometers in case of silicon cells.

Commonly used methods for spectral irradiance measurements of sun
simulators are listed in Table 3. For precision measurements,
narrow-band spectro-radiometers are normally used, whereas for
routine checks of spectral stability, simple measurements with
broad band receivers are usually preferred.

Equally important as the sun simulator is the reference cell used
to set the illumination intensity (irradiance).

1.	SPECTRO-RADIOMETER MEASUREMENTS (300 NM TO 1200 NM).
2.	NARROW BAND FILTER SET PLUS RADIOMETER.
3.	SET OF THREE CALIBRATED SOLAR CELLS WITH WIDEBAND FILTERS (UV - VISIBLE - INFRARED).
4.	SET OF PRIMARY STANDARDS WITH DIFFERENT SPECTRAL RESPONSE.
5.	"RED - BLUE RATIO" CONTROL.

Table 3:   Sun Simulator Spectral Irradiance Control
           Methods

3.2 Pulsed Sun Simulator with Reference Cells

Generally, modules and arrays are too large to test using conventional steady-state solar simulators, hence other methods for
obtaining performance (Isc 1, I-V) measurements must be used.
The approach used at ESTEC is to measure modules using a long-arc
xenon-lamp pulsed solar simulator. The indoor pulsed simulator
method has the advantages of easy availability and, because of the
short duration of the irradiance pulse (2 msec), there is no heating of the solar cell modules. Flash intensity is controlled and
data are taken at room temperature. Only minor corrections are
done to obtain data at standard conditions (e.g. 25°C., AMO
equivalent irradiance).

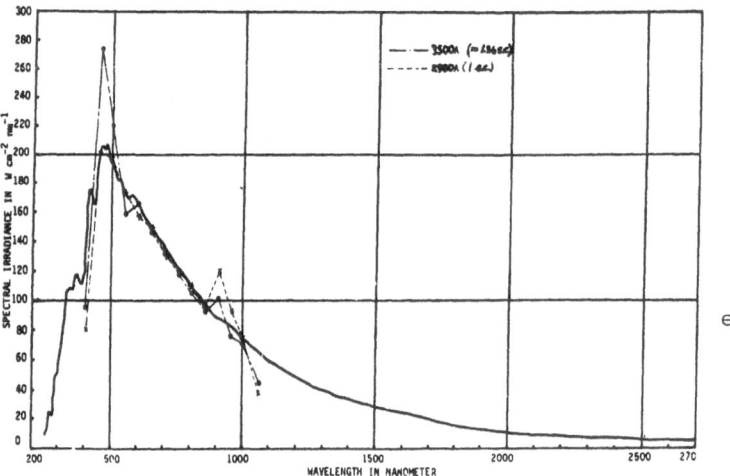

Figure 5:   LAPSS Spectrum

- 202 -

Pulsed sun simulators have been developed and manufactured mainly
for large area irradiance in satellite solar array testing. The
commonly used "LAPPS" (large area pulsed simulator) from Spectrolab is a complex and versatile facility consisting mainly of a
high energy pulsed xenon lamp in a suitable lamp-house, a regulated power supply and a computerised data handling unit. The
system does not require optical filtering, since the spectral
match of the pulsed xenon light to AMO sunlight is already adequate for all applications where appropriate reference cells are
available. The good spectral match is due to self-absorption of
the xenon emission lines in the high current density arc (see
Figure 5).

Photographs of the two flasher systems available at ESTEC are
shown in Figures 6 and 7. The major difference between the
Spectrolab and the TRW pulsed simulator concept lies in the number of flashers used to produce a complete I-V-curve: the Spectrolab LAPPS provides up to 40 points of an I-V-curve within one
single flash, whereas the TRW system produces one point per flash
only.

As in the case of stead-state simulator measurements, the accuracy
of pulsed simulation tests depends fully on the accuracy and
spectral equivalence of the reference standards which have to be
used to adjust the irradiance level of the light pulse. For
flasher tests, it is common practice to use reference cells which
have the same spectral response distribution as the cells on a
module or panel under test. Therefore, no correction for spectral
mismatch is usually required.

3.3 Spectral Response Measurements

Solar cell spectral response measurements provide a powerful tool
in solar cell diagnostics.

In addition, spectral response measurements are equally important
in performance measurements and calibration. Spectral Response data
can be used to directly compute the short-circuit current of a
solar cell by folding the spectral response with the solar irradiance, according to equations (i) and (ii) of paragraph 1. In this
case, the absolute spectral response curve of the test cell must
be known precisely.

In a second method, the spectral response data are only needed as
relative values for second order correction of spectral errors
occurring in sun simulator calibration measurements.

Both methods have been considerably improved during previous years
and they represent today a valuable supplement to simulated sunlight calibration methods.

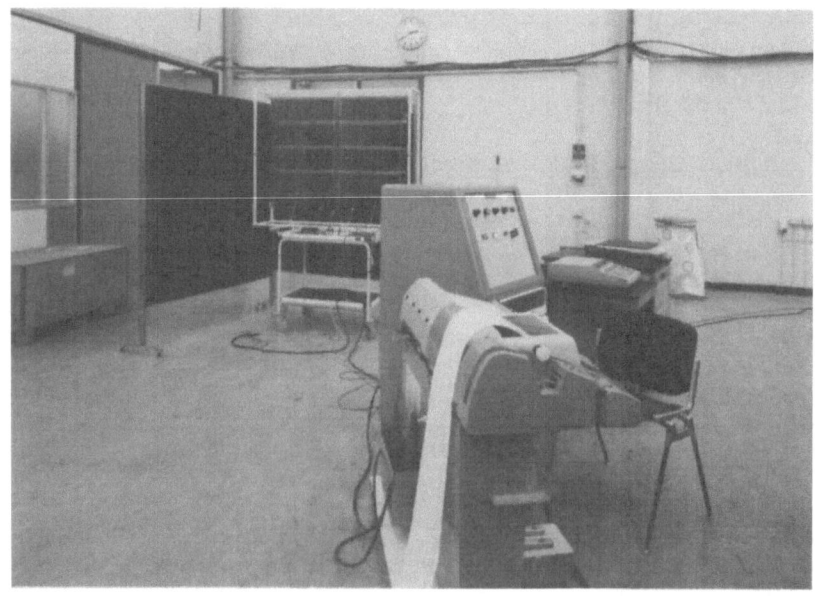

Figure 6: Photographs of Spectrolab LAPPS

Figure 7: Photograph of TRW LAPPS

Monochromators and narrow band pass monochromatic interference
filters are the most common sources of light. Monochromators can
range continuously but are limited in area coverage, uniformity
and intensity. Interference filters do not suffer from these
limitations, but their response generally cannot be continuously
varied over a wide range, since solar cells are generally devices
with smooth, broadband spectral sensitivities. Therefore, continuous wavelength coverage is not very important and filter methods
are normally preferred. Figure 8 shows a typical filter wheel
spectral response test set-up which in steady-state operation is
equipped with two 1000W Quartz-Iodine lamps.

The major disadvantage of spectral response measurements is the
low irradiance available at the cell surface. Typically, monochromatic irradiance values behind interference filters are 5 to 6
orders of magnitude lower than sunlight integral irradiance. This
can lead to significant errors in cases where:-

a) the solar cell sensitivity is intensity dependent;

b) the solar cell spectral response shows spectral quenching or
   enhancement effects.

In these cases, high-intensity DC-bias light can be used, on
which the modulated monochromatic light is superimposed. The
spectral response signal can then be extracted from the total
solar cell output by standard lock-in amplifier techniques.

Figure 8: Photograph of Spectral Response Measurement Set-up

In the past, absolute accuracy has been a substantial problem in
spectral response measurements. In practice, errors of +20% were
not unusual.

Recently, however, spectral response calibrations on solar cells
have been performed for ESTEC by the Physikalisch-Technische
Bundesanstalt (PTB) in Brannschweig (Germany) which are claimed to
have an absolute accuracy of +2% at a reference wavelength of 436um
and a relative accuracy of the other points of the spectrum with
respect to the reference point of +1%.

In order to avoid the potential problems in connection with low-
level monochromatic irradiation, it is advisable to use a high-
irradiance pulsed light source, which, in spite of the high attenuation through the filter, still provides an irradiance close to
actual sunlight conditions at the solar cell surface without overheating the filters (Reference 6). A schematic diagram of such an
arrangement is shown in Figure 9.

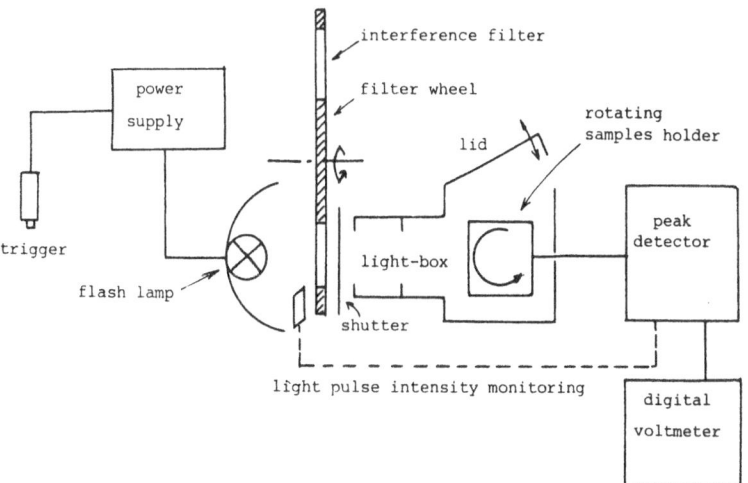

Figure 9: Pulsed Spectral Response Measurement Principle

3.4 Solar Cell Reference Standards

Calibrated solar cell standards are needed for most indoor performance measurements, as described above.

Cells which have been carefully calibrated by a recognised institution following established procedures, are usually called primary standards. Accepted methods for calibration of primary standards at AMO include the following:

- Balloon flight at high altitudes (CNES, JPL);

- Calibration in global sunlight (sun and sky), using a
  horizontal pyranometer (RAE-Malta/Cyprus);

- Calibration in collimated direct sunlight (CNES: Table-
  Mountain);

- Calibration with High-altitude Aircraft (NASA-Lewis);

- Indoor calibration, using a precision sun-simulator and a set
  of accepted calibrated solar cells.

For routine operations in daily use in cell production, secondary working standards are normally used which are spectrally identical to the production cells and which are derived from primary standards by the indoor calibration method mentioned above.

A detailed review of ESTEC's experience with standards is given in paragraph 5.

3.5 Other Measurement Methods

If the short circuit current of a solar cell module or array is known from simulated sunlight tests of the components or sub-assemblies at an earlier stage, it is possible to determine the J-V-characteristic of the arrays in the dark without sun-simulation.

The testing of the dark solar cell array forward characteristics has received considerable attention for two reasons:

a) solar simulators which illuminate very large areas or volumes sufficiently uniformly so that arrays mounted to spacecraft can be meaningfully tested may not be available, and

b) large, deployable, oriented arrays, already integrated to a spacecraft and mounted in a stowed condition, cannot always be readily unfolded for testing.

Dark forward-testing consists of connecting an external DC bias source to the terminals of a non-illuminated solar cell array such that the array becomes forward biased. During the test, the dark forward characteristics of the solar cell array are measured as a function of the bus voltage.

Accurate results are obtained when the maximum forward bias current exceeds 50% of the short-circuit current which would be obtained under one solar constant illumination.

The accuracy of this test method depends upon the temperature uniformity that can be achieved over the entire array. During the test, the forward bias causes a temperature rise of the array under test, and this, under most practical circumstances, leads to an undesirable error. Therefore, pulsed bias methods have been developed.

Another source of error lies in the difference between a photo-voltaic I-V-characteristic and the dark current forward character-istic caused by the equivalent series resistance of the array which leads to an artificial increase of the array voltage in the dark characteristic.

A new area of indoor electrical measurements on cell modules and panels deals with the AC-characteristics of a photovoltaic power source. The AC-properties are of importance in application where the subsequent power conditioning circuit operates in a fast switch-ing mode. They are also to be considered as potential error sources in pulsed light measurements.

The AC-properties of an array can be described in terms of classical network line theories (Reference 7), where the individual cell in the chain is characterised by its AC impedance properties.

Lastly, an important area of electrical measurements is the assessment of solar array performance in partial shadowing. This operation mode is potentially hazardous and can lead to permanent damage, due to hot spot formation or second breakdown of reverse biased solar cells (Reference 8).

4. ERROR SOURCES IN INDOOR MEASUREMENTS

The electrical performance of a solar cell or a solar array is usually measured by comparing its output against the short circuit current of a calibrated reference cell (standard) under simulated sunlight. At first sight, the errors in these performance measurements should only be related to one or a combination of the following three main factors:

- a difference between the spectral composition of AMO sunlight and the simulated light;

- a difference in the spectral response of reference cell and the cell/array to be measured;

- an error in the calibration level of the reference cell.

The electrical performance measurement should therefore be a straightforward operation which should lead to reproducible results within the accuracy limitations of phometric measurements. Considerable experience has been gained during the past decade from comparison of ground calibrations with orbital flight data and cross-checks between the various ground calibration laboratories in Europe and the US.

However, in spite of the increased experience and refined measurement procedures, it is found that little progress has been made in reducing calibration discrepancies. Presently, discrepancies in the order of 3-5% are being discussed, which exceed the bandwidth of any single calibration error source by a factor of more than two.

The reasons for this contradictory situation are mainly that:

- the rapid evolution of solar cell technology has led to the introduction of new solar cells which are spectrally and geometrically different and for which no sufficient calibration history exists, and

- the assessment of calibration error sources is often confused and incomplete because it is overlooked that many error sources

(each of which is in the noise of the overall measurement accuracy)
can add up to a significant total error if no proper distinction
is made between random deviations, for which a certain tolerance
band is allocated and systematic deviations which shift the calibration levels into a "preferred" direction and which should not
be tolerated at all.

4.1 Potential Error Sources

The final purpose of solar cell calibrations and electrical performance tests is to predetermine the operational performance of
solar arrays by indoor measurements. In the case of space applications, this usually includes three steps:

(i) Final Performance Measurement of Flight Panels

These are usually performed by using a LAPSS (Large Area Pulsed
Sun Simulator). The light intensity of the LAPSS is adjusted to
the equivalent one solar constant AMO level by means of a Secondary Working Standard (SWS) which should have the same spectral
response as the cells on the panel to be measured, in order to
minimise spectral mismatch errors resulting from the imperfect
spectral energy distribution of the LPSS.

The potential error sources of this measurement and their corresponding error bandwidths are listed in Table (4). The resulting
total root mean square (RMS) error of this measurement is about
+2% if it is assumed that the SWS has been calibrated accurately,
i.e. its calibrated short-circuit current value is the true current
at one solar constant AMO irradiation.

NO.	ERROR SOURCE	UNCERTAINTY
1.	ELECTRICAL MEASUREMENT REPRODUCIBILITY	± 1 %
2.	OPTICAL AND THERMAL IMPERFECTIONS IN TEST ARRANGEMENT	± 1 %
3.	TOLERABLE FLIGHT PANEL PERFORMANCE CHANGE AFTER TEST	± 1 %
4.	SPECTRAL RESPONSE DIFFERENCE BETWEEN SWS AND TEST PANEL	± 0.5%
5.	SWS LONG TERM STABILITY	± 0.5%
	RMS ERROR BANDWIDTH	± 1.9%
	MAXIMUM ERROR BANDWIDTH	± 4 %

Table 4: Error Sources in Final Acceptance Tests

(ii) Secondary Working Standard (SWS) Calibration

The requirements for the calibration of SWSs are described in
ESA-PSS-25. The accuracy of this calibration depends strongly on
the spectral match of the sun simulator used and on the accuracy
and the spectral match of the primary solar cell standard used for
setting the intensity of the simulator. Moreover, the optical and
geometrical arrangement of the calibration set-up are of importance.

The potential SWS calibration error sources and their corresponding
error bandwidths are given in Table (5) for the case of using the
ESTEC calibration system and for the case of using a simpler system
as it is used on line production control, for example.

The total RMS error with the ESTEC system is about $\pm1\%$ whereas it
is $\pm2\%$ in the case of the production control method.

NO.	ERROR SOURCE	ESTEC REFERENCE SYSTEM		SINGLE STANDARD CALIBRATION	
		CHARACTERISTICS	UNCERTAINTY	CHARACTERISTICS	UNCERTAINTY
1	PRIMARY STANDARD ISC-CALIBRATION	MULTIPLE STANDARDS FROM MALTA, JPL, CNES + FLIGHT DATA	$\pm 0.3\%$	SINGLE STANDARD, EG JPL - BALLOON	$\pm 1.0\%$
2	PRIMARY STANDARD LONG-TERM STABILITY	FREQUENT PRIMARY STANDARD CROSS-CHECKS	$\pm 0.3\%$	NO DIRECT CROSS-CHECKS WITH STANDARDS FROM OTHER SOURCES	$\pm 1.0\%$
3	SECONDARY STANDARD SPECTRAL MISMATCH	SPECTRAL MATCH OF ESTEC SUN SIMULATOR GUARANTEES $\pm1\%$ EVEN FOR WORST CASE SPECTRAL MISMATCH OF IRRADIATED CELLS	$\pm 0.5\%$	SPECTRAL MATCH OF SUN SIMULATOR LEADS TO $\pm2\%$ ERROR FOR IRRADIATED CELL CALIBRATION (SEE OTS CALIBRATION HISTORY)	$\pm 1.0\%$
4	SECONDARY STANDARD ELECTRICAL MEASUREMENT REPRODUCTION	UNCRITICAL SINCE ONLY THE RATIO OF TWO CURRENT VALUES ENTERS	$\pm 0.5\%$	SAME AS ESTEC; SOMETIMES TWO DIFFERENT INSTRUMENTS ARE USED FOR STANDARD AND TEST CELL, RESP.	$\pm 0.5\%$
5	DIFFERENCES IN OPTICAL CONFIGURATION BETWEEN PRIMARY & SECONDARY STANDARD	SOMETIMES CELL SIZE IS DIFFERENT (E.G. 2 X 2 STANDARD CELL & 2 X 4 TEST CELL)	$\pm 0.5\%$	PRIMARY STANDARD MOUNTED ON REFLECTING GROUND PLATE	$\pm 1.0\%$
	RMS ERROR BANDWIDTH		$\pm 0.9\%$		$\pm 2.1\%$
	MAXIMUM ERROR BANDWIDTH		$\pm 2.1\%$		$\pm 4.5\%$

Table 5: Error Sources in Secondary Standard Calibration

(iii) Primary Standard Calibration

Primary solar cell standards are calibrated by methods as summarised in paragraph 3.4. Generally, these methods have an accuracy
of about $\pm1\%$, i.e. the calibrated short circuit current of a primary standard corresponds to its current at 1 solar constant, AMO
conditions within $\pm1\%$.

This value has been used for the assessment of error sources in the calibration of SWSS shown in Table (5). Since the ESTEC Calibration Base system relies on the average of a large number of primary standards calibrated by different methods, the uncertainty from primary standard calibration enters with the RMS of the individual calibrations only which is about ±0.3%.

4.2 Tolerable Deviations

Solar array performance specifications for ESA projects generally allow for ±2% calibration uncertainty which is included in the power calculations as an additional loss factor. As shown in paragraph 4.1 (i), this covers the errors of the LAPSS measurement used for final acceptance tests, assuming that these errors occur at random and are not of a systematic nature.

This leaves no further margin for additional error sources which come in during primary and secondary standard calibration. Consequently, these calibrations should be performed with a maximum of accuracy and systematic deviations from the in-orbit situation should not be tolerated at all.

When comparing the results from two different calibration methods, it is important to distinguish between systematic and random errors. Systematic errors are those which produce a systematic, reproducible offset between the results obtained by the two methods. By repeating the measurements, this offset will not be reduced but will become more pronounced above the "background noise" of random errors. Deviations caused by random errors, however, will be reduced by repeated measurements.

From Tables (4) and (5) of the previous paragraph, it is apparent that the total calibration error is the product of ten different error sources, each of which produce an uncertainty of 1% or less. It is useless to argue on further reductions of any of these uncertainty-bands. One has to accept a reasonable overall error bandwidth of about ±1% for the calibration of SWS and about 2% for LAPPS tests of flight panels.

5. PAST SOLAR CELL CALIBRATION EXPERIENCE AT ESTEC

5.1 Verification of Primary Standards

During 1976, ESA and COMSAT collaborated in a series of "Round Robin" measurements on fifteen solar cells at nine test facilities throughout Europe and the USA, to establish the size of the "creditibility gap" (Reference 9). Short circuit current values are depicted in Figure 10 and show a spread of 5% for cells rejecting UV below 0.35 um and a 13% spread for cells with no UV

rejection filters. These results gave rise to a more general
investigation of primary standards of various types obtained
from the calibration methods described in paragraph 3.4.

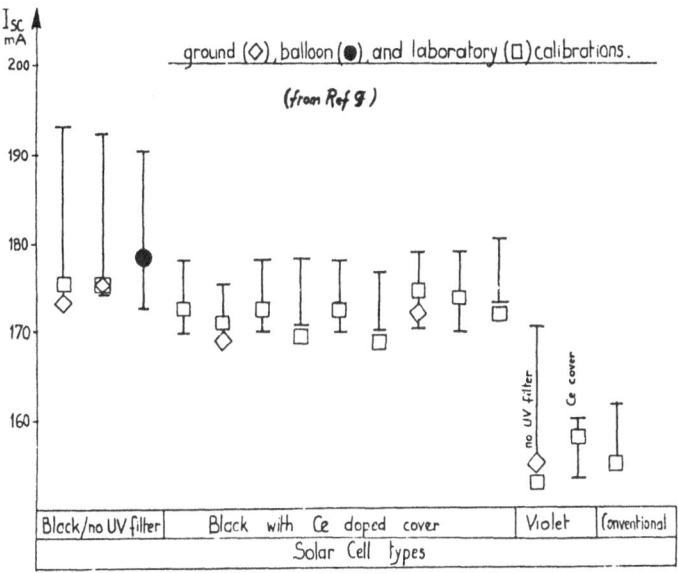

Figure 10: Measurement Discrepancies observed in the Past

A comparison between different calibration methods is often
hampered by the different (permanent) mounting of the standards
which makes it difficult to use the same solar cell for subsequent
primary calibrations with different methods. A parallel calibration of various solar cells of the same spectral response group
with different calibration methods, on the other hand, is usually
too expensive and provides only a reduced data base.

The comparability of different primary standards calibrated in
different years and with different methods has been greatly
improved by the availability of the Bosch sun simulator. The
spectrum match of this simulator is such that spectral mismatch
errors are less than +1% for all the primary standards which had
been compiled at ESTEC in the previous years.

Figure 11 shows the deviations of the calibrated short circuit
current value from the values measured with the Bosch simulator
at AMO equivalent intensity for various primary standards. The
majority of the standards lies within the tolerance band of +1%.
A considerable number of standards, however, show deviations of

up to +5%. In some cases, it was possible to identify the reasons for the anomalous deviation: one group of standards, for example, was originally mounted on white, reflecting base plates. The scattered light was reflected from the base plate onto the solar cell during the calibration. This obviously resulted in optimistic calibration. For several standards, however, no explanation could be found for the anomalous deviation. Moreover, the anomalies were not restricted to a particular calibration method but were found in all methods.

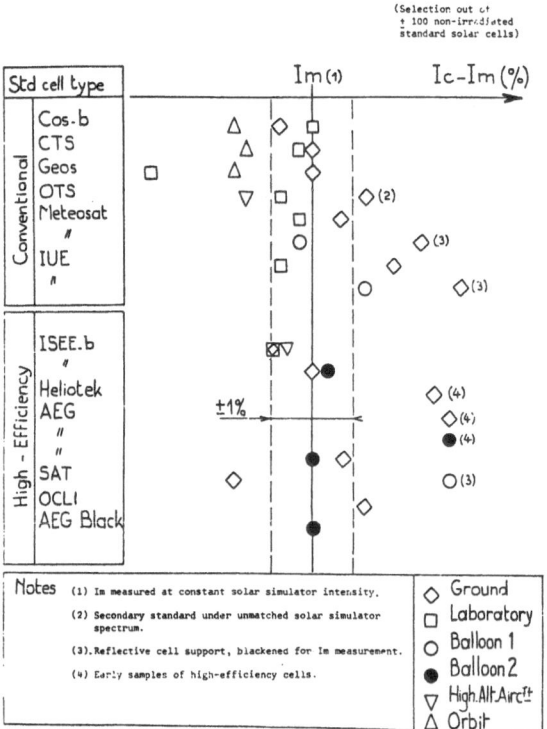

Figure 11: Comparison between Calibrated & Remeasured Isc-Data

Two explanations have to be considered for the standards with anomalous deviation:

- The primary calibrations might have taken place under unfavourable conditions which were not recognised (in fact, the records of most of the primary calibration methods show examples for the invalidation of calibration results due to unfavourable conditions that were recognised only afterwards).

- The primary calibration might have been accurate but the performance of the standards changed in the time between the calibration and the comparison. Damage or inherent instabilities
  of the standards could cause this change.

The results presented in Figure 11 led to the following conclusions:

a) A primary standard is only acceptable if a cross-check with a
large number of other standards under close spectral match sun
simulation has been passed with success.

b) It is necessary to keep a large number of primary standards
which constitute a reliable base.

c) Accurate (and inaccurate!) standards can be obtained with
any of the methods described.

An updated summary of the recalibration data history of frequently
recalibrated standard cells in 1981 showed maximum deviations of
about 5%, confirming the earlier findings described above.

5.2 Comparison with In-Orbit Flight Data

During recent years, several satellites were launched which were
equipped with instrumentation to measure solar cell performance in
orbit: CTS, GEOS-1, GEOS-2, JSEE-B, Meteosat and OTS. A summary
of the solar array performance of these satellites is found in
References (10) and (11).

The spinning satellites with body-mounted cylindrical arrays,
generally yield less accurate data, since data are retrieved
only indirectly from data measured at the main bus or the battery
charger and the effect of spacecraft orientation with respect to
the sun has a large impact. Nevertheless, for Meteosat, good
agreement with the pre-launch indoor measurement was found; for
GEOS-2, the initial data were within the +2% tolerance band of
the ground tests and only JSEE-B deviated by about -4% from the
ground measurement data.

For three-axis stabilised satellites with sun-oriented solar array
wings, more accurate data can be expected, in particular, if special
instrumentation is available.

On CTS (launched in January 1976), a special solar cell test patch
of 3 x 9 solar cells formed part of a solar array experiment.
The test patch was mounted on one of the two sun-oriented deployable solar array wings. The current-voltage curve of the test
match, including a short circuit measurement, was continuously
monitored. The data were transmitted to the ground station via

telemetry, with a resolution of 0.7%.

The measured short circuit current of the test patch was found
to be 1.5% lower than predicted by the ground calibration.
Considering the fact that the solar cell type used for CTS
(1Ω cm cells made from crucible grown silicon) can show a 1%-2%
photon degradation at beginning-of-life, the agreement between
the ground calibration and the flight measurement can be called
very good.

For OTS, which was launched in May 1978, the agreement between
ground measurement and orbit performance was excellent: the
short circuit current of one of the eight array sections was
reproduced within 0.5%.

In December 1981, MARECS was launched and its flight data confirmed the previous findings. Again, the performance prediction
had been accurate within 0.5%.

5.3 The ESTEC Calibration Base (ECB)

The experiences gained in previous years with primary solar cell
standards of various origin, as described in the previous paragraph, have considerably increased the confidence in most of the
standards. Several standards which showed an anomalous behaviour,
as discussed in paragraph 5.1, were excluded.

Only those standards are approved which have been verified by
cross-checking with other reliable standards, i.e. with standards
which lie in the +1% range of Figure 11.

Various procedures for the verification of standard calibration
were studied. The simplest is the direct comparison with accepted
standards under a sun simulator with an AMO spectrum, accompanied
by a spectral response measurement against an accepted spectral
response standard and a calculation of the short circuit current
derived from folding the spectral response curve into the AMO
solar spectrum. From the spectral response measurement and the
spectral distribution of the simulator light, the spectral mismatch error induced in the sun simulator measurement can thus be
calculated and if it exceeds 1.0%, it can be included as a correction into the comparison. A standard is accepted if the short
circuit current derived from the spectral response measurement,
the primary calibration value of the short circuit current and
the short circuit current measured in the comparison under the
sun simulator, agree within +1%. This procedure can be applied
only to solar cell standards which have a linear intensity dependence and do not show spectral quenching or enhancement. If non-
linearities occur, the spectral response measurement has to be

performed in a way that no additional errors are introduced.

An alternative method of standard verification is the comparison of the results of two different primary standard calibration methods. In practice, this is often found very time-consuming and expensive.

The ESTEC solar cell standard base consists at present of about 40 verified primary standards of various silicon solar cell types, covering a wide range of spectral response characteristics. This standard base is used for:

- the periodically repeated control of the accepted standards in order to detect and eliminate standards which have degraded with time;

- the calibration of secondary standards;

- the verification of new primary standards.

5.4 Combined Space/Terrestrial Calibration Procedure

In contrast to the "normal incidence" and the "global" methods, which aim at approaching standard sunlight conditions by using actual terrestrial sunlight, this method requires a precisely matched sun simulator. Therefore, this method can be called the sun simulator method.

Moreover, the sun simulator method uses a set of spectrally different AMO reference solar cells out of an accepted reference standard base (e.g. the ESTEC reference standard base, described in Reference 2), instead of a pyrheliometer or pyranometer. The requirements on a reference standard base are given in Reference 2.

The requirements on a sun simulator for this application have been described in Reference 1 and are summarised as follows:

- The spectral match of the sun simulator to extra-terrestrial sunlight (AMO) shall be so close that spectral mismatch errors resulting from the different spectral response of the full variety of cells to be measured at constant intensity setting of the simulator are less than 1%.

- The short- and long-term stability of the simulator shall be such that a reproducibility of cell calibration of better than +0.3% is guaranteed.

For the calibration of reference cells at standard terrestrial sunlight conditions (e.e. AM 1.5), the following steps are required:

1) Set the simulator to AMO sunlight intensity and verify the spectral match to AMO by:

   a) measuring the relative spectral energy distribution;
   b) comparing the short circuit current values of spectrally different accepted AMO reference standards measured at constant intensity setting of the simulator with the original calibration data.

2) Measure the AMO short circuit of the cells to be calibrated.

3) Measure the relative spectral response of the cells to be calibrated. Make sure that errors due to non-linear intensity dependence, and enhancement/quenching effects are eliminated.

4) Place an AM 1.5 filter which simulates the atmospheric absorption/scattering in the simulator light beam and measure the AM 1.5 short circuit current of the cells to be calibrated.

5) For verification, compute the difference between AMO and AM 1.5 short circuit current of the cells to be calibrated:

$$I_{SC\ AMO} - I_{SC\ AM\ 1.5} = \int S(\lambda) \left\{ E_{AMO}(\lambda) - E_{AM\ 1.5}(\lambda) \right\} d\lambda$$

$$= \int S(\lambda) \cdot A(\lambda) \quad d\lambda$$

($S(\lambda)$ is the absolute spectral response derived from the measured relative spectral response $\xi(\lambda)$: $S(\lambda) = K \cdot \xi(\lambda)$ with $K = I_{SC\ AMO} / \int \xi(\lambda) \cdot E_{AMO}(\lambda) d\lambda$; $E(\lambda)$ is the spectral irradiance; $A(\lambda)$ is the atmospheric attenuation used for the specification of AM 1.5 standard sunlight, as derived from AMO sunlight).

From this, the AM 1.5 short circuit current can be derived and compared with the value measured in (4). A calibration is accepted only if the two values differ by less than 1%.

The advantages of the sun simulator method are:

-   It is independent of weather conditions, seasons, location;

-   It is continuously available under controlled laboratory conditions;

-   Spectral measurements of the light source and of the solar cells are necessary for control purposes only, whereas the actual calibration is based on a simple short circuit current measurement, thus reducing the complexity of the measurement and the influence of spectral errors.

As mentioned above, however, these advantages come into being only
if the rather extensive requirements on the sun simulator and the
reference standard base are fully met.

References

1. "Spectral Match of the Sun Simulators Required for Measuring
   Today's Solar Cells", Bogus, K.- Proc. of the Photovoltaic
   Power Generation Conference, September 1974, Hamburg.

2. "Solar Cell Calibration": Recent Experiences at ESTEC and
   and Proposal of a Combined Space and Terrestrial Calibration
   Procedure", K. Bogus, J. C. Larue and R. L. Crabb - Proceedings of the Photovoltaic Solar Energy Conference, September,
   1977, Luxemburg.

3. Distribution of Energy in the Solar Spectrum - Y. A. Makarova
   and A. V. Kharitonov, NASA TT.F.803, June 1974.

4. "Terrestrial Photovoltaic Measurement Procedures", ERDA/NASA
   1022-77/16; NASA TM 73702, June 1977.

5. Specification No. 101 of the CEC, Issue 2, 1980.

6. "Pulsed Measurement of Solar Cell Spectral Response",
   J. C. Larue - Proceedings: 2nd Photovoltaic Solar Energy
   Conference, April 1979, Berlin.

7. Solar Cell Arrays in Unsteady Electrical Operation -
   O. H. Gruber, Proceedings of 1st Symposium on Photovoltaic
   Generators in Space, Noordwijk, September 1978.

8. "Effect of Partial Shadowing on Solar Panels..", J. C. Larue
   and E. Du Trieu; Proceedings of 3rd Photovoltaic Solar Energy
   Conference, October 1980, Cannes.

9. "What is Simulated AMO?" - J. F. Allison and R. L. Crabb;
   Proceedings of the 12th Photovoltaic Specialists' Conference,
   Baton Rouge, 1976.

10. "The Flight Performance of the CTS ("Hermes") Power Subsystem
    and Solar Array", J. V. Gore and K. Bogus; Proceedings of the
    20th Symposium, Royal Society of Canada, Ottawa, November 1977.

11. Review of In-orbit Performance of ESA's Spacecraft Solar Arrays,
    E. G. Suppa and R. L. Crabb; Proceedings of the 14th PVSC,
    January 1980, San Diego.

METHOD FOR A RELIABILITY STUDY ON PHOTOVOLTAIC MODULES
APPLICATION FOR THE QUALIFICATION OF CELLS AND MODULES

Antony DESOMBRE

QUALITY DEPARTMENT
PHOTOWATT INTERNATIONAL S.A.
6, rue de la Girafe
14000 CAEN
(31) 95 09 46

ABSTRACT

The photovoltaic solar modules for terrestrial applications must
be designed for operation with only power loss during 5, 10 or
20 years in very severe environmental conditions.

   With the view of assuring user satisfaction, high resistance
to the above mentioned stresses must lead to a well adapted design of the solar module and a good reproducibility of its manufacture. For these reasons, it is necessary to analyse the mechanisms in use, through a functional reliability diagram of a solar
module in order to reduce the risk of degradation with an acceleration of these stresses in view of forecasting the life expectancy in use.

   The results are used to determine the qualification plan of
cells and modules with the aim of reducing the qualification
cost. Examples of acceleration effects are given for the temperature, the humidity and the elimination influence of technological
choices.

1. INTRODUCTION

1.1. Definition

The quality is the ability of a product or a service to satisfy
the user's needs. One of the quality aspects of this product is
its reliability which is to assure the intended function during
a sufficient time and under definite environmental conditions,
and this reliability will only be assured after a number of tests
or selected samples.

G. Beghi (ed.), Performance of Solar Energy Converters: Thermal Collectors and Photovoltaic Cells, 221–234.

1.2. Main purposes in a reliability study

This definition involves 4 main purposes in a reliability study.

The first one is to assure a good function. This is the reason why it is necessary to define what can be called "a good function" and to analyse the failure modes in relation to the technological choices and to determine the influence of each part on the system's reliability.

The second one implies the duration notion. It is necessary to specify how long the device can work and give satisfaction to the user.

The third one includes the climatic, electrical or mechanical stresses depending on site in the world and on the conditions of use.

The last one is the probability. In practice, it is impossible to test all the devices in each test. This is the reason why the reliability tests are applied to a sampling.

1.3. Design survey

In view of assuring user satisfaction the reliability target must lead to a well adapted design of the photovoltaic solar module and a good reproducibility of its manufacture.

Before speaking of the method of reliability applied to the photovoltaic module, il would be interesting to conduct a brief survey on the composition of a module.

A module consist of placing a number of cells in series to obtain the wanted voltage. A cell is a photo sensitive diode generally using silicon material. The current is collected on the rear face by a large contact area and transmitted to the following cell by a connection. The front contact area must be designed with a compromise between the minimum occulted area and the minimum series resistance.

When the cells are interconnected they are embedded in a plastic material (like resin or film). The front face protection is assured by a glass plate, plastic film or directly by the potting material. The back face protection consists of a glass plate, metallic substrate or plastic film. The mechanical peripheral protection is generally assured by a rubber and an aluminium frame which allows a fixation on the structure. The electrical outputs consist of cables, connectors, studs, clips, etc...

2. DESCRIPTION OF METHOD

The reliability requirements are very important to ensure a good adaptation to user needs. In the case of the photovoltaic solar modules for terrestrial applications, the life expectancy is more than 5 years (with a goal of 20 years). Moreover a solar module must operate under severe conditions with a combination of environmental stresses met in different parts of the world and different types of applications.

2.1. First step

The first step of this reliability study is to determine the possible stresses in relation to use. The main types of stresses are
- climatic stresses such as temperature, humidity, rapid change of temperature, sunlight,
- electrical stresses such as polarization and module coupling mode,
- mechanical stresses due to the mounting conditions of the modules on array structure, such as twisting,
- mechanical stresses due to the environment such as wind, sand wind, contamination and pollution, abrasion etc...
- mechanical stresses due to an accident, such as impact, hail, stones, etc...

2.2. Second step

In the second step, the possible failure mechanisms are studied in relation to the functionnal reliability diagram (see fig. 1) In this diagram there is a distinction between the catastrophic failure and the drift failure. The catastrophic failure is a sudden and complete lack of function for instance an open-circuit, an intermittent contact or a short circuit.

The drift failure of a part introduces a progressive decrease of the output characteristics and its influence on this characteristic depends on the failure mode (for instance : increase of the contact series resistance, decrease of the encapsulation transparency etc...).
This is the reason why in this diagramm there is one series line for catastrophic failure, including the output connection (+ and -) and n units with the cell and its interconnection. If there is an open circuit of a part, the whole network is opened and it is the same for an intermittent contact.

The drift failure can affect only one part of this series line. This is the reason why there are 3 kinds of attenuators :
. an increase of the series resistance due to a contact corrosion
. a decrease of the sunlight flux up to the junction level, specifically due to a reduction of the transmission factor of the encapsulants,
. an increase of the leakage currents, mainly in relation to the cell junction degradation.

The figures 2 and 3 give examples of evolution of the electrical characteristics due to respective decrease of the transmission of the sunlight up to the junction and an increase of the series resistance. The shape of the electrical characteristic dégradation guides the failure analysis to determine the possible cause of this evolution.

On the other hand a well adapted visual inspection at each step can give failure prediction or confirmation that a visual evolution has no influence on the future electrical characteristics.

2.3. Third step

The third step deals with a research of the accelerated testing and the determination of the acceleration laws and factors. The main parameters are :
. the temperature accelerated in high temperature storage,
. the humidity, in combination with the temperature, accelerated by storage in humidity,
. the sunlight combined with the temperature and humidity with the acceleration of sunlight level, the number of hours of daylight, the cell temperature, the thermal stress and the humidity percentage (during the dark phase),
. the changes of temperature to simulate the day/night cycle, a cloud passage, and the rain during sunlight, by rapid change of temperature (thermal cycle) with acceleration of the transfer times and by increasing the difference of extreme temperatures.
. the mechanical stresses accelerated by a number of different tests such as twist, static and cyclic pressure load, hail or impact, abrasion and pollution etc...
    The interest for accelerating these stresses is to obtain data as quickly as possible in order to assess the life expectancy in different working conditions and to verify whether the technological choices are well adapted to satisfy the application.

2.4.

Finally, when these three steps are completed, it is important to review the design reliability to establish the possible interaction between the technological choices, the type of failure mechanism and the accelerated stresses used to simulate the normal use conditions.
    Then, during a module development phase, each technological choice must be submitted to a suitable test programme. The main relationship between these choices and their influence on the module reliability can be summarized as follows :
. influence of the cell material on the mechanical and thermal stress behaviour,
. influence of the contact technology on the corrosion risk in humidity in association with temperature,
. influence of the interconnection method on the thermal cycle behaviour,
. influence of the encapsulation method and fixing frame on the mechanical behaviour specifically due to mounting structure, static and cyclic pressure load.

. influence of the encapsulation method on the cell protection
  against humidity penetration. The types of front face, rear
  face and side protection must be taken into account.
. influence of the encapsulant on the transmission stability specially under sunlight and at high temperature,
. influence of the front face protection on the resistance to the
  environment such as contamination, sand wind, birds, abrasion
  etc..., on the cleanability and on the mechanical resistance
  due to an accident such as impact, hail, stones etc...,
. influence of the electrical output method on the risk of humidity corrosion.

3. PRACTICAL ASSESSMENT

For a complete reliability study the first approach consists of
ascertaining how the solar module is able to under successfully
all the mechanical and climatic tests described in the available
specifications.
    In the second phase, it would seem necessary to accelerate
these stresses to obtain results as quickly as possible. These
results, at different levels of stress permit assessment of the
main acceleration laws and factors to forecast the life expen-
tancy in different conditions of use. These estimations are used
to verify if the technological choices are well adapted to user
needs and also to set up regular test in order to permit rapid
correction action in processing.

3.1. Quality programme

In general, a customer wants to buy modules which are qualified
by himself or by an authorized organization. This is the reason
why it is necessary to define a quality test programme.
    From the manufacturer's point of view it is more convenient
to have a harmonized system to avoid multiplication of different
quality test procedures depending on the customer.
    For the customer it is more economic to use modules quali-
fied by an authorized organization, especially if his needs are
lower in comparison to the production level of the supplier.
    These quality test programmes are defined by a national,
european or international organization after preparation by an
had-oc working group composed of customers'representatives, manufacturers and public organizations. At the request of the commission of the european communities the joint research centre of the
Ispra establishment was given the task of coordinating this
action inside a European Working Group on photovoltaic testing
and to apply the defined quality test programme. This test programme is studied by this working group in a proposed specification "test recommendation for quality testing of terrestrial
photovoltaic modules".

This specification describes a lot of requirements about test description, failure criteria and test procedure. When a manufacturer intends to introduce a new product to the market he must prove that this module can successfully pass this quality test. This is the reason why the first step is to verify this point. But this is insufficient to know fully the module behaviour on any site and the life expectancy in each type of climate depending on the location of use.

3.2. Accelerated test programme

In view of obtaining precise informations about the module reliability in use it is too long to wait for the results of a number of modules on different sites. This is the reason why during the development phase of a technology or a module, PHOTOWATT International S.A. would want to have a number of accelerated stress tests conducted to know the failure mechanism of each technology and the life expectancy of the module.
This programme is based on acceleration of climatic, electrical and mechanical stresses. In fact, it is possible to design a module to endure the mechanical stresses with a sufficient safety margin. This is more complicated for its climatic stress behaviour. This is the reason why the first reliability survey set up by PHOTOWATT International S.A. has consisted in studying how the temperature (in storage and in thermal cycling), the humidity in association with temperature, the solar radiation in association with temperature and humidity may influence the technological choices, mainly for contacts, interconnections and encapsulation. In this type of study it is necessary to keep in mind that the acceleration laws are only valid for one failure mechanism and the acceleration factors depend on the tested technology.Moreover each technology has its own limit and this involves avoiding to generate phenomenon which have no relation to its us. In view of obtaining significant results at several acceleration levels, even relatively near the maximum conditions of use, the first approach consists of studying a known sensitive technology to the concerned stress. When the acceleration law and factor are determined these results are verified on the selected technology. The first phase of this programme was to define the test conditions,the intermediate and end point parameters and the failure criteria.
        - temperature influence
There are 2 kinds of influences : the storage at high temperature and the rapid change of temperature. For the high temperature storage it is necessary to test the technologies at the max. published temperature and if we want a more accelerated test we must avoid to generate secondary effects without relation to the use.The specification give 100°C as a good compromise or a more realistic approach consists of taking into account the nominal operating cell temperature and adding + 50°C to be in the worst case (the result is between 85°C and 100°C).

An example of this approach is given in figure 4 where we have plotted the time before observing the same level of colouration of the encapsulant in function of the ambiant temperature. It appeared that the results were aligned on a straight line when we used a logarithmic scale for the time duration and a line scale for the inverse of temperature expressed in KELVIN degrees. So, we can assimilate that to an ARRHENIUS law type :

$$\nu = A \, \exp \left[- \frac{B}{T} \right]$$

Where ν is the reaction speed
      T is the temperature in KELVIN degrees
      B is a constant depending on the analysed technologies
          (B is a function of the activation energy)

In practice we have observed that there is an acceleration factor of about 1,5 by step of 10°C.

The rapid change of temperature test intends to simulate the variations of temperature in use due to the day / night cycle, the thermal cycling when it rains on an illuminated module and when there is a cloud passage. We can accelerate these three phenomenons by increasing the temperature range up to the maximum published limits (in general -40°C, +85°C) or more, and by reducing the transfer time between the 2 extreme temperatures (in general the total time of a cycle is around 2 or 3 hours depending on the equipment capacities).

But, it appeared that it is very important to verify during the whole cycles the electrical continuity of the module to prevent open circuit or intermittent contact at a specific temperature.

The experience indicates that 200 cycles in these conditions are equivalent to 5 or 10 years in use.

- humidity influence

The photovoltaic modules are often used in tropical regions where there is a high percentage of relative humidity with a high temperature. Due to the variability of these two parameters from one region to another, it is necessary to determine the life expectancy in relation with them.

In order to study this humidity influence on the contact corrosion, it appears necessary to combine this effect with the module temperature. Three types of tests can be used : 55°C with 95% RH, 85°C with 85% RH and the boiling water steam test (only for investigation).

In the first step, PHOTOWATT International S.A. has embedded humidity sensitive cells in modules to study the contact corrosion in these three test conditions. For each test we observed the time to detect the same series resistance evolution.

These results are plotted on a graph given in figure 5. If we use
a logarithmic scale for the time duration and a linear scale for
the sum of temperature in degrees CELSIUS and the percentage of
relative humidity the results are aligned and we can use the experimental law :

$$\delta = \exp \left[A + B(\theta\,°C + \%\ RH) \right]$$

where A and B are two technological factors.

In fact it is important to distinguish two successive phenomena. First the penetration velocity of the humidity inside the
module depends on the resistance of the encapsulation (choice of
encapsulant, type of front, back and side protection, distance
between the edge and the sensitive zones). Secondly the reaction
speed of the humidity on these contacts leads to a power loss.
    The humidity penetration way depends on the encapsulation
technologies. The best protection for the cells are two laminated
glass plates and the design must be made taking into account a
sufficient distance between the edge and the contact area.

- sunlight influence

This parameter is difficult to accelerate due to the fact that
it is necessary to reproduce the solar spectrum and to increase
the sunlight level especially in the UV wavelength.
But, if we increase the sunlight level there is simultaneously
an increase of the cell temperature especially by the infra-red
radiation.
    This is the reason why the light source can be a xenon arc
lamp with a sufficient UV radiation. In fact if we want to test
a big size module, it is necessary to put a lot of xenon arc
lamps to have a sufficient homogeneity on the tested module.
Moreover the ambient temperature must be chosen to limit the cell
temperature at a maximum level with the aim to have a sufficient
acceleration factor.
    To effect this test in pratice we illuminate the module for
a significant number of hours per day (more than 20 days) with
an illumination acceleration factor by bringing the module
nearer the light source, and by bringing an ambient temperature
such as the module temperature up to the maximum value. In this
case there are 3 kinds of acceleration factors : the illumination
level (specially in UV), the duration of this illumination by
day, and the temperature effect. During this illumination phase,
we can project a stream of water to create a thermal shock.
During the off-phase (less than 4 hours per day) we stop the
illumination, the module temperature is reduced at 25°C and we
introduce 100% relative humidity. An other type of sunlight
simulation is the UV test but experiments show that it is interesting to combine this UV effect with the temperature.

3.3. Application by PHOTOWATT of this method of reliability for the qualification of cells and modules

The life cycle of a module can be divided into the following phases :

STUDY OF TECHNOLOGY

From the long experience from the reliability studies on cells and modules, PHOTOWATT International S.A. defines for each technology the well adapted test programme. During the R & D phase, it is necessary to have a good cooperation between the development team and the quality department. The quality department participates in the design from an engineering point of view and studies the possible technologies which can be divided into two parts :
. technology of the cells = material, contact technology,inter connection method,
. technology of the encapsulation = type of front face, rear face and side protection, encapsulant, frame design, electric output.

For each technological choice we set up the suitable test programme and the results are very useful to determine the good technological choice.

STUDY OF A NEW MODULE

When PHOTOWATT International S.A. wants to introduce a new module onto the market we study the influence of each technological choice on a complete module integrating the mutual effect of each part and taking into account the influence of the module size and the design.

This qualification plan is set up to determine the reliability of the module (ie the life time in function of the different types of climates).

If the results show that the module is well adapted to the user's need, PHOTOWATT International S.A. starts the production phase.

PRODUCTION OF THE MODULE IN LARGE SCALE

During this production phase the quality department of PHOTOWATT International S.A. assures the reproducibility of the process by a number of controls during the process, by a quality assurance programme on the module and by long term reliability tests which are compared to the results in site. All the informations are compiled to consolidate the results obtained during the development phase.

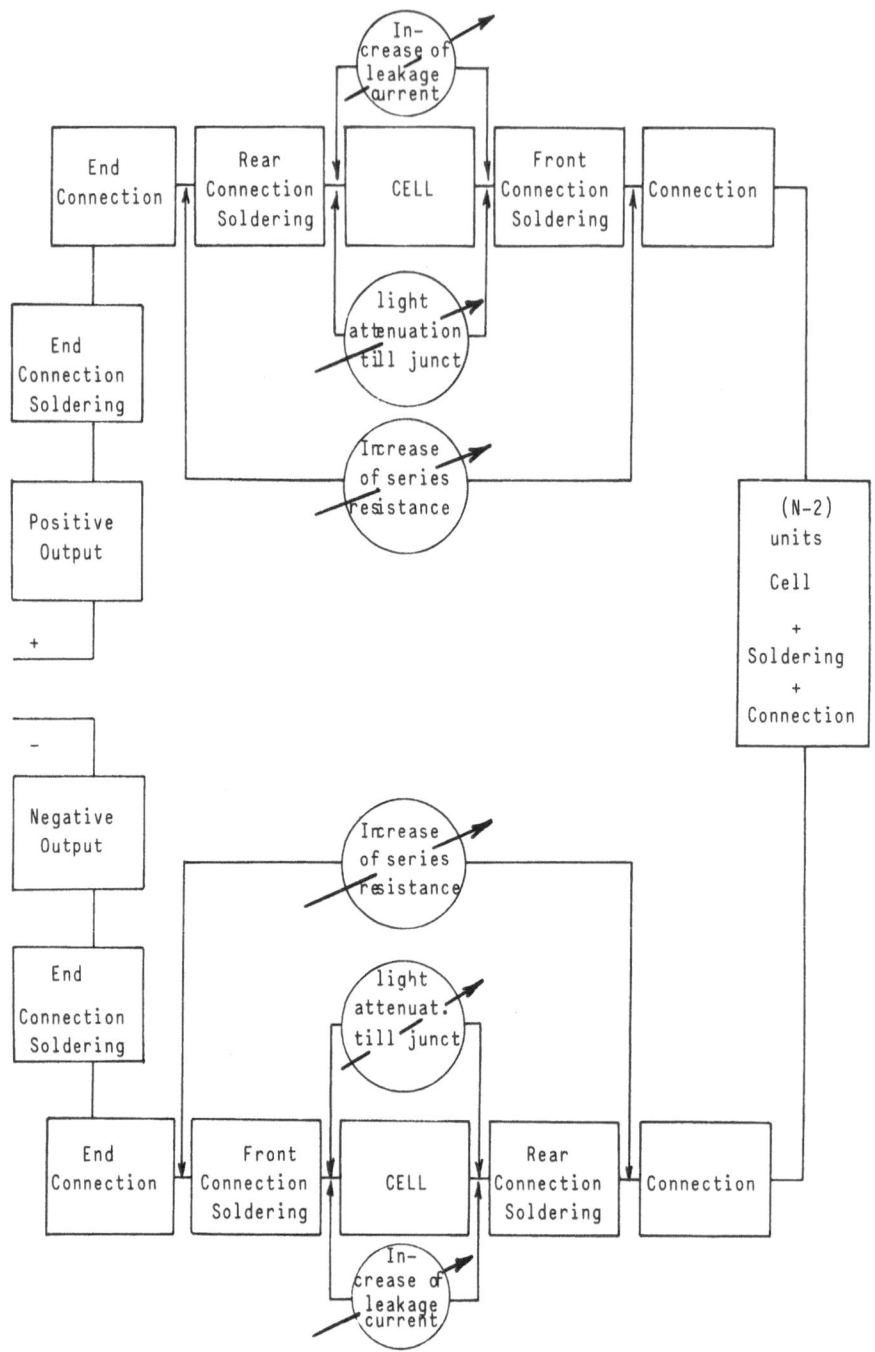

Figure 1. FONCTIONNAL RELIABILITY DIAGRAM

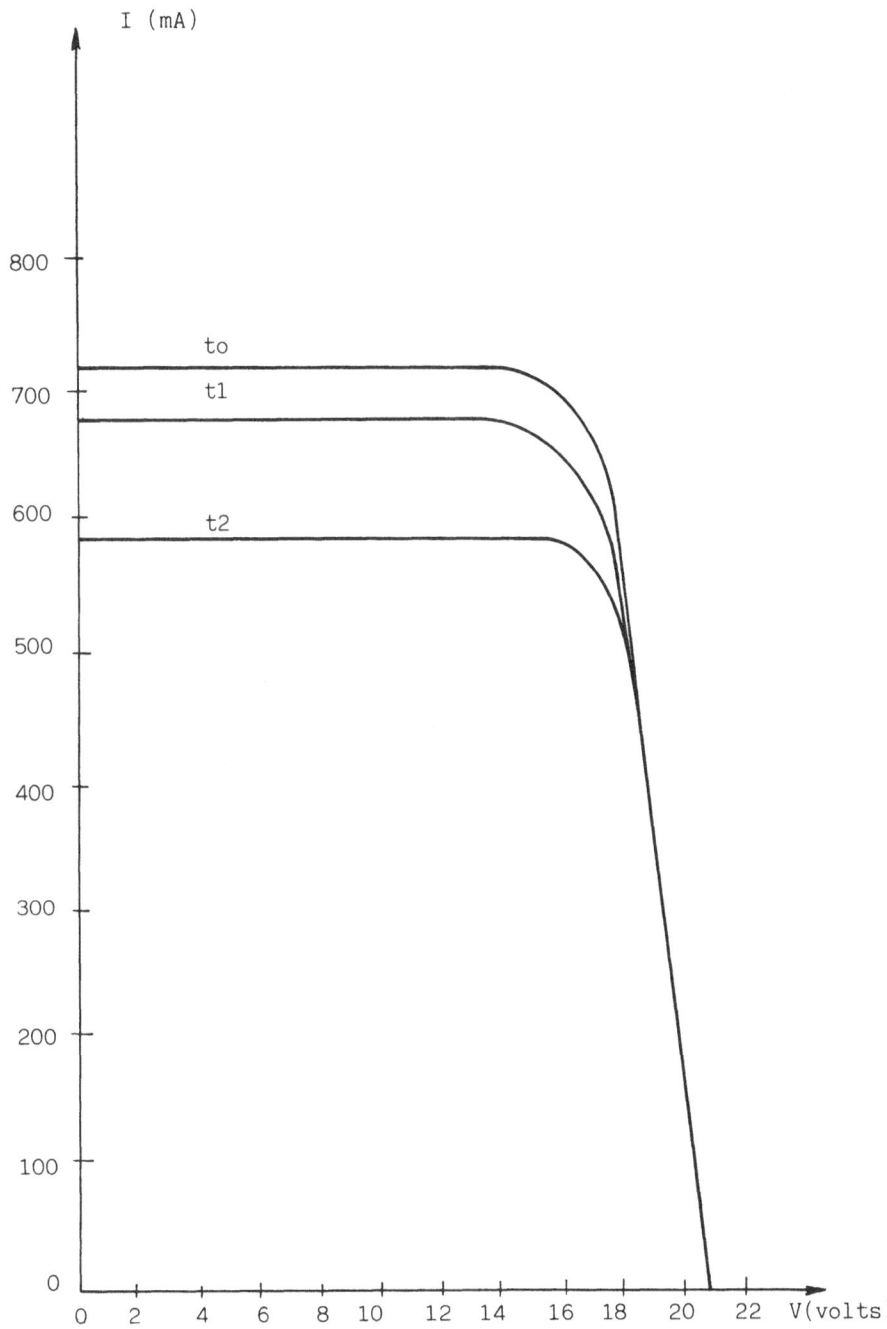

Figure 2. EVOLUTION OF THE ELECTRICAL CHARACTERISTICS
DUE TO A DECREASE OF THE SUNLIGHT TRANSMISSION

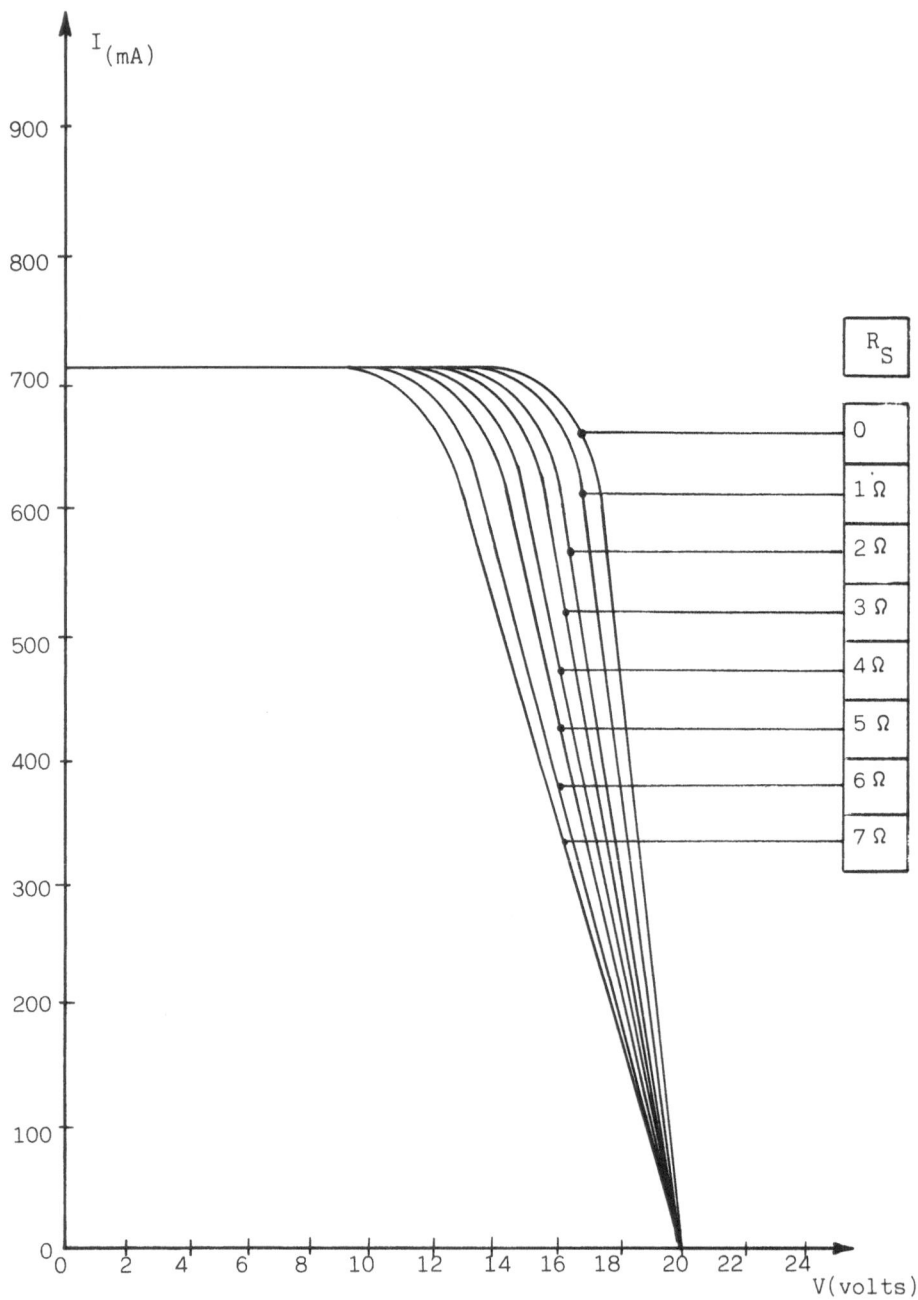

Figure 3. EVOLUTION OF THE ELECTRICAL CHARACTERISTICS
DUE TO AN INCREASE OF THE SERIE RESISTANCE

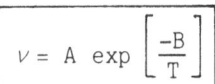

$$\nu = A \exp\left[\frac{-B}{T}\right]$$

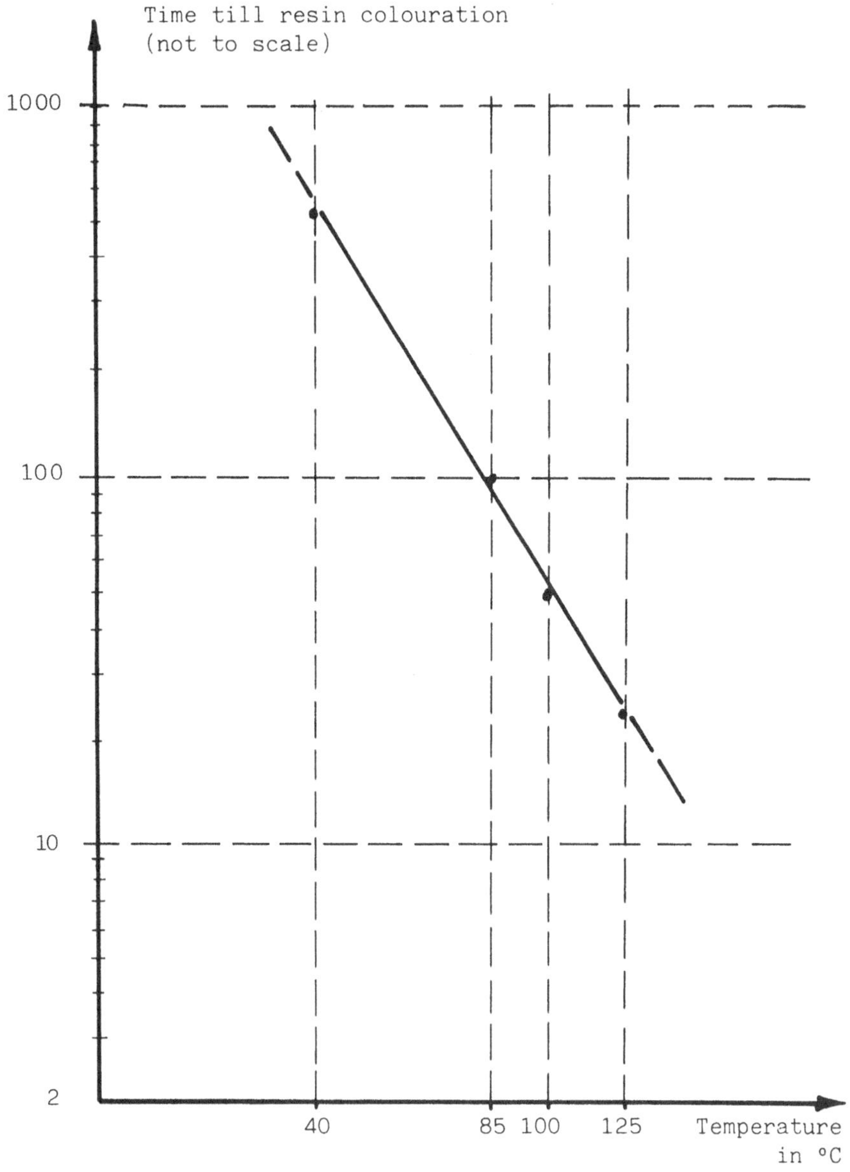

Time till resin colouration
(not to scale)

1000

100

10

2

40   85 100   125   Temperature
in °C

Figure 4. EXAMPLE OF ACCELERATION IN TEMPERATURE

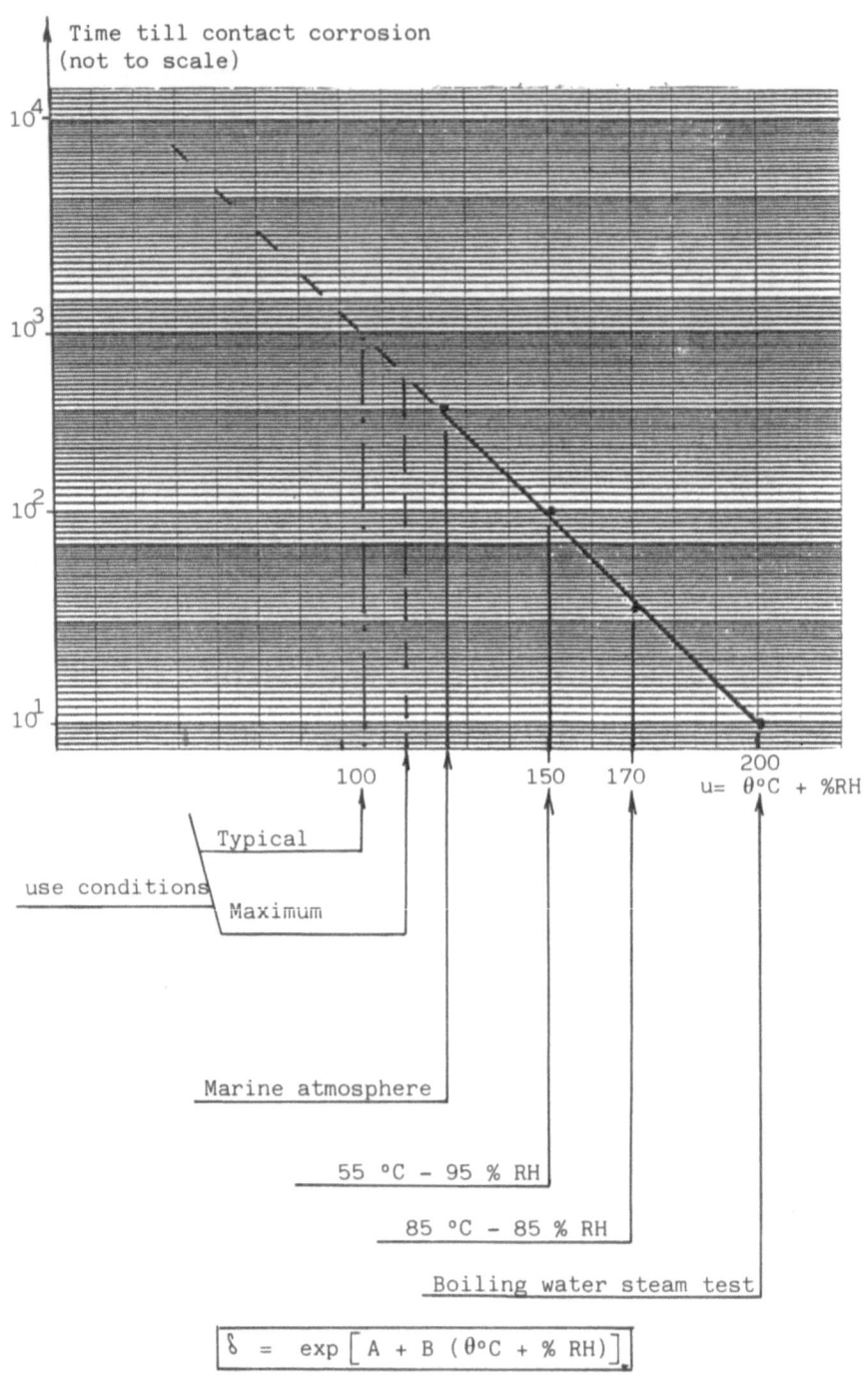

Figure 5. EXAMPLE OF ACCELERATION IN HUMIDITY + TEMPERATURE

PHOTOVOLTAIC INDOOR TEST FACILITIES AT THE JOINT
RESEARCH CENTRE - ISPRA ESTABLISHMENT
SOLAR SIMULATORS

E. Rossi-Gianoli and S. Pace

Commission of the European Communities
Joint Research Centre - Ispra Establishment
I-21020 Ispra (Va), Italy

1. INTRODUCTION

The most expensive but also most critical components of any
solar indoor test facility are devices to simulate natural sunlight.
Their quality determines to a large extent the accuracy and the
significance of test results. In the past large solar simulators
were only used for space technology research problems, a field
where the cost of test objects or their later use justify rather
large investments. In the field of terrestrial applications, say
for domestic water heating or for small photoelectric devices,
we have at first sight the opposite tendency, solar simulators
tend to be much more expensive than the objects under test.
However, this at first obvious discrepancy is somewhat different
if we consider the number of objects to be tested. If solar energy
will have the expected large impact, costs for testing will become
negligible against the actual production and sales volume. On the
contrary, the risk to produce or invest large amounts of money
in this new industry requires even the most careful and thus
expensive testing, since even minor improvements in performance
or in device lifetime would have very substantial financial conse-
quences. Against this background we consider it entirely justified
to make large investments in test facilities. This should be done
at least in certain test agencies, if not also in any large company.
Indoor stations could have a substantial economical impact in as
far as these facilities would provide useful and repeatable guarantee conditions. This would allow not only to verify if material

G. Beghi (ed.), Performance of Solar Energy Converters: Thermal Collectors and Photovoltaic Cells, 235–258.
Copyright © 1983 ECSC, EEC, EAEC, Brussels and Luxembourg

corresponds to specifications, but also to differentiate between meteorological and material risks and to eliminate additional costs due to too large safety margins.

At present the main use of our facilities is to make performance and qualification tests of prototype and production modules which will be installed at 18 pilot plants in Europe.

In the following sections we would like to describe briefly some fundamentals on solar simulation, then the ESTI (European Solar Test Installation) facilities and some typical results.

2. SOLAR SIMULATION

A complex problem lies in the production of radiation with the qualities of natural sunlight. The basic characteristics of natural sunlight are:

- intensity (max. about $1.0-1.2 \text{ kW/m}^2$)
- uniformity (perfect)
- quality (direct and/or diffuse)
- spectral irradiance (characterized by various air mass values)
- solar subtense angle ($\pm$ 16 min.)
- daily variation (dependent on latitude and declination angle).

To construct and operate a single solar simulator with all these properties would be difficult and expensive and not even necessary for all measurements. The analysis of the needs for a solar simulator reveals essentially two fields for applications:

a) For many tests of thermal and photovoltaic systems one requires uniform irradiation across an extended test area, the incident energy should be variable between 100 and 1200 W/m^2 and the spectrum should be daylight-like.

b) For other tests on angular incidence effects and on concentrator devices, the parallelism of the incident rays will be essential.

These arguments lead to the definition of two simulator types:

1) A simulator with a rather large collimation angle and a uniformity in a test plane.

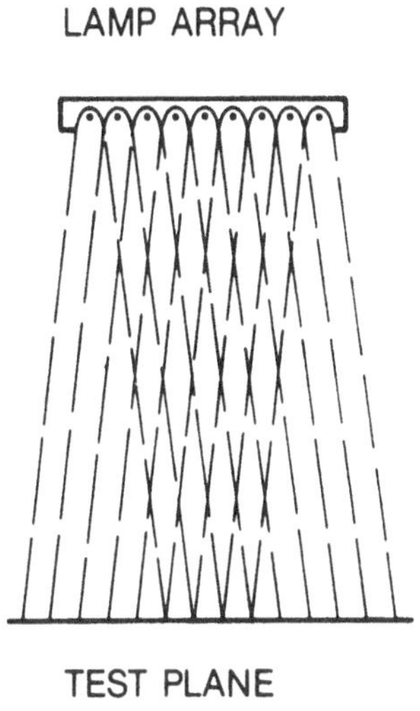

LAMP ARRAY

TEST PLANE

Figure 1. Multiple source simulator

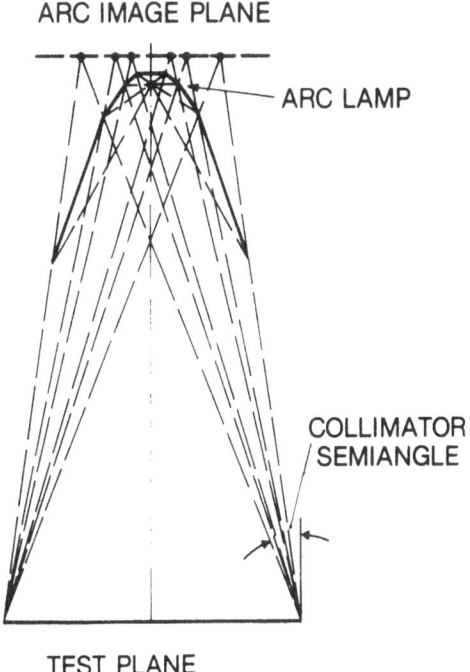

ARC IMAGE PLANE

ARC LAMP

COLLIMATOR
SEMIANGLE

TEST PLANE

Figure 2. Multiple mirror simulator

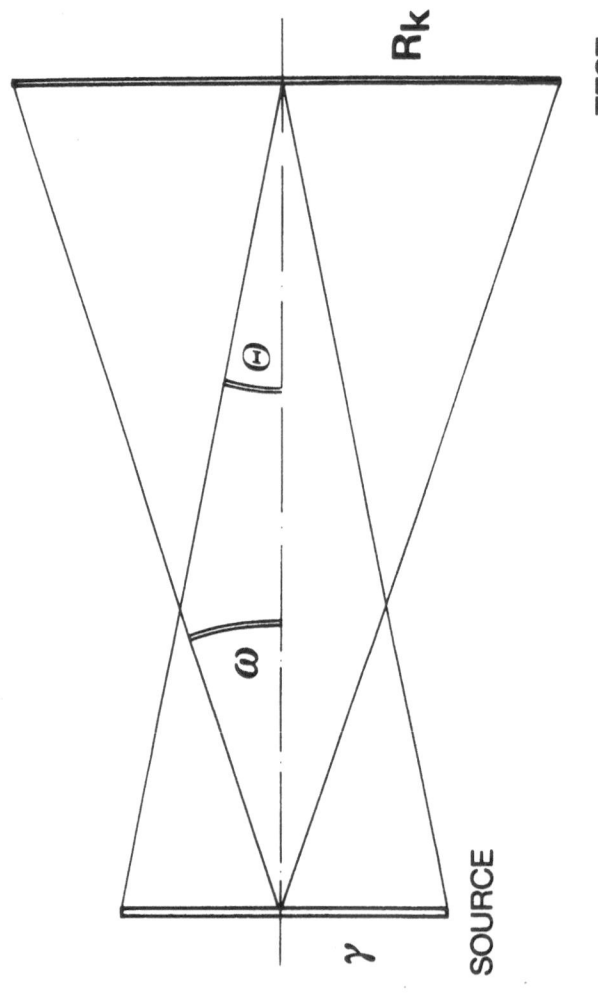

ω : DIVERGENCE ANGLE
Θ : COLLIMATION ANGLE

Figure 3. Collimation angle

2) A simulator with a small collimation angle and a uniformity in a test volume.

The most economic solution for a type 1) simulator (uniform but uncollimated light) is an array of many lamps (Fig. 1) with a sun-like spectrum; such lamps have been developed in recent years for photographic purposes, they are gas discharge lamps with special rare earths additives.

Because of the rather large emission angle of the lamps ($\alpha \simeq 40^{\circ}$) this source cannot produce collimated light. The lamps should be distributed in such a way that the superposition of the individual light beams produces a uniform intensity distribution.

Another solution is a system of multiple mirrors around the single source (Fig. 2); however, this requires a very high intensity lamp. (A system of this type has been installed at Canada's National Solar Test Facility).

For the type 2) simulator (uniform and parallel light) high power xenon short arc lamps are needed, only such high power lamps are able to produce sufficient intensity at small collimation angles. This might be understood from the following:
The basic problem of solar simulation is the generation of uniform irradiation with an intensity and a subtense angle corresponding to natural sunlight within the whole test volume. Small collimation angles are obtained by putting the apparent source to infinity using a suitable optical device.

The irradiance at a point in the volume is given by

$$E = t\pi B\Omega_o \sin^2\theta.$$

Here t is the transmittance factor of the optical system, B the irradiance of the source, Ω_o is the unit solid angle and θ is the half-angle subtended by the source as seen from a point in the test volume.

If we want to simulate natural sunlight conditions, we have to generate $E \simeq 1$ kW/m^2 with a θ-value of 16 min. This means that the radiance (luminance) should be:

$$B = 46.2/\pi \ (MW/m^2 \ sr) = 1.44 \times 10^9 \ (cd/m^2).$$

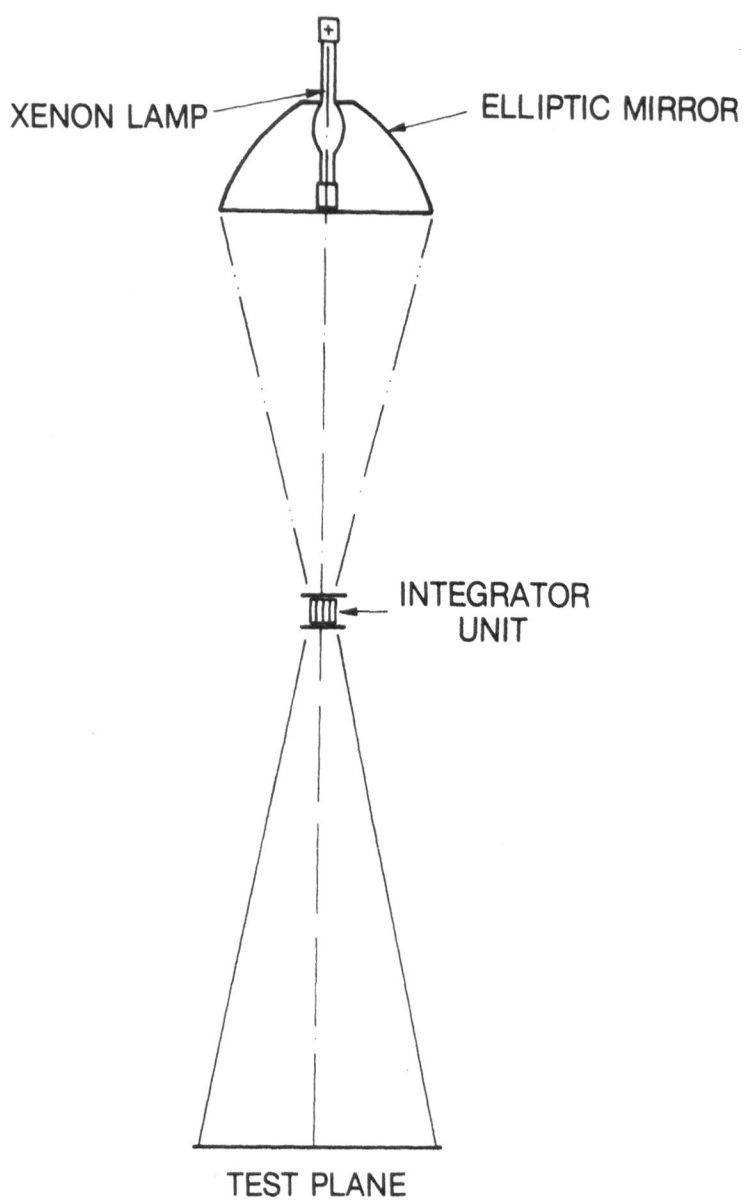

XENON LAMP   ELLIPTIC MIRROR

INTEGRATOR
UNIT

TEST PLANE

Figure 4. Light simulator with integrator unit

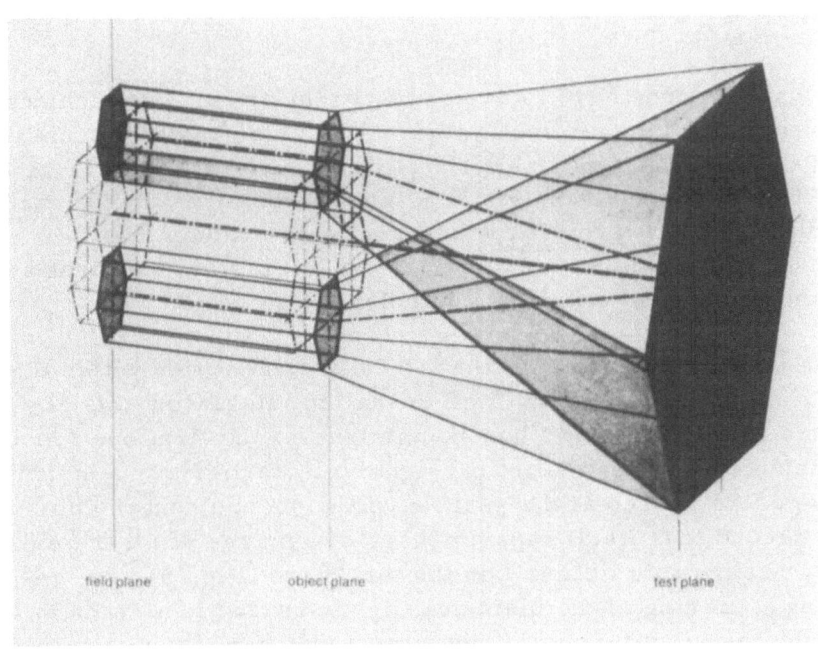

field plane   object plane   test plane

Figure 5. Integrator

Such high luminance values can be obtained using high pressure short arc xenon lamps (e. g. a 30 kW lamp produces 7.5×10^9 cd/m^2).

If we need not only a high intensity but at the same time a small collimation angle _and_ parallel rays we have to image the apparent source at infinity using a collimation system. If r is the radius of the apparent source, R_k the radius of the collimator ($\simeq$ radius R of the test plane), θ the half collimation angle, then the divergence angle ω is given by (Fig. 3):

$$\omega = \frac{R}{r} \theta .$$

On the other hand, since the apparent source has to be imaged at infinity, we have to put it into the focal distance of the collimator, thus

$$\omega = \frac{R}{F}$$

or

$$r = \frac{\theta}{\omega} R = F \ tg \ \theta.$$

The last relations have to be used to fix ω and r. The technical problem arises then if such values can also be realized from the point of view of the available light source, with the additional condition that the irradiation in R must be uniform. In the case of solar simulation for space research this problem has been attacked in various ways, the two most interesting solutions are shown schematically in Figs. 4 and 6.

In the first system (Fig. 4) the intensity distribution within the test plane is made uniform by a so-called integrator: this is a system consisting of two honeycomb lenses; the first one produces in conjunction with an elliptical light collector mirror a multiple image of the source at the position of the second honeycomb lens. This second lens itself superimposes the images of all the elementary segments of lens 1 in the test plane (Fig. 5). The resulting averaging of the individual light contributions leads to a high degree of uniformity. (Systems of this type have been installed at the European Space Research and Technology Centre, Noordwijk, the Netherlands).

In the second system (Fig. 6) the averaging process is based on

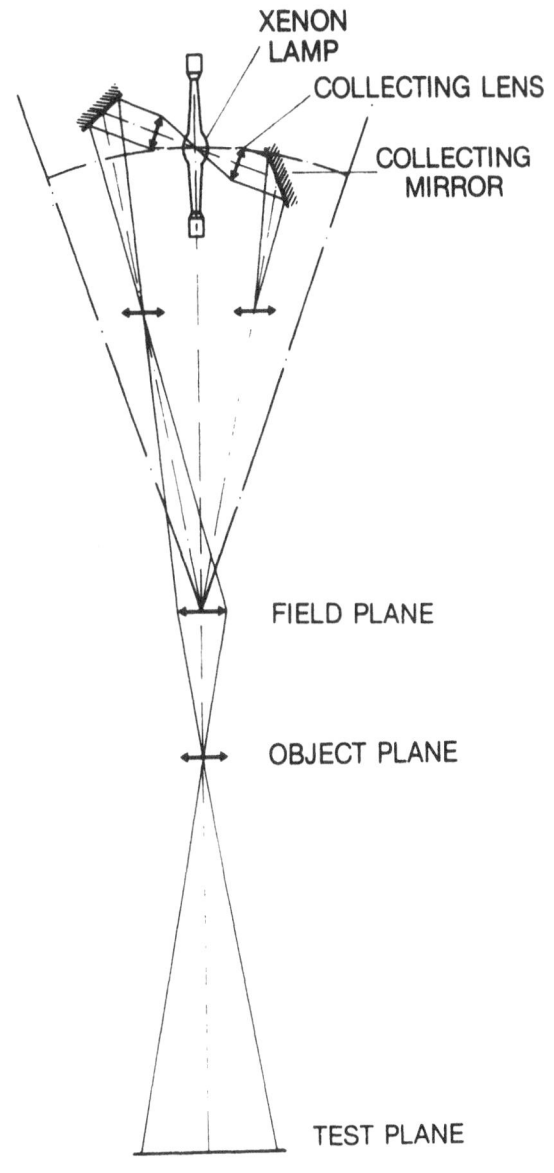

Figure 6. Light simulator with mirrors and lenses

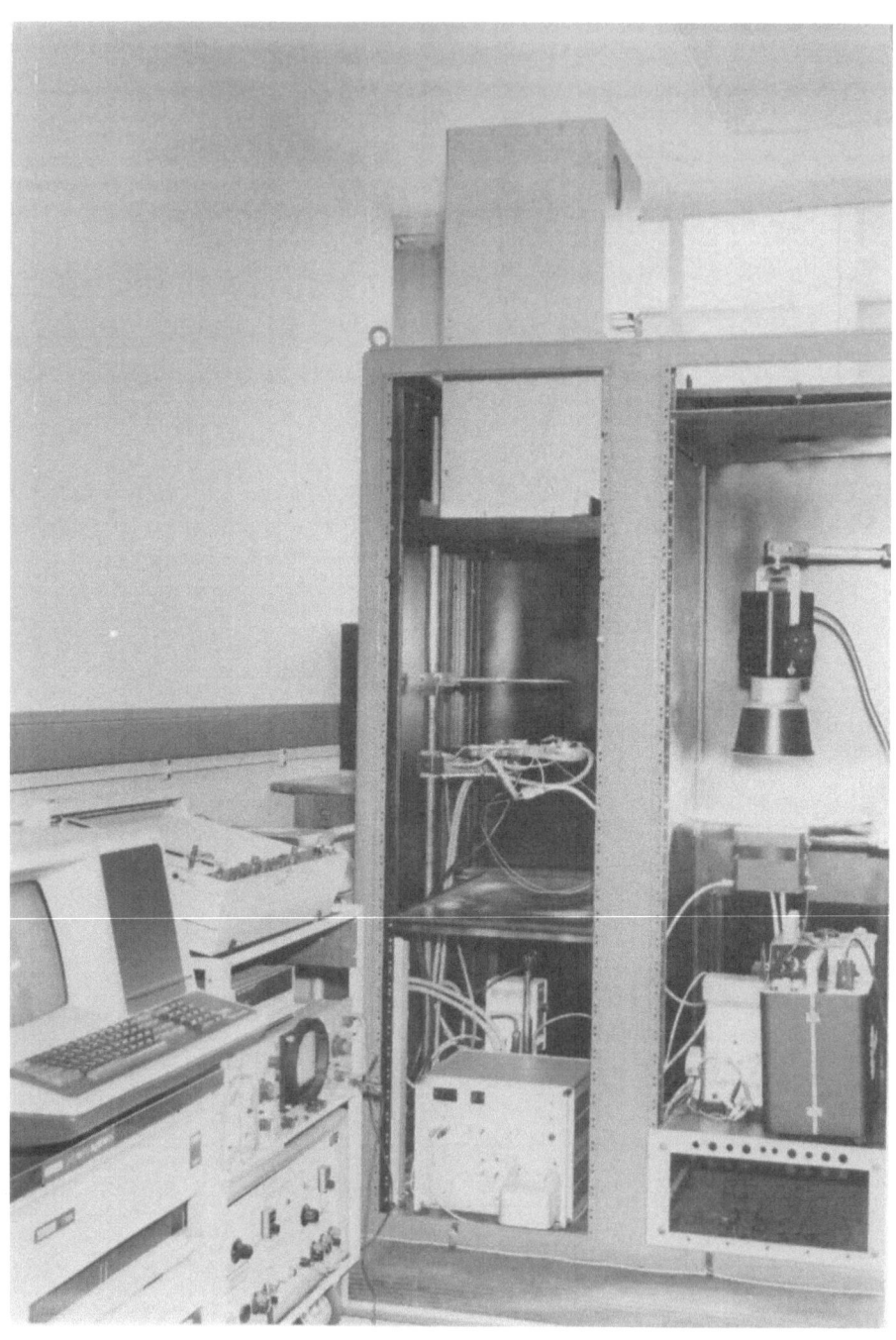

Figure 7. LS-0 and SPR facilities

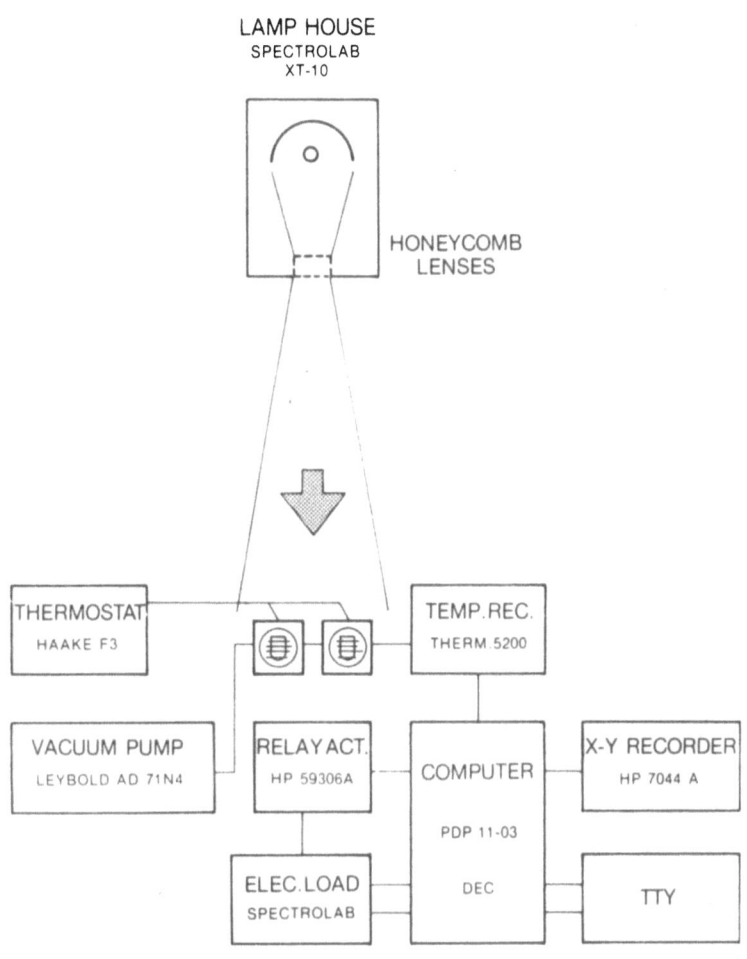

Figure 8. Block diagram of LS-0

the direct superposition of uniform sections of the light emitted
by the arc of each lamp, thus light collection and superposition
are made in one step. This second system should have an advantage if light with a very small collimation angle ($< 1^o$) is needed.
(Such a system has been installed at the Centre National d'Etudes
Spatiales, Toulouse, France).

3. ESTI FACILITIES

3.1 LS-O

The light source is a 1 kW xenon arc lamp (HANOVIA) which
irradiates a 20 x 20 cm^2 test area with an intensity of 1000 -
1400 W/m^2. The uniformity of the light intensity distribution
($\pm$ 3%) is obtained with an integrator comprising 70 lenses. A
set of 35 interference filters inside the integrator allows a spectral match close to the AMO distribution.

This facility (Fig. 7) is used for calibration of solar cells, in
the sense of secondary standard production and for determination
of their temperature coefficients. The reference cell and the
test cell are held on a special support with a vacuum system.
The cell temperature is kept constant at 25 ± 2^oC using a thermostat. The I-V curve is measured with an Electronic Load which
is under control of a PDP 11/03; curves can be plotted on an
X-Y plotter and are stored on a floppy disk (Fig. 8).

The I-V curves of the reference solar cell (AEG No. 166, 5 x 5
cm^2) and of the test cell (Ferranti No. 127, 5.7 cm $\emptyset$) executed
at 1000 W/m^2 according to the CEC Specification 101, are plotted
in Fig. 9.

Figure 7 shows near the LS-O simulator a facility for solar cell
spectral response measurements. The facility consists of a
1500 W$_s$ Multiblitz flasher, a filter wheel actuated by stepping
motor and the test plane with a vacuum hold-down system for
the solar cells. The short circuit current values corresponding
to the maximum light intensity, are measured by two peak detectors.

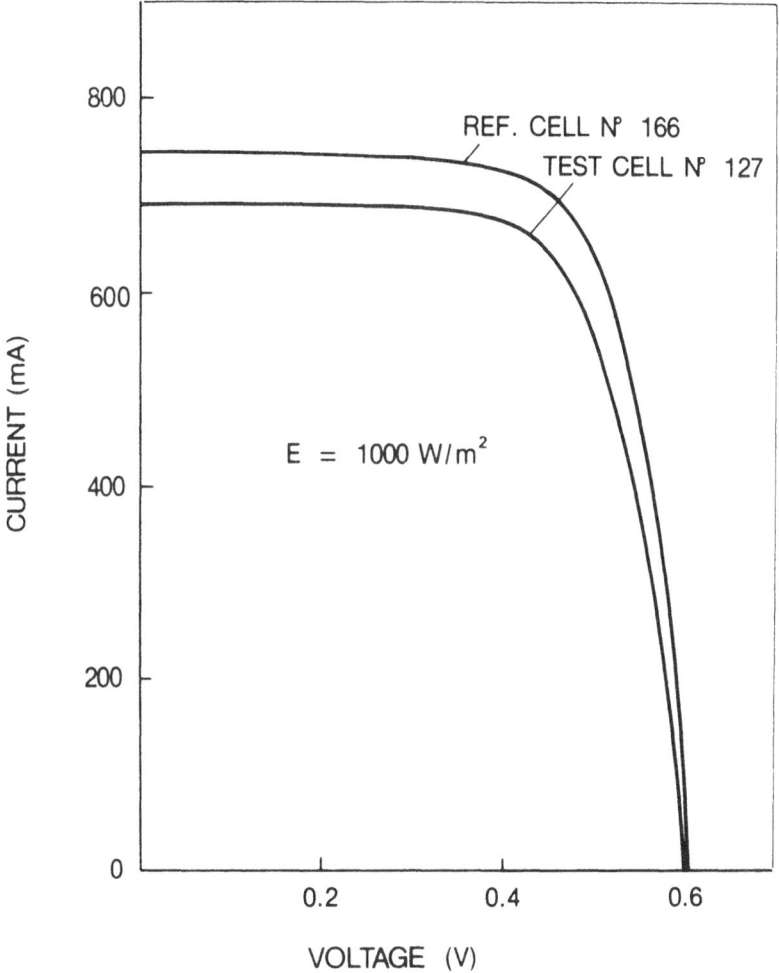

Figure 9. Results of solar cell calibration

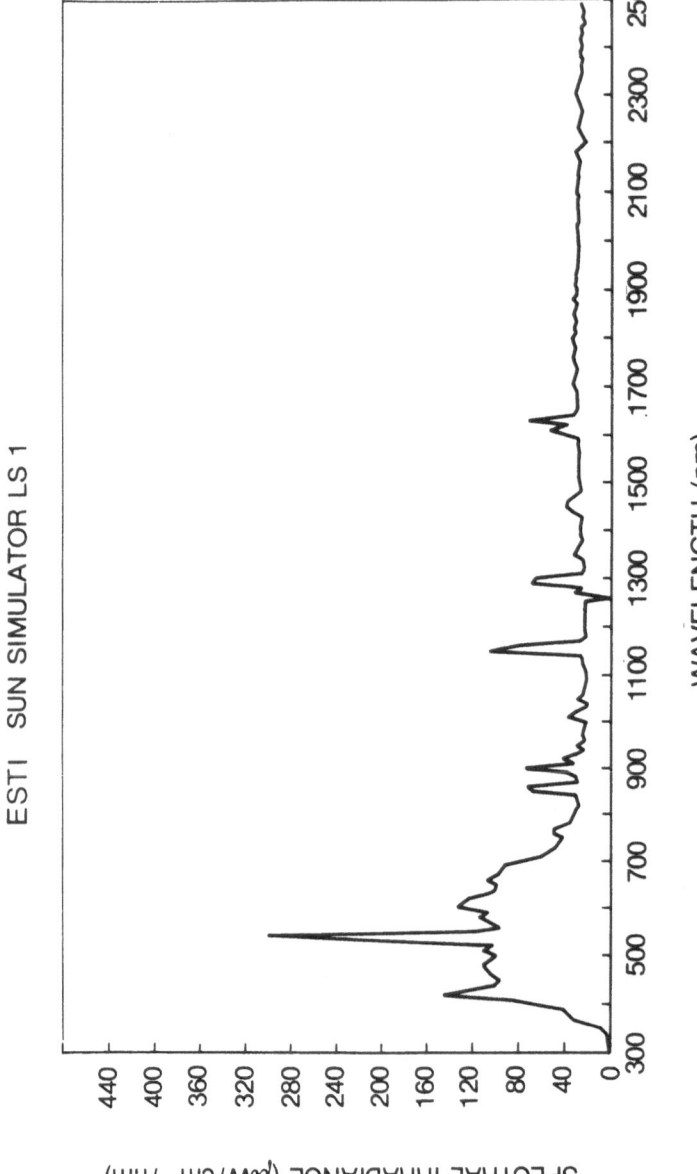

Figure 10. LS-1 spectrum

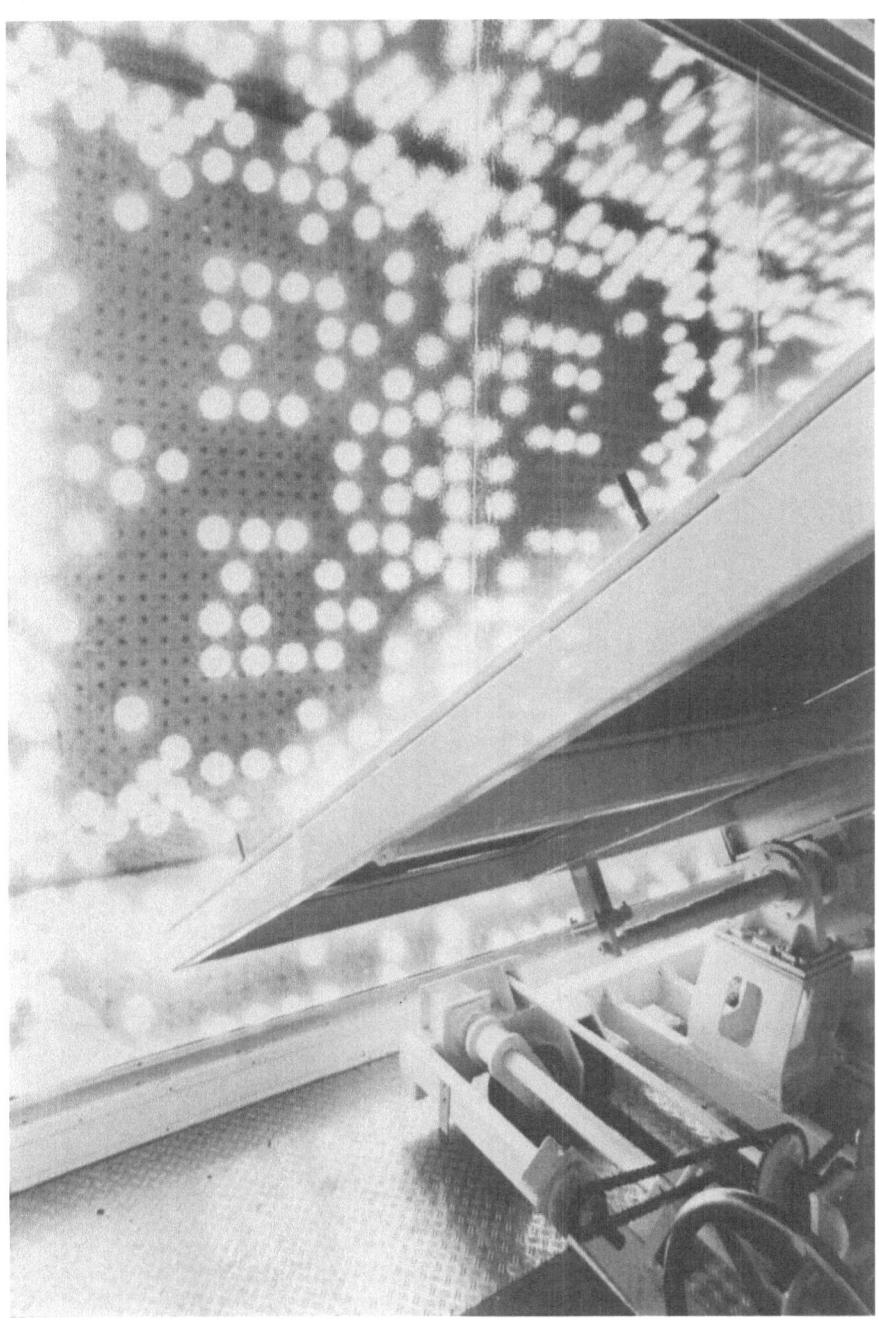

Figure 11. LS-1 lamp distribution

Figure 12. LS-1 facility

3.2 LS-1

The radiation is produced by 295 HQI-R discharge lamps having the spectrum, as measured with a spectroradiometer, shown in Fig. 10. The lamps are distributed in such a way that the superposition of the individual light beams produces a uniform intensity distribution ($\pm$ 3%) on a test plane of 3 x 4 m^2 (Fig. 11). In order to reduce boundary losses a reflecting mirror channel of 1.4 m depth has been placed between source and test plane.

A variation of the light intensity from 200 to 1200 W/m^2 is possible by switching on different lamp patterns. The light source is separated from the test chamber by a low absorption glass window. Within the test chamber it is possible to change the air temperature between -40 and +60°C. The temporal temperature stability amounts to $\pm$ 1°C, heating and cooling rates are about 0.3°C/min (Fig. 12). Data acquisition is made automatically by means of a PDP-11/03 station computer.

This facility is mainly used for thermal collector performance measurements; in case of photovoltaic testing we use it for the temperature cycling test and the humidity-freezing test according to the CEC Specification 501.

3.3 LS-2

The light source is a high pressure xenon lamp (Ushio Type 25 kW/UXW 25000F) which irradiates a hexagonal test plane area with a width of 1.7 m. The uniformity of the light intensity distribution ($\pm$ 5%) is obtained with an integrator comprising 18 quartz rods. The spectrum of this simulator, as measured by a spectroradiometer, is shown in Fig. 13. Since in this case no filters are used, we see the characteristic xenon peak.

This facility is used for indoor measurements of NOCT (Nominal Operating Cell Temperature) value of modules. The test procedure essentially consists in measuring the temperature of the module under 800 W/m^2 using a calibrated module for obtaining the required standard conditions (Fig. 14a, b).

A collimation mirror combined with the simulator also allows to measure angular incidence effects and the efficiency of concentrator cells.

ESTI   SUN SIMULATOR LS 2

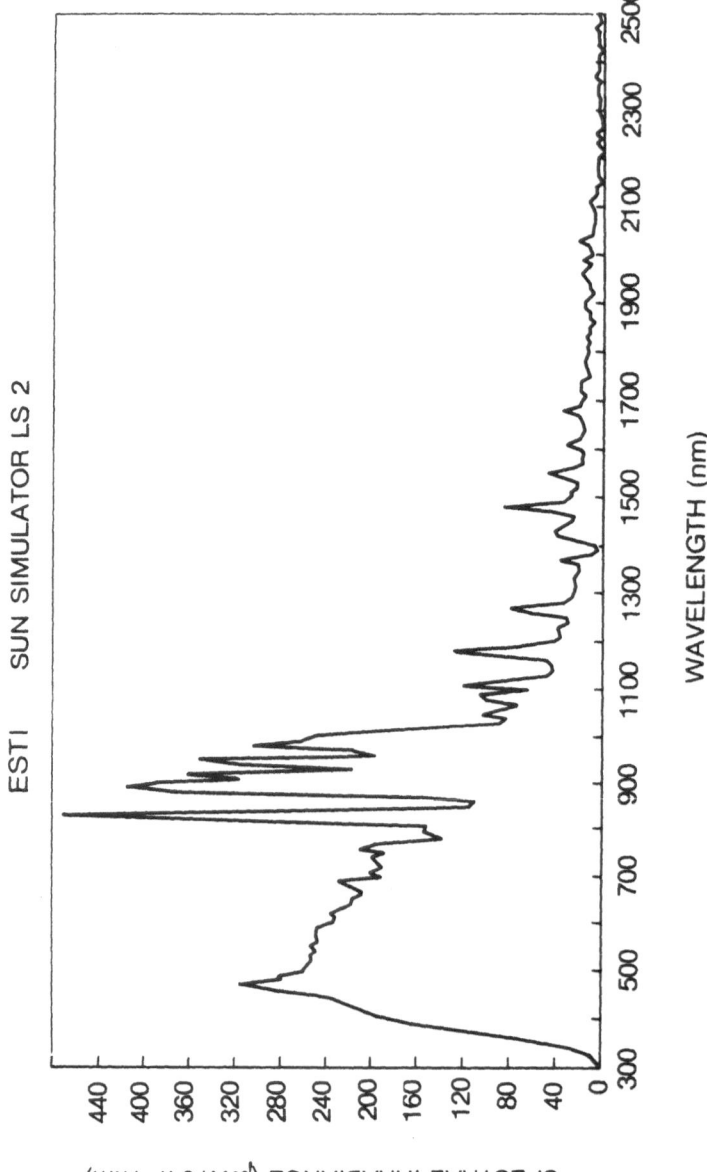

WAVELENGTH (nm)

SPECTRAL IRRADIANCE ($\mu W/cm^2/nm$)

Figure 13. LS-2 spectrum

Figure 14a. NOCT installation

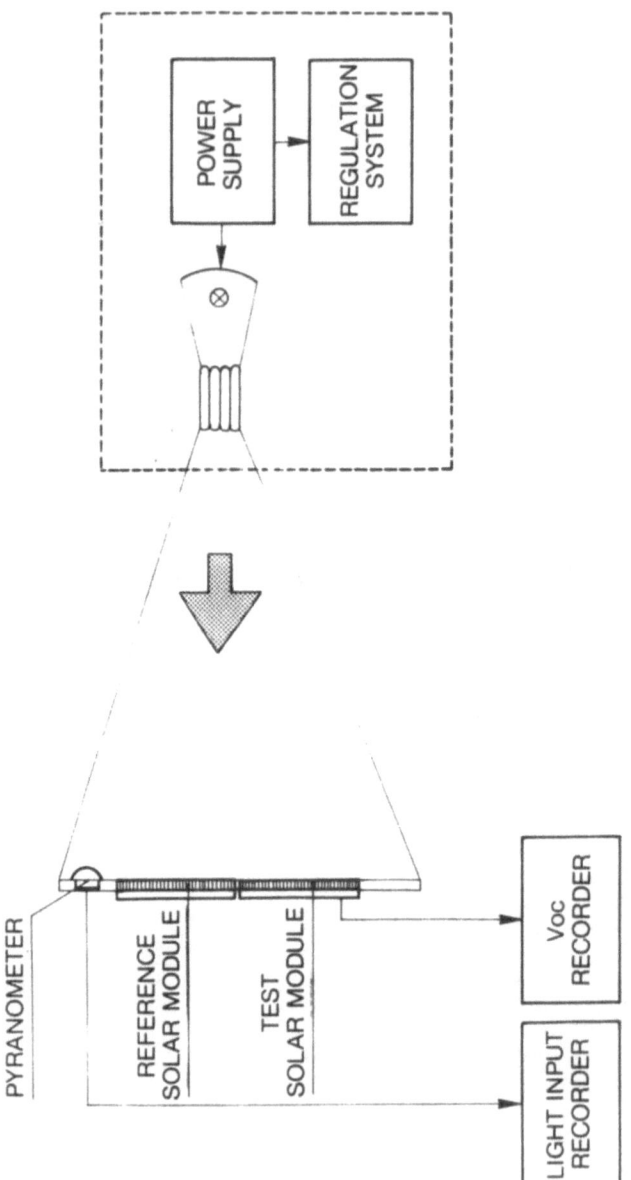

Figure 14b. Block diagram of NOCT measurement

Figure 15. LS-3 test plane

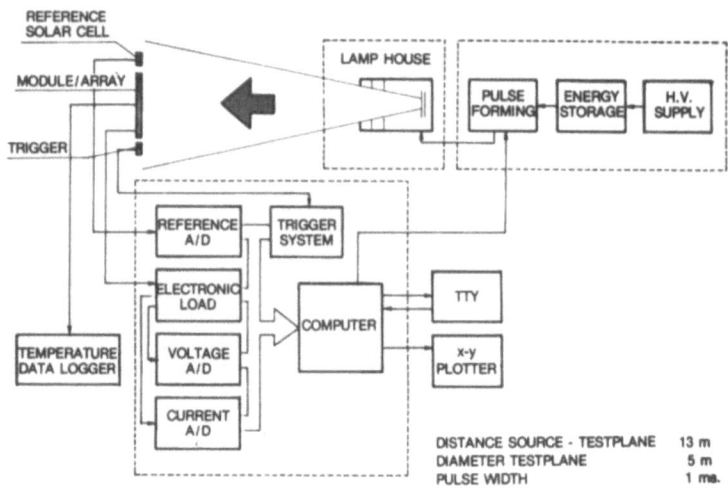

Figure 16. Block diagram of LS-3

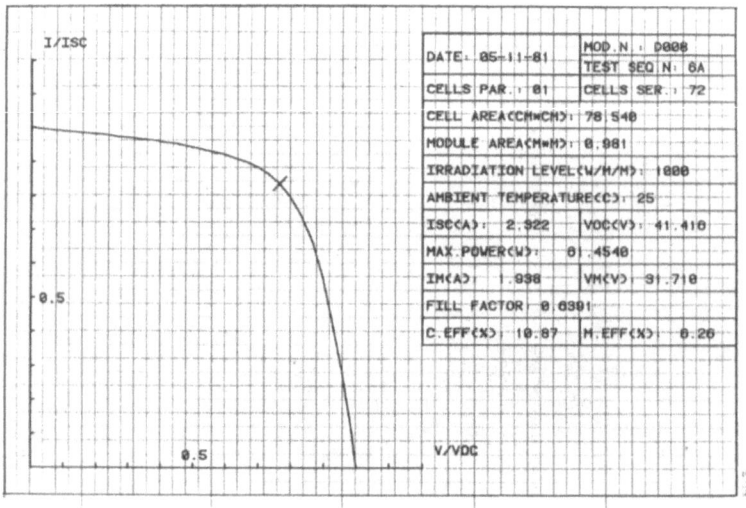

Figure 17. Performance measurement of a 35 W solar module

3.4 LS-3

The large area pulsed solar simulator (LAPSS) consists of two major sub-systems, the pulse illuminator system and the data acquisition and processing system.

The light source, consisting of two xenon pulse flash tubes EG&G, irradiates the test plane area (2.5 x 2.5 m^2, 5 m diameter) at a distance of 13 m with an irradiance intensity which is adjustable between 0.5 and 1.4 kW/m^2. The pulse duration is about 1 msec with a repeatability of $\pm$ 2%. The spectrum is a close match to the AMO distribution.

For this simulator a large light-tight enclosure has been constructed. The room is black inside and is equipped with four sets of curtains, in order to eliminate reflections and to reach the required irradiance uniformity on the test plane ($\pm$ 1.5% over 2.5 x 2.5 m^2, $\pm$ 3% over a 5 m diameter test plane) (Fig. 15).

The other major sub-system of this simulator is the data acquisition and processing system. This system, synchronized with the operating of the pulsed simulator, measures the I-V performance characteristic of the test object; performs an analog to digital conversion on all data; stores the digital information; provides mathematical corrections and prints out the corrected I-V characteristic of the test object. Major components of this system are: a standard solar cell, a fast electronic load, the Data General Nova 3 computer and the HP 7221A X-Y plotter (Fig. 16). Data are stored on paper tape and on floppy disk.

Since light pulses of 1 msec duration are sufficient to take measurements, it is possible to build a very intense flash-light source and keep a large distance between test plane and light source with a consequently high uniformity over large areas. Such a simulator is especially useful for testing extended photovoltaic arrays.

Another major advantage of a pulsed system is the fact that the irradiation itself does not change the temperature of the test object. It facilitates measurements at a constant or standard temperature. We are using this simulator for performance measurements of solar modules according to the CEC Specification

101 and especially also for all performance control tests during the execution of the 501 test programme.

Figure 17 shows the result of such a measurement on a 35 W module.

THE SOLAR TEST FACILITY LS-1 - TECHNICAL DESCRIPTION

G. Blaesser, H. Hettinger, K. Krebs, S. Pace,
A. J. Prins, and E. Rossi-Gianoli

Commission of the European Communities
Joint Research Centre - Ispra Establishment
I-21020 Ispra (Va), Italy

1. INTRODUCTION

The outdoor testing of solar equipment shows a number of intrinsic difficulties: irradiation and environmental conditions,
e. g. temperature or wind, cannot be chosen arbitrarily nor independently one of the other. It is difficult, therefore, to repeat
measurements under the same conditions. Moreover, it may last
very long to conduct statistically significant experiments for a
chosen type of weather or irradiance. Since an exact reproduction of a specific performance measurement is only possible at
very few test sites, it is questionable whether outdoor test
stations are sufficient to compare or optimize converters.

Most of these disadvantages may be avoided using indoor testing.
There it is, in principle, possible to create defined and reproducible test conditions, not only for a wide range of input parameters including also extreme values, but also for arbitrary
combinations of these parameters. Moreover, indoor tests may
be done at any time and for conditions corresponding to arbitrary geographical sites.

The main benefits to be expected from the operation of a larger
indoor testing facility are related to the following two areas:

G. Beghi (ed.), Performance of Solar Energy Converters: Thermal Collectors and Photovoltaic Cells, 259–295.

1. The technology of thermal and photovoltaic converters, and
2. questions related to guarantee conditions.

Concerning the first argument the most important properties of
a converter are its conversion efficiency, i.e. the amount of
useful energy it delivers, and its reliability, i.e. the time it
runs without failure. Amongst the conditions which influence the
efficiency are some which are difficult to simulate by simple
means, e.g. various incidence angles, concentrated radiation,
panel temperatures under natural or forced cooling, or integration of incident energy over a day for a given site. Nevertheless,
it would be very useful to know the influence of these factors in
function of different construction technologies or in function of
various utilization methods. This would obviously allow the
manufacturer of converters or systems to define better the performance of his products. He could then compare the advantages
between different technological approaches and find the best
compromise between cost and performance or reliability.

As concerns the second argument, the essential problem for the
user is the question how much energy his installation delivers
under real climatic conditions. It is possible to extrapolate these
conditions from past values, but often these values will not be
available for a chosen site, i.e. one would have to wait several
years to verify that a given installation corresponds to the initial
request. Of course, one may be content with partial tests and
then propose a warranty only in function of these screening tests,
however, experience shows that actually such guarantees are
often insufficient to convince a user who wants to invest in this
field.

With these problems in mind it has been decided at the JRC to
set up test facilities for performance and reliability assessment
of solar technology products. A major part of this activity concerns indoor testing. Two aspects will be considered; one is
actual testing of products, the other to arrive at a clear definition
where indoor methods are appropriate and where not.

In the present report it is intended to give a brief description of
the indoor test facility LS-1. A major component of this facility
is a multiple lamp system to produce a uniform but uncollimated
light distribution across a test plane which is inside a climatic

chamber.

For the light source we have selected a new but now commercially available type of lamp with a long life time, in order to
reduce test costs and to provide experience for the later use of
this or similar lamps in smaller simulators.

The aspects of collimated light, i.e. angular incidence effects
and a closer spectral match to sunlight will be studied in another
simulation system.

In the following sections we will describe briefly the illumination
system, the thermal control facilities and the thermal collector
test loops.

2. ILLUMINATION SYSTEM

The most economical solution to illuminate uniformly a large
test area is a lighting system consisting of many lamps with an
appropriate spectrum. In order to realize such a source, we
have chosen a non-uniform distribution of 295 OSRAM HQI-R
250 W/D lamps, which are distributed on a grid structure of
3.4×4.4 m^2. To reduce edge losses and to increase the irradiance on the test plane we have added a frame of mirrors in a
distance of 10 cm from the lamps on the edge. In order to compute the light distribution of this arrangement, a computer program was developed.

The calculation of the irradiance L in the test point with coordinates x, y relative to the centre of the test plane is based on the
formula

$$L(x, y) = \sum_{i=1}^{N} D(\tan \nu_i(x, y)) / R_i^2(x, y)$$

where N is the number of lamps ("lamp" = light source + reflector), $R_i(x, y)$ the distance from the test point to the lamp i, and
the "indicatrix" D (angular distribution of the light emitted from
a single lamp) is numerically given as function of the tangent of
the angle $\nu_i(x, y)$ between the axis of symmetry of the lamp i and
the vector $R_i(x, y)$ pointing from that lamp to the test point.
We assume that the lamps are oriented with their axes of sym-

metry normal to the "lamp plane" (plane of the lamp distribution).

The actual calculations - performed by the program "LUX" - use additionally to the test plane coordinates (x, y) a coordinate system x', y', z' with origin in the lower left-hand corner of the lamp plane. In this coordinate system the lamp plane is the plane $z' = 0$, while the centre of the test plane has the coordinates $x' = x_o$, $y' = y_o$; $z' = a_1$. If lamp plane and test plane are parallel to each other, the x', y', z'-coordinates of the point (x, y) are found by the translation

$$x' = x + x_o$$
$$y' = y + y_o$$
$$z' = a_1$$

If the test plane is inclined with respect to the lamp plane its position can be described in terms of the Euler angles (e_1, e_2, e_3) of rotation around the plane centre, and the coordinates x', y', z' of the test point are given in terms of the well-known transformation matrix $Q_{ik}(e_1, e_2, e_3)$ by

$$x'(x, y) = Q_{11}x + Q_{12}y + x_o$$
$$y'(x, y) = Q_{21}x + Q_{22}y + y_o$$
$$z'(x, y) = Q_{31}x + Q_{32}y + a_1 \ .$$

After having determined these primed coordinates the test point distance $R_i(x, y)$ from the lamp i with coordinates $x'(i), y'(i)$, which are contained in the "lamp file", is calculated as

$$R_i^2(x, y) = (x'(x, y) - x'(i))^2 + (y'(x, y) - y'(i))^2 + (z'(x, y))^2$$

The tangent of the angle v_i is determined from

$$\tan v_i = \sqrt{R_i^2(x, y) - (z'(x, y))^2} / z'(x, y)$$

and the corresponding value of the function D is found by interpolation from a table (the "indicatrix file").
In this way the value of $L_i = D(\tan v_i)/R_i^2$ is obtained for all lamps i, and the sum of these values gives the total irradiance L in the point (x, y).

Additional programs are provided for the generation and revision
of the indicatrix and lamp files. The program "FILGEN" requests
the lamp coordinates for the lamps contained in the lower left-
hand quarter of the lamp plane and automatically generates the
other lamp positions necessary for a symmetry of 4 (plane symmetry group "mm"). It also provides for a symmetrical change
of positions; in particular, lamps might be switched off in symmetrical sets by giving negative values for its coordinates. The
program "FILPRT" gives an ordered printout of all switched-on
lamps. Finally, the program "IND" is used for the generation
of the indicatrix file: measured values of the irradiance distribution on a plane at a standard distance from a single lamp are
read in as a function of the distance from the centre of the distribution.

The FORTRAN listings of these programs are given in Appendix 1.
By means of these programs (given in Appendix 1) we have selected 6 sets of lamps which give a uniform irradiance distribution
at a distance of 220 cm from the source. Figures 1 to 7 show the
lamp positions which produce the most uniform distribution for
the following intensities: 250, 470, 640, 1050 and 1170 W/m^2.
The degree of uniformity obtained can be seen in Figs. 8-11,
where we have plotted the irradiance for three characteristic
directions in the test plane.
The actual switching of the lamps will be done according to the
scheme indicated in Fig. 12.

The light source described above does, of course, not produce
parallel or collimated light. This source is only intended for
measurements under uniform irradiation and simulates as such
average global irradiation. The aspects of direct light will be
treated with the facility LS-2.

3. TEMPERATURE CONTROL SYSTEM

3.1 Summary

The light simulator LS-1 consists of two climatic chambers,
one contains the light source, while the second contains a table
on which solar collectors and/or photovoltaic modules are
mounted for testing. Both chambers are separated by a common
transparent window (see Figs. 13a, b, c). The outer dimensions

of the facility are 9.4 x 5.4 x 4.6 m^3, while the inner dimensions
of the useful space are 7.9 x 5.2 x 4.4 m^3. This space is then
equally divided into the light source and the test chamber.

A testing table is installed at a distance of about 2.6 m from the
light source; it has a surface of 3.65 x 4.65 m^2 and may thus
carry up to 4 thermal collectors. The light source allows a
uniform irradiation between 250 and 1170 W/m^2 in steps of about
200 W/m^2. It consists of 295 lamps with a power of 250 W each
and with an average life time of 4000 h.

The cooling aggregates for both chambers are located in a sound-
proof room in a building adjacent to the experimental hall. The
power supply for the light source is installed close to the LS-1,
all control and registration systems are assembled in a separate
control room. Special attention is paid to facilitate maintenance
and repair of the equipment.

A more detailed description of the installation is given in the
following sections.

3.2 Light Source Chamber

Chamber. The chamber is composed of standardized, easily
removable sandwich panels and elements covered on both sides
with non-reflective, baking enamelled 1 mm thick aluminium
sheets. Panels and elements ensure a heat isolation with a K
value of 0.2 kcal/m^{2}h°C. Wall, ceiling and floor thicknesses
are 120 mm, while the entire chamber is supported on 20 mm
thick wooden strips in labyrinth form to assure a good aeration
under the chamber. The materials are fire-proof and temperature and corrosion-resistant. The floor is covered with 3.5 mm
thick aluminium plates and can take a maximum load of 3000 kg/m^2.
Two 1.5 m long rails for the light source carriage are embedded
in the chamber floor. The chamber is accessible through two
insulated doors: one door gives access to the heat exchanger and
the rear side of the light source; the other to the front of the
light source, the mirror channel and the separation window.
Multipane glass windows are installed on eye-level in both doors.

An about 1.3 m deep mirror channel reflects the three outer
rows of lamps of the light source onto the test table in the test
chamber. A 50 mm large opening between the upper mirror part

and the window ensures sufficient heat removal.

The air duct is installed in the rear wall containing the necessary equipment for air cooling. Radial fans circulate the air
through the duct in which the heat exchanger and other air conditioning elements are installed. The (cooled) air enters the
room at the opening between the light source and the mirror
channel, flows at first partially over the window and then over
the lamps, passing the air ducts and sectional aeration axial
fans of the light source and returns at the rear side into the
radial fans. Approximately 8000 m^3 air per hour can be circulated, adjustable within a range of 2000 - 8000 m^3/h.

The temperature is controlled within a range of $50 \pm 20^{o}C$ by a
3-point PID controller with resistance thermometer Pt 100 Ω
mounted in the air duct. The air is dried down to a dew point of
-5°C, thus preventing condensate at the window during low
temperature operations in the test chamber.
The chamber interior can be illuminated by fluorescent tubes.

Light Source. The light source is installed at a distance of 1.5 m
from the separation window. It consists mainly of a rectangular
metallic frame with a surface of 4.65 x 3.65 m^2 and is composed
of a system of 35 double surface protected steel U-profiles.
Each 120 mm large U-beam has 45 sleeves for easily fixing the
lamp sockets. 295 lamp sockets are distributed over the 1575
possible fixing points in a pattern shown in Fig. 7. The distance
between the fixing points is 100 mm; the minimum socket distance is 141 mm. Igniters and starters of the lamps are fixed at
the rear side of the U-profiles.

The lamps are of the type HQI-R 250 W, produced by OSRAM
(mercury arc lamps with dysprosium additives and integrated
reflector). It is clear that the total power of the lamp source
must be switched on in steps, maximum 18 lamps per phase at
about 55 A in each step. The necessary electric equipment (relais,
induction coils and starter circuits) is installed in cabinets (see
Fig. 13a) along the side of the light source chamber.

The lamps are cooled by circulating air, flowing over the lamp
bulbs from the top to the sockets. The fixing points which are
not occupied by lamp sockets are covered by plates in order to

assure that the maximum cooling air flow passes along the lamps.
The entire light source is installed on a carriage with wheels and
can be moved over a distance of 0. 7 m towards the rear wall of
the chamber, for inspection, maintenance and repair.
A retractable ladder gives access to the lamps, mirrors and
window.

3.3 The Separating Window

This window serves as a separating wall between the atmosphere
in the test chamber $(-40^{\circ}C \leqslant t \leqslant 60^{\circ}C)$ and in the light source
chamber $(t = 50 \pm 20^{\circ}C)$. A difference of pressure between both
sides of the window is avoided. The entire window of $3.5 \times 4.5 \ m^2$
is supported by a rigid frame. A trap door in the ceiling between
both chambers gives access for replacement of the window using
the main crane of the experimental hall.
The side walls and the floor of the chambers are provided with
a guiding groove of low thermal conductive material into which
the pane can be inserted and pressed on.

3.4 Test Chamber

Chamber. This chamber is composed, in the same way as the
light source chamber, of standardized easily removable sandwich panels and elements covered on both sides with non-reflective 1 mm thick white enamelled aluminium sheets. Panel elements and connections ensure a heat insulation with a K-value of
$0.2 \ kcal/m^2h^{\circ}C$.
The thickness of the wall, ceiling and floor is 120 mm, while the
chamber is supported by a labyrinth frame made of 20 mm thick
wooden strips to assure a good aeration and to protect against
freezing. The materials are fire-proof and temperature-resis-
tant. The inner surfaces are corrosion-proof and washable.
The floor is covered with 2.5 mm thick aluminium plates and can
withstand a maximum load of 3000 kg/m^2 and 300 kg per (rubber)
wheel. The internal dimensions are $4 \times 4.8 \times 4 \ m^3$.
Rails are provided in the chamber floor in order to be able to
move the testing table within a range of about 1.3 m.

The heavy duty walk-in type doublewing door has a clear opening
of $3.6 \times 2 \ m^2$. The door panel corresponds to the design of the
chamber walls. The door can be locked from outside, while a
safety turn latch permits the opening from inside even if pad-

locked on the outside. A special door-frame heating element
prevents freezing and condensation. A multipane inspection
window is installed in the chamber door at eye-level.

The chamber can be illuminated by water-proof incandescent
lamps with external switch.
The chamber space temperature can be regulated between $-40^{\circ}C$
and $+60^{\circ}C$ with an accuracy (with radiation load) in time of $\pm 1^{\circ}C$
and in space of $\pm 5^{\circ}C$. The average cooling and heating rates
without radiation load are $0.33^{\circ}C/min$, thus the chamber temperature can be changed between $+60^{\circ}C$ and $-40^{\circ}C$ within 5 hours.

The air duct is installed in the rear wall containing the necessary equipment for air conditioning. Corrosion-proof radial fans
circulate the air through the duct in which the copper aluminium
lamella heat exchanger and the corrosion-proof electric heater
are located. A drip, respectively humidification, try of stainless steel with overflow pipe collects surplus condensate. The
air enters the test chamber through a corrosion-proof perforated
ceiling, thus ensuring a regular and draught-free air distribution
in the entire chamber. After passing the test chamber the air is
equally taken up and recirculated through the air duct. About
$9000 \ m^3$ air per hour is circulated adjustable within a range of
$2000 - 9000 \ m^3/h$.

The chamber temperature is regulated by a 3-point controller
with PID behaviour as well as by feedback having two sensitive
resistance thermometers Pt 100Ω, of which one is installed
near the test table and the other one in the air duct.
Eleven 100 mm diameter wall passages are provided in the rear
wall for test loops and measurement leads.
The panels of the inner wall on the rear side are easily dismount-
able in order to give access for maintenance.

Test Table. The test table consists mainly of a steel frame, a
trolley and a rocking table. The table itself is a rigid frame
construction which has a useful surface of $3.2 \times 4 \ m^2$, and is
suitable to hold four solar collectors fixed by rapid fixture devices. The table can be inclined (with a speed of $10^{\circ}/min$) between its vertical and horizontal position. The lower part of
the table is provided with a positioning device by means of which
the table can be moved (with a speed of $0.1 \ m/min$) within a
range between 1.9 and 3.2 m from the light source, respectively

0. 4 and 1. 7 m from the separating window. Both movements are motorised and remotely controlled either from the control room or from a desk near the walk-in door.

The table positions are indicated by a digital display. Four retractable (rubber) wheels permit to withdraw the test table through the walk-in door. The solar collectors are fixed by rapid fixing devices on the table outside the chamber. The double wing doors are shut immediately after the test table has passed in order to maintain clean conditions in the chamber.

If motors or control would fail, the movements can be carried out manually by hand-driven wheels. Movements of the table inside the chamber beyond the indicated ranges are prevented by electric limit switches and mechanical stops in order to avoid damage to the chamber walls or the window.

The water in- and outlets of the collectors are connected to the test loop outside the chamber by eight flexible insulation tubes, which leave the chamber through the rear wall.

3. 5 Cooling Units

The cooling units of both chambers are located in a sound-proof room (Engine Room I, see Fig. 14). The cooling is based on the evaporation of Frigen refrigerants, which remove the heat out of the chamber via heat exchangers. The refrigerating units are cooled by industrial water (max. 14 m^3/h).

3.6 Electric Power Supply

The electric power (380 V, 3-phase, 50 Hz) is supplied from the power distribution room. For the light source 95 kVA and for the refrigerating units 135 kVA are needed. The light source electric equipment is divided into a low voltage circuit (control room) and a power circuit (cabinet close to the light source chamber).

3. 7 Control and Operation of the Installation

The operation and supervision of the installation as well as the collection and recording of experimental data are done in the control room. All information received in the control room can be introduced into the central computer for treatment, control and storage. The following control functions for the light simulator are available:

- light indicators for each lamp (configuration, failures) of the light source,
- temperature regulation of both chambers,
- operation of the air circulation fans,
- movements and positions of the test table (also from a switch board near the door of the test chamber),
- emergency stop.

Further some safety measures are built in, such as:

- limitation of the movement of the test table,
- automatic stop of the cooling units due to the lack of water,
- automatic stop of the light source due to insufficient ventilation,
- automatic stop if one of the chamber doors is open.

The lamps of the light source can be pre-selected either manually or automatically in one of the 6 power ranges corresponding to the irradiance values (250, 470, 640, 850, 1050, 1170 W/m^2) by switching the number of lamps as indicated in Fig. 12. Each power level is achieved in successive steps, the selection is possible in increasing or decreasing order. The automatic control of the light source is realized by means of a computer program while the manual operation is done by push buttons on the control rack.

3.8 Maintenance and Repair

Maintenance is carried out according to the instructions of the manufacturers. The replacement of lamps or the cleaning of the window and the mirrors is facilitated by retracting the light source and by using a special ladder between mirror and light source.
The elements of the air conditioning system are accessible for maintenance by removing the panels of the inner rear walls of both chambers.
The replacement of the window (4 x 5 m^2) of the LS-1 light source housing may be carried out via a removable door in the top of LS-1 using special brackets installed on the main crane.

4. THERMAL COLLECTOR TEST LOOPS

4.1 Indoor Test Loops

The LS-1 facility will be mainly used for performance measurements on thermal collectors. The lay-out of the test loops is flexible enough, so that different test procedures can easily be applied. The size of the test area allows to perform measurements with up to four collectors simultaneously. The collectors are fixed on a table which can be moved out of the chamber and rotated for mounting. The irradiation is generally performed with the collectors in a vertical position.

The flow sheet of the four test loops is shown in Fig. 15. The four solar panels can be investigated in parallel or in series. The circuit contains two thermostatically controlled tanks, one for hot water and one for cold water.
For temperatures below 0°C, the water in the circuit is replaced by a mixture of water and antifreeze. In each of the four sub-circuits the inlet water temperature of a collector can be adjusted and controlled independently by step motor-driven mixing valves. A constant head is provided by an elevated tank with overflow-return to the supply tank. This arrangement allows to choose different flow rates for each of the four collectors running in a parallel circuit. During irradiation the water enters into the collector from below. For thermal loss measurements without irradiation, the flow can be inverted in order to maintain the direction of the temperature gradient as in the operation with light.

The determination of the collector efficiency requires very precise measurements of the water flow rate through the solar collector as well as of the difference between the inlet and outlet temperatures of the water. The flow rate is measured with an axial-flow turbine flow meter with an accuracy of about 0.5%. High precision temperature measurements are obtained by using quartz thermometers with a resolution of about 10^{-4} °C and an absolute accuracy of 0.04°C.

In order to simulate wind conditions a fan is used to blow air across the collectors in the plane of the glass covers. The wind speed is controlled by an anemometer. The ambient temperature as well as the surface temperatures of the various collector

components are measured with resistance thermometers Pt-100 Ω.
A high precision pyranometer is used to measure the incident
radiation; the instrument is fixed to the collector stand within
the collector test plane.

4.2 Outdoor Correlation Test Loop

For an absolute calibration of the indoor facility it is necessary
to perform correlation experiments outdoors, in our case we
are using for this purpose the BSE method.

The general expression defining the instantaneous efficiency is:

$$\eta = \frac{q_u}{A.I_n} = F_R \left[(\tau\alpha)_e - \frac{U_L(T_F - T_A)}{I_n} \right]$$

where the symbols have their usual meaning.

Since the loss coefficient $\left(U_o = F_R U_L\right)$ is approximately linearly
related to the temperature difference

$$U_o = a + b(T_F - T_A),$$

the efficiency of the collector can be expressed as a second
order function of the temperature difference

$$\eta = F_R \left[(\tau\alpha)_e - a \frac{T_F - T_A}{I_n} - b I_n \left(\frac{T_F - T_A}{I_n} \right)^2 \right] \tag{1}$$

An efficiency test involves thus the experimental determination
of the coefficients $F_R(\tau\alpha)_e$, $F_R a$, $F_R b I_n$. Following the BSE
method this can be done by two independent tests:

1. "No heat-loss test" for the determination of $F_R(\tau\alpha)_e$.

   In this case $T_F = T_A$, and eq. (1) becomes

   $$q_u = F_R(\tau\alpha)_e I_n$$

   where I_n and $q_u = A m c_{tf}(T_{f,e} - T_{f,i})$ are obtained experimentally.

2. "No insolation test" for measuring the heat losses as a function of $T_F - T_A$.

In this case $I_n = 0$, and from eq. (1) follows

$$\frac{q_u}{A} = - F_R a(T_F - T_A) + b(T_F - T_A)^2,$$

which enables us to determine the quantities $F_R a$ and $F_R b$.

Our experimental test loop is shown in Fig. 16.

A water mixing device controls the fluid temperature $T_{f,i}$ at the input of the collector by two regulating linear precision valves, one for setting the flow of hot water ($\sim 80°C$) and one for setting the flow of cold water ($\sim 10°C$).

The hot water source is a water heater for domestic use (capacity about 300 l), while the cold water is available (at a temperature of about 8-12°C) from a heat exchanger. A pump provides the fluid flow in the closed loop.

A Digital Equipment Corporation PDP-11 computer controls the operational procedure and all measuring instruments. A feedback system, under computer control, drives by means of stepping motors the two regulating valves (actuated in push-pull for a constant total flow rate) in order to get the correct hot/cold water ratio in function of the reference value $T_{f,i}$ (temperature of the fluid entering the collector). Details of the regulating valves are shown in Fig. 17. The facility can deliver water at any temperature between 10 and 90°C with a high degree of stability and reproducibility.
A unique feature of this device is the advantage that one is not obliged to wait the time a big storage tank usually needs to reach a new temperature.

A stationary mount is provided for testing at selectable fixed angles.

In Fig. 18 a block diagram of the data acquisition system is shown. The computer, via a control interface and a logic conversion and pulse-shaping circuit, allows random access channel selection of the Leeds and Northrup 2740 Scanner which transfers the selected input signal i_n $(n = 1, 2, \ldots, 20)$ to a Numatron multi-range digital millivoltmeter which, in turn, sends the signal to the computer memory.

A Canberra computer interface establishes system organization directly through the DEC Unibus and simultaneously transfers information to the Canberra Databus which is a parallel bidirectional bus shared by all Canberra Datanim units, e. g. the axis positioners. Each axis positioner, in turn, drives a stepping motor.

In Fig. 19 a typical plot of the measured heat loss versus the difference between mean fluid temperature (T_F) and air temperature (T_A) is reported. Figure 20 shows a plot of efficiencies as calculated from the experimental points of Fig. 19.

The collector (model HI-F) was obtained from the DRU Company, Ulft, the Netherlands. Efficiency measurements for another collector of known performance have been made: the comparison with data obtained by other authors has shown good agreement.

ANNEX I

```
        PROGRAM LUX
C       A PROGRAM TO CALCULATE THE LUMINOSITY DISTRIBUTION
        DIMENSION V(20,2),R(600,2),Q(3,3)
        PRINT 1000
        TYPE 1001
        CALL ASSIGN(3,'DUMMY',-1,'OLD')
        TYPE 1002
        CALL ASSIGN(4,'DUMMY',-1,'OLD')
        READ(3)((R(I,K),K=1,2),I=1,600)
        READ(4)((V(I,K),K=1,2),I=1,20)
        TYPE 1003
        ACCEPT 1004,X1,Y1
        PRINT 1005,X1,Y1
        TYPE 1006
        ACCEPT 1004,X0,Y0
        PRINT 1007,X0,Y0
        X1=X1/2
        Y1=Y1/2
        TYPE 1008
        ACCEPT 1009,B1
        PRINT 1010,B1
        TYPE 1011
        ACCEPT 1009,A1
        TYPE 1012
        ACCEPT 1013,E1,E2,E3
        Q(1,1)=COS(E3)*COS(E1)-SIN(E3)*SIN(E1)*COS(E2)
        Q(1,2)=COS(E3)*SIN(E1)+SIN(E3)*COS(E1)*COS(E2)
        Q(1,3)=SIN(E3)*SIN(E2)
        Q(2,1)=-COS(E3)*SIN(E1)*COS(E2)-SIN(E3)*COS(E1)
        Q(2,2)=COS(E3)*COS(E1)*COS(E2)-SIN(E3)*SIN(E1)
        Q(2,3)=COS(E3)*SIN(E2)
        Q(3,1)=SIN(E1)*SIN(E2)
        Q(3,2)=-COS(E1)*SIN(E2)
        Q(3,3)=COS(E3)
        Z=1
        PRINT 1014,A1
        PRINT 1015,E1,E2,E3
        PRINT 1016
        PRINT 1017,((Q(I,K),K=1,3),I=1,3)
        TYPE 1018
        PRINT 1018
2       TYPE 1019
        ACCEPT 1004,X,Y
        S1=0
        X2=Q(1,1)*X+Q(1,2)*Y+X0
        Y2=Q(2,1)*X+Q(2,2)*Y+Y0
        R2=A1+Q(3,1)*X+Q(3,2)*Y
        IF(R2.LT.1) GOTO 50
        DO 10 I=1,600
        IF(R(I,1).LT.0) GOTO 10
        R1=SQRT((R(I,1)-X2)**2+(R(I,2)-Y2)**2)
        IF(R1.GT.R2) GOTO 10
        A=R1*B1/R2
        U2=B1*B1/(R1*R1+R2*R2)
        DO 5 K=1,20
        IF(V(K,1).GT.A) GOTO 6
5       CONTINUE
        K=20
6       D1=V(K,1)-V(K-1,1)
        D2=V(K,2)-V(K-1,2)
        S=V(K-1,2)+(A-V(K-1,1))*D2/D1
        S1=S1+S*U2
10      CONTINUE
```

```
        TYPE 1020,X,Y,S1
        PRINT 1020,X,Y,S1
        TYPE 1021
        ACCEPT 1022,I
        IF(I.GT.0) GOTO 2
        TYPE 1023
        ACCEPT 1022,I
        IF(I.GT.0) GOTO 51
50      STOP
51      PRINT 1024
        PRINT 1025
        PRINT 1026,((V(I,K),K=1,2),I=1,20)
        GOTO 50
1000    FORMAT(T30,'ILLUMINATION DISTRIBUTION ON THE TEST PLANE'//)
1001    FORMAT(' GIVE NAME OF LAMP FILE'/)
1002    FORMAT(' GIVE NAME OF INDICATRIX FILE'/)
1003    FORMAT('$DIMENSIONS OF TEST PLANE X,Y?')
1004    FORMAT(2F8.0)
1005    FORMAT(' THE DIMENSIONS OF THE TEST PLANE ARE',F8.2,
       C'X',F8.2,'CM')
1006    FORMAT('$GIVE X,Y-COORDINATES OF TEST PLANE CENTER!')
1007    FORMAT(' THE X,Y-COORDINATES OF TEST PLANE CENTER ARE!',2F8.2)
1008    FORMAT('$NORMAL DISTANCE OF INDICATRIX?')
1009    FORMAT(F8.0)
1010    FORMAT(' NORMAL DISTANCE OF INDICATRIX!',F8.2)
1011    FORMAT('$DISTANCE OF TEST PLANE CENTER?')
1012    FORMAT('$GIVE EULERS ANGLES E1,E2,E3:')
1013    FORMAT(3F8.0)
1014    FORMAT(' DISTANCE OF TEST PLANE CENTER!',F8.2)
1015    FORMAT('OORIENTATION OF THE TEST PLANE(EULERS ANGLES IN DEGR.):
       C',3F8.2)
1016    FORMAT('0',T20,'ROTATION MATRIX Q(I,K)!'/)
1017    FORMAT(3F15.2)
1018    FORMAT(T15,'ILLUMINATION INTENSITY IN TEST POINTS'/)
1019    FORMAT('$GIVE TEST POINT COORDINATES X AND Y!')
1020    FORMAT(T10,'X=',F8.2,T25,'Y=',F8.2,T50,'I=',F8.2)
1021    FORMAT('$WANT MORE TEST POINT CALCULATIONS?')
1022    FORMAT(I2)
1023    FORMAT('$DO YOU WANT A LISTING OF THE INDICATRIX?')
1024    FORMAT('0',T20,'INDICATRIX!'/)
1025    FORMAT(T18,'X',T28,'Y'/)
1026    FORMAT(T10,2F10.2)        ,
        END




        PROGRAM FILGEN
C       A PROGRAM FOR LAMP FILE GENERATION, EXTENSION AND CORRECTION (SYMMETRY44)
        DIMENSION R(600,2)
        COMMON R,X1,Y1,X0,Y0
        TYPE 1000
        TYPE 1001
        TYPE 1002
        TYPE 1003
1       TYPE 1004
        ACCEPT 1005,I
        IF(I.GT.3) GO TO 1
        TYPE 1006
        ACCEPT 1007,X0
        TYPE 1008
        ACCEPT 1007,Y0
        X1=X0/2.
        Y1=Y0/2.
        GO TO (2,30,40),I
2       TYPE 1009
        CALL ASSIGN(3,'DUMMY',-1,'NEW')
3       TYPE 1010
        M=1
        ACCEPT 1005,N
        IF(N.LE.150) GOTO 4
        TYPE 1011
```

```
        GOTO 3
4       DO 5  I=M,N
        CALL SYM(I)
5       CONTINUE
        DO 6 I=N+1,150
        DO 6 I1=1,2
        R(I,I1)=-1
        R(I+150,I1)=-1
        R(I+300,I1)=-1
        R(I+450,I1)=-1
6       CONTINUE
7       TYPE 1012
        ACCEPT 1005,I
        IF(I.EQ.0) GO TO 8
        PRINT 1013
        PRINT 1014
        PRINT 1015,(I,((R(I+K-1,J),J=1,2),K=1,4) ,I=1,600,4)
8       REWIND 3
        WRITE(3)((R(I,K),K=1,2),I=1,600)
        TYPE 1016
        STOP
30      TYPE 1017
        CALL ASSIGN(3,'DUMMY',-1,'OLD')
        REWIND 3
        READ(3)((R(I,K),K=1,2),I=1,600)
        TYPE 1018
        ACCEPT 1005,M
        TYPE 1019
31      ACCEPT 1005,N
        IF(N.LE.150) GOTO 4
        TYPE 1011
        GOTO 31
60      TYPE 1020
        CALL ASSIGN(3,'DUMMY',-1,'OLD')
        REWIND 3
        READ(3)((R(I,K),K=1,2),I=1,600)
61      TYPE 1021
        ACCEPT 1005,I
        IF(I.EQ.0) GOTO 7
        TYPE 1022
        ACCEPT 1005,IL
        IF(IL.GT.150) GOTO 65
        TYPE 1023,R(IL,1),R(IL,2)
        CALL SYM(IL)
        GOTO 61
65      TYPE 1026
        GOTO 61
1000    FORMAT(' THIS PROGRAM OFFERS YOU THE FOLLOWING OPTIONS:')
1001    FORMAT('        -TO GENERATE THE LAMP FILE [I=1]')
1002    FORMAT('        -TO EXTEND THE LAMP FILE   [I=2]')
1003    FORMAT('        -TO CORRECT THE LAMP FILE  [I=3]')
1004    FORMAT('$INDICATE THE DESIRED OPTION BY CHOOSING I:')
1005    FORMAT(I3)
1006    FORMAT('$MAXIMUM X-VALUE?')
1007    FORMAT(F7.0)
1008    FORMAT('$MAXIMUM Y-VALUE?')
1009    FORMAT(' GIVE NAME OF NEW LAMP FILE'/)
1010    FORMAT('$NUMBER OF LAMPS IN FIRST QUADRANT?')
1011    FORMAT('??NUMBER OF LAMPS IS TOO LARGE!!')
1012    FORMAT('$PRINT LAMP FILE?')
1013    FORMAT(T31,'COORDINATES OF LAMP POSITIONS'//)
1014    FORMAT(T4,'I',T10,'M(I,1)  M(I,2)',T30,'M(I+1,1) M(I+1,2)',
     C  T54,'M(I+2,1) M(I+2,2)',T78,'M(I+3,1) M(I+3,2)')
1015    FORMAT(I4,F10.2,F8.2,F15.2,F9.2,F15.2,F9.2,F15.2,F9.2)
1016    FORMAT(' LAMP FILE GENERATED')
1017    FORMAT(' GIVE NAME OF LAMP FILE TO BE EXTENDED'/)
1018    FORMAT('$AT WHAT LAMP NR. CHANGE STARTS?')
1019    FORMAT('$GIVE NEW TOTAL NUMBER OF LAMPS IN QUADRANT:')
1020    FORMAT(' GIVE NAME OF FILE TO BE CHANGED'/)
1021    FORMAT('$DO YOU WANT MORE CORRECTIONS?')
1022    FORMAT('$LAMP NR.?')
1023    FORMAT(' OLD LAMP COORDINATES ARE',2F8.1)
1024    FORMAT('$GIVE NEW LAMP COORDINATES:')
1025    FORMAT(2F7.0)
1026    FORMAT(' LAMP NR. TOO LARGE')
        END
        SUBROUTINE SYM(I)
```

```fortran
        DIMENSION R(600,2)
        COMMON R,X1,Y1,X0,Y0
        J=I+150
        K=I+300
        L=I+450
101     TYPE 1100,I
        ACCEPT 1101, R(I,1),R(I,2)
        IF ((R(I,1).LT.0).OR.(R(I,2).LT.0)) GOTO 155
        IF  (R(I,1)-X1) 102,140,110
102     IF  (R(I,2)-Y1) 103,150,120
103     R(J,1)=R(I,1)
        R(J,2)=Y0-R(I,2)
        R(K,1)=X0-R(I,1)
        R(K,2)=R(I,2)
        R(L,1)=R(K,1)
        R(L,2)=R(J,2)
        GO TO 160
110     TYPE 1102
        GO TO 101
120     TYPE 1103
        GO TO 101
140     IF(R(I,2).EQ.Y1) GO TO 155
        R(J,1)=R(I,1)
        R(J,2)=Y0-R(I,2)
        R(K,1)=-1
        R(K,2)=-1
        R(L,1)=-1
        R(L,2)=-1
        GOTO 160
150     R(J,1)=X0-R(I,1)
        R(J,2)=R(I,2)
        R(K,1)=-1
        R(K,2)=-1
        R(L,1)=-1
        R(L,2)=-1
        GOTO 160
155     R(J,1)=-1
        R(J,2)=-1
        R(L,1)=-1
        R(K,2)=-1
        R(L,1)=-1
        R(L,2)=-1
160     RETURN
1100    FORMAT('4X,Y-VALUES FOR LAMP NR.',I3,'?')
1101    FORMAT(2F8.0)
1102    FORMAT(' ?X-VALUE TOO LARGE!!')
1103    FORMAT(' ?Y-VALUE TOO LARGE!!')
        END


        PROGRAM FILPRT
C       A PROGRAM TO PRINT THE LAMP FILE IN ORDERED FORM
        DIMENSION R(600,2)
        TYPE 100
        CALL ASSIGN(3,'DUMMY',-1,'OLD')
        READ(3) ((R(I,K),K=1,2),I=1,600)
        I=1
        J=2
1       IF(J.GT.600) GOTO 10
        IF(R(I,1).LT.0) GOTO 5
        I=I+1
        J=J+1
        GOTO 1
5       IF(R(J,1).LT.0) GOTO 7
        R(I,1)=R(J,1)
        R(I,2)=R(J,2)
        R(J,1)=-1
        I=I+1
        J=J+1
        GOTO 1
```

```
7          J=J+1
           IF(J.LT.600) GOTO 5
           I=J+1
10         CONTINUE
           DO 16 IP=1,I-2
           TO=R(IP,1)
           RO=R(IP,2)
           IF(IP.EQ.(I-2)) GOTO 15
           DO 14 K=IP+1,I-2
           IF(TO.GT.R(K,1)) GOTO 12
           GOTO 14
12         T1=TO
           R1=RO
           TO=R(K,1)
           RO=R(K,2)
           R(K,1)=T1
           R(K,2)=R1
14         CONTINUE
15         R(IP,1)=TO
           R(IP,2)=RO
16         CONTINUE
           A=1
           DO 18  N=IA,I-2
           GOTO 20
18         CONTINUE
20         N=N-1
           IF(N.EQ.IA) GOTO 27
           DO 26  IP=IA,N
           RO=R(IP,2)
           IF(IP.EQ.N) GOTO 25
           DO 24  K=IP+1,N
           IF(RO.GT.R(K,2))  GOTO 22
           GOTO 24
22         R1=RO
           RO=R(K,2)
           R(K,2)=R1
24         CONTINUE
25         R(IP,2)=RO
26         CONTINUE
27         IA=N+1
28         CONTINUE
           PRINT 103
           PRINT 104
           PRINT 105,(K,R(K,1),R(K,2),K=1,I-2)
           STOP
100        FORMAT(' GIVE NAME OF LOOP FILE TO BE PRINTED')
101        FORMAT('$GIVE MAXIMUM X-VALUE OF LOOP DISTRIBUTION')
102        FORMAT(F7.0)
103        FORMAT('0',T30,'POSITIONS OF SWITCHED-ON LOOPS')
104        FORMAT(I10,'NR',I26,'X',I11,'Y')
105        FORMAT(I5,I6,2F15.2)
           END
```

```
           PROGRAM LOO
C          A PROGRAM TO GENERATE OR CHANGE THE LOOP DISTRIBUTIONS
           DIMENSION V(20,2)
1          TYPE 100
           TYPE 101
           TYPE 102
           ACCEPT 103,I
           GOTO(2,10),I
           GOTO 1
2          DO 4 I=1,20
           TYPE 104,I
4          ACCEPT 105,V(I,1),V(I,2)
           TYPE 106
           CALL ASSIGN(3,'DUMMY',-1,'NEW')
           WRITE(3)((V(I,K),K=1,2),I=1,20)
           DO 6 I=1,20
```

```
6        TYPE 107,I,V(I,1),V(I,2)
         STOP
10       TYPE 108
         CALL ASSIGN(3,'DUMMY',-1,'OLD')
         REWIND 3
         READ(3)((V(I,K),K=1,2),I=1,20)
12       TYPE 109
         ACCEPT 103,I
         IF(I.EQ.0) GOTO 15
         TYPE 110
         ACCEPT 103,L
         TYPE 111,L,V(L,1),V(L,2)
         TYPE 112
         ACCEPT 105,V(L,1),V(L,2)
         GOTO 12
15       REWIND 3
         WRITE(3)((V(I,K),K=1,2),I=1,20)
         DO 16 I=1,20
16       TYPE 107 ,I,V(I,1),V(I,2)
         STOP
100      FORMAT(' IF YOU WANT')
101      FORMAT('        -TO GENERATE THE INDICATRIX, TYPE: 1')
102      FORMAT('        -TO CHANGE THE INDICATRIX, TYPE: 2')
103      FORMAT(I2)
104      FORMAT('$GIVE X,Y-VALUES FOR INDICATRIX ELEMENT',I3,':')
105      FORMAT(2F7.0)
106      FORMAT(' GIVE NAME FOR NEW INDICATRIX FILE')
107      FORMAT(T4,I3,T15,F8.2,T30,F8.2)
108      FORMAT(' GIVE NAME OF OLD INDICATRIX FILE')
109      FORMAT('$MORE CORRECTIONS?')
110      FORMAT('$INDICATRIX ELEMENT NR.?')
111      FORMAT(' OLD X,Y-VALUES OF INDICATRIX ELEMENT ',I3,' ARE: ',2F8.2)
112      FORMAT('$GIVE NEW X,Y-VALUES:')
         END
```

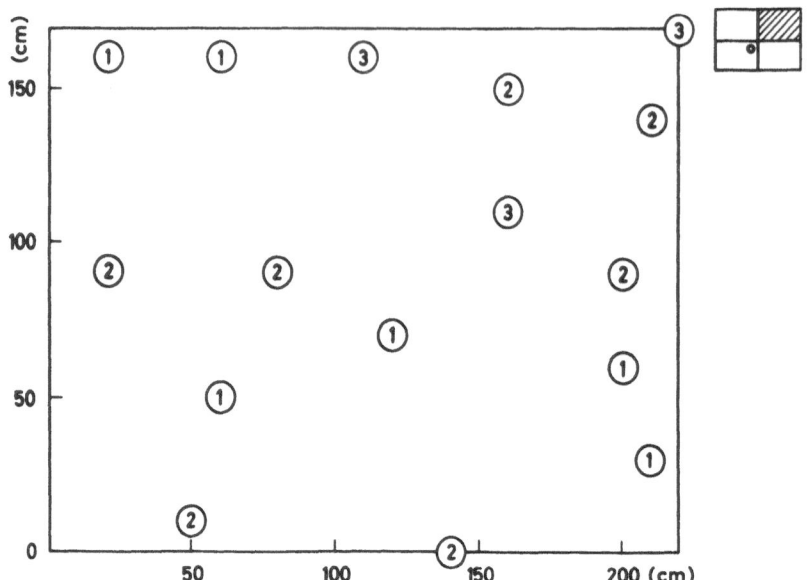

Fig. 1: Lamp distribution at 250 W/m² (total number of lamps: 62)

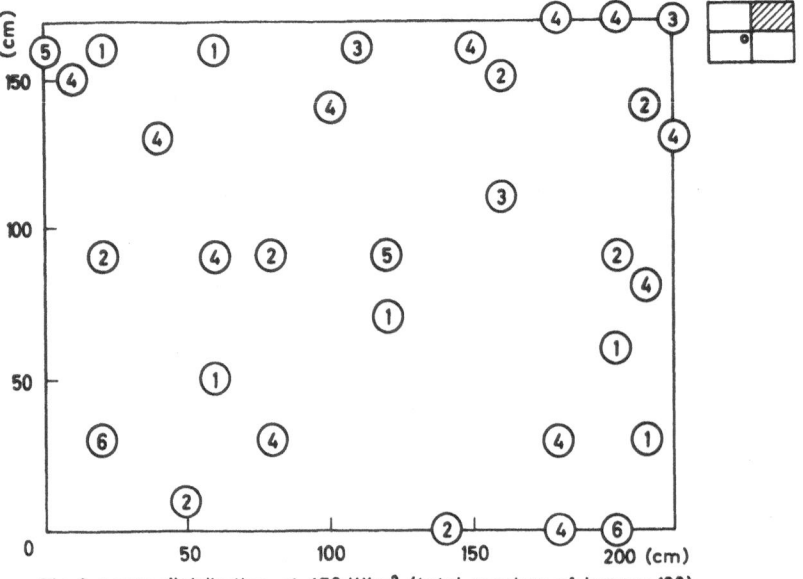

Fig. 2: Lamp distribution at 450 W/m² (total number of lamps: 120)

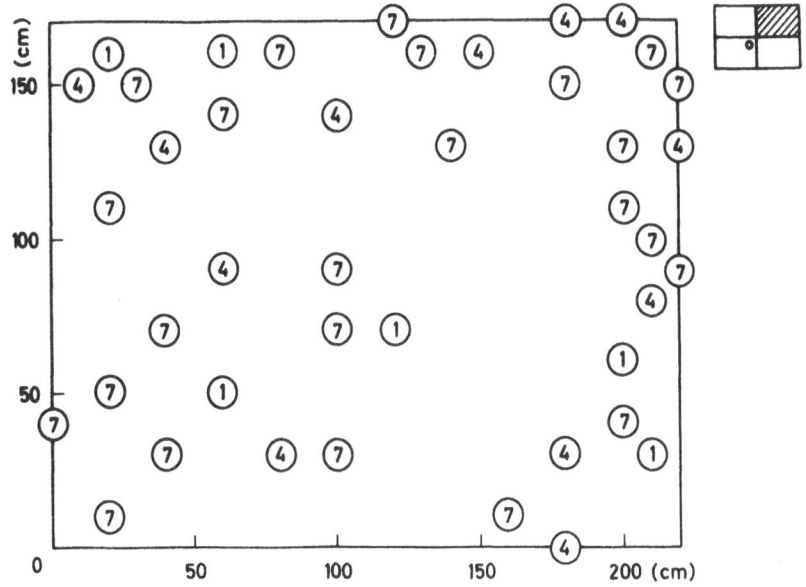

Fig. 3: Lamp distribution at 650 W/m² (total number of lamps: 164)

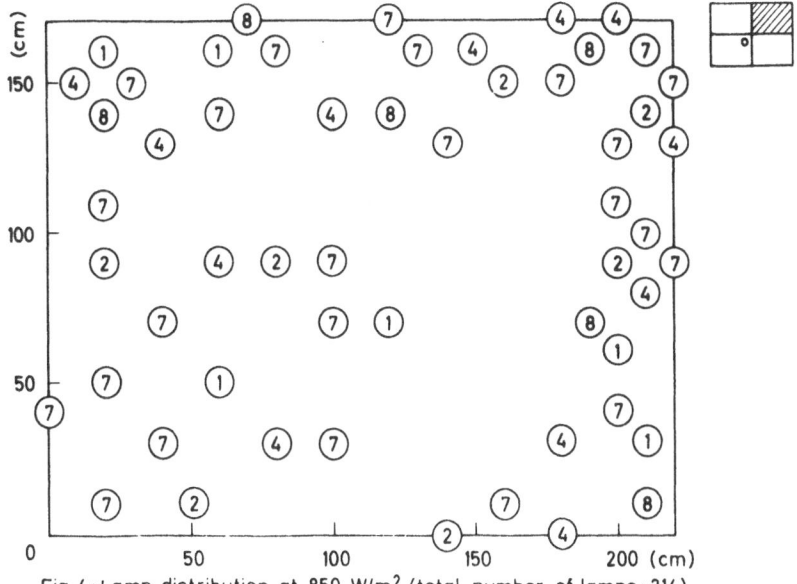

Fig. 4: Lamp distribution at 850 W/m² (total number of lamps: 214)

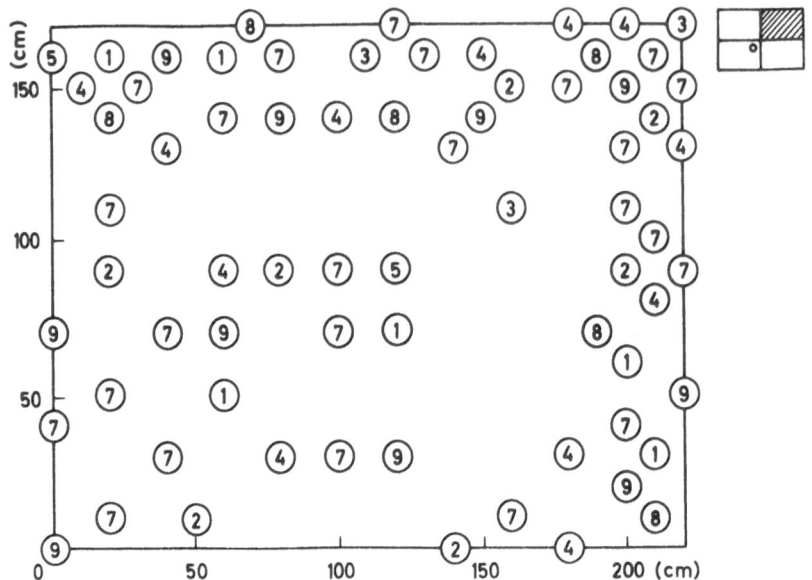

Fig. 5 : Lamp distribution at 1050 W/m² (total number of lamps : 267)

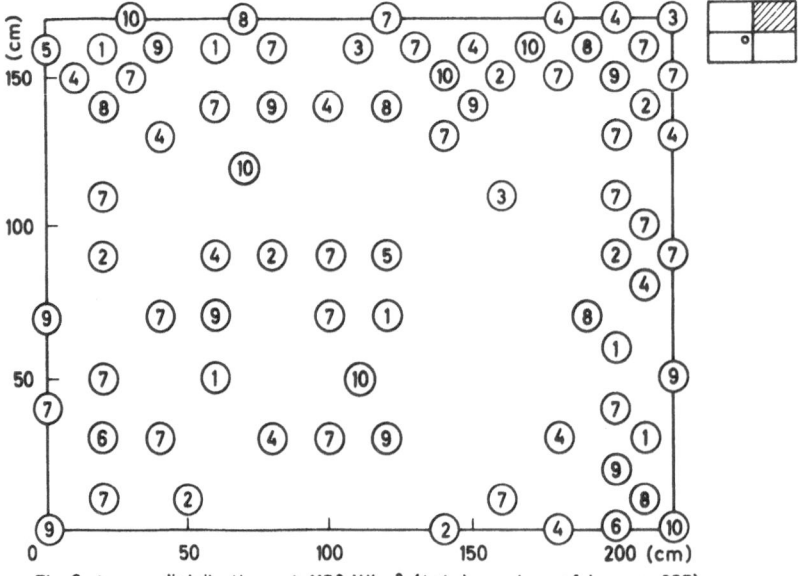

Fig. 6 : Lamp distribution at 1150 W/m² (total number of lamps: 295)

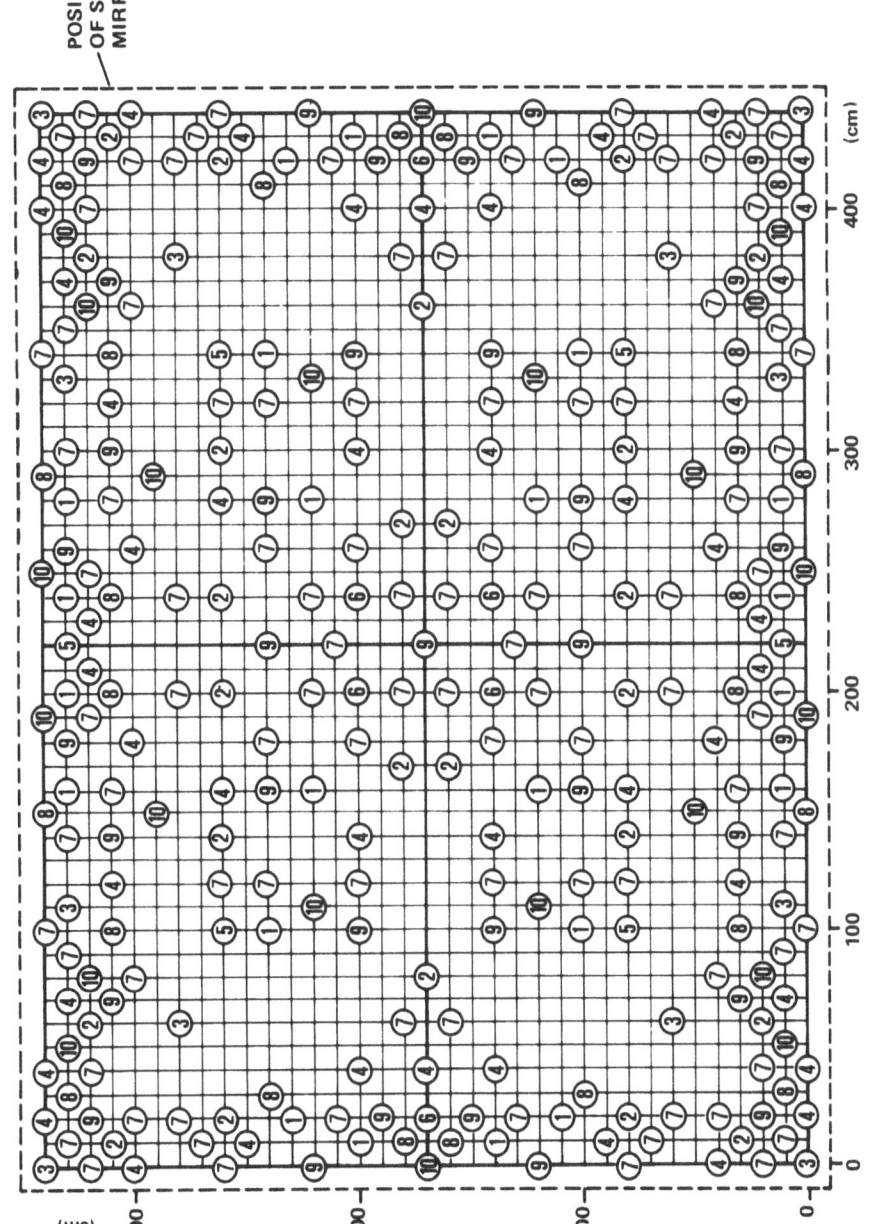

POSITION OF SIDE MIRRORS

Fig. 7. Light source of system LS-1: lamp distribution according to group number

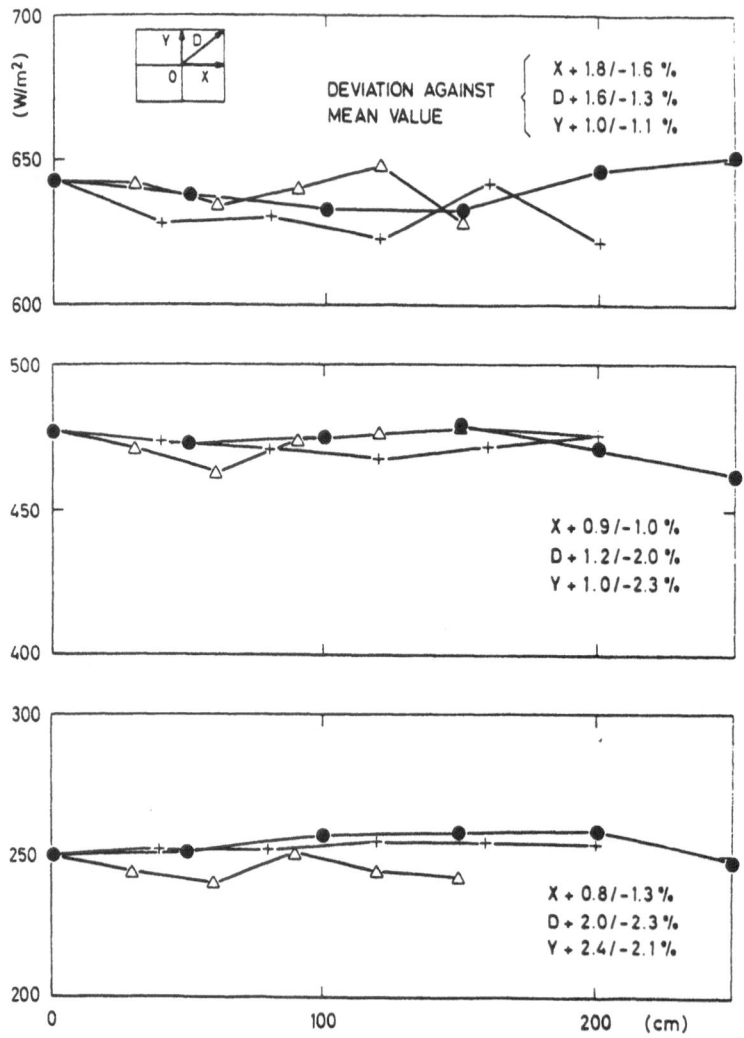

Fig. 8 : VARIATION OF LIGHT INTENSITY AT A DISTANCE OF 220 cm
BETWEEN SOURCE AND TEST PLANE ALONG DIREC-
TIONS X (+), D (●), Y (△)

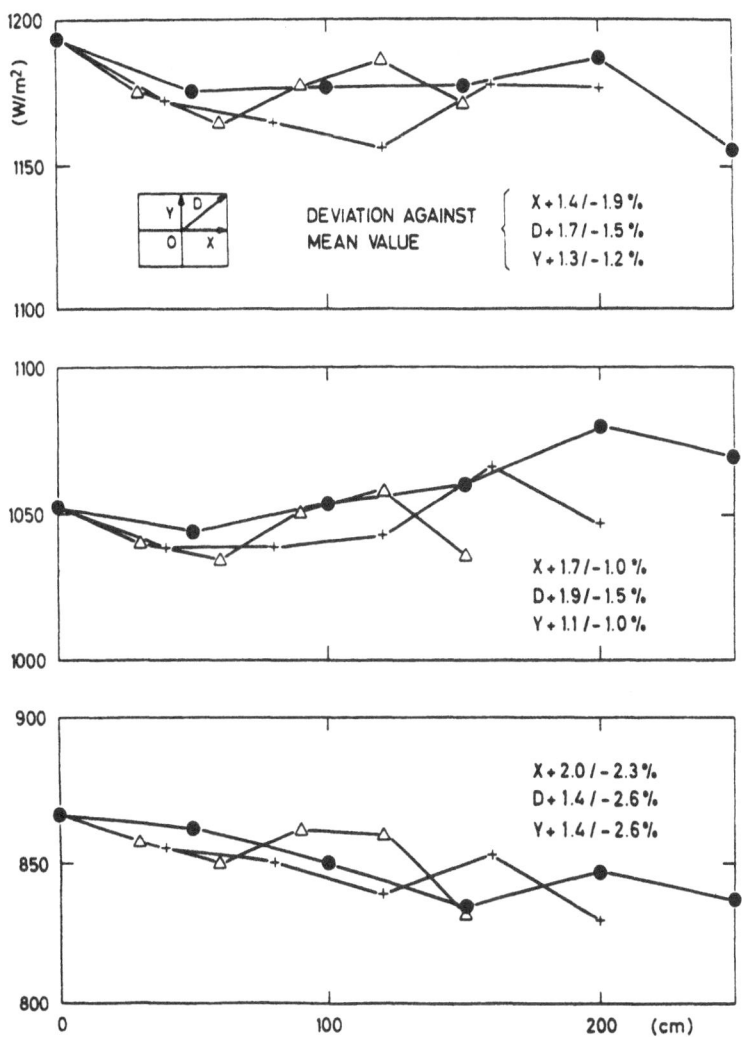

Fig.9: VARIATION OF LIGHT INTENSITY AT A DISTANCE OF 220 cm BETWEEN SOURCE AND TEST PLANE ALONG DIREC-TIONS X (+), D (●), Y (△)

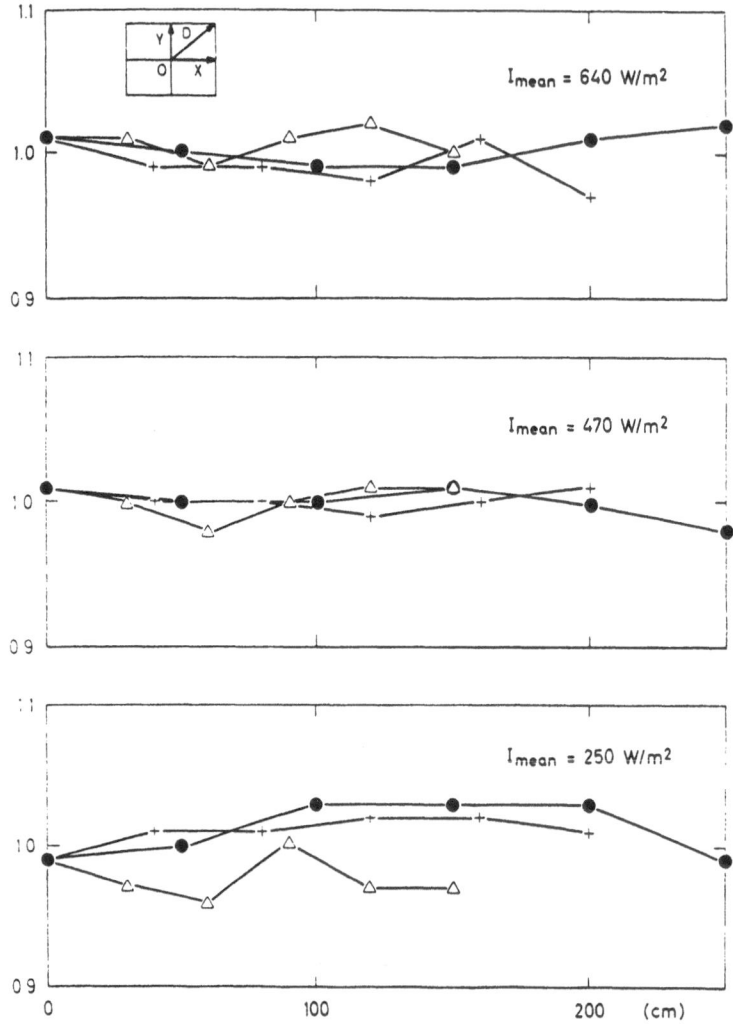

Fig.10 : RELATIVE INTENSITIES AT A DISTANCE OF 220 cm
BETWEEN SOURCE AND TEST PLANE ALONG DIREC-
TIONS X (+), D (●), Y (△)

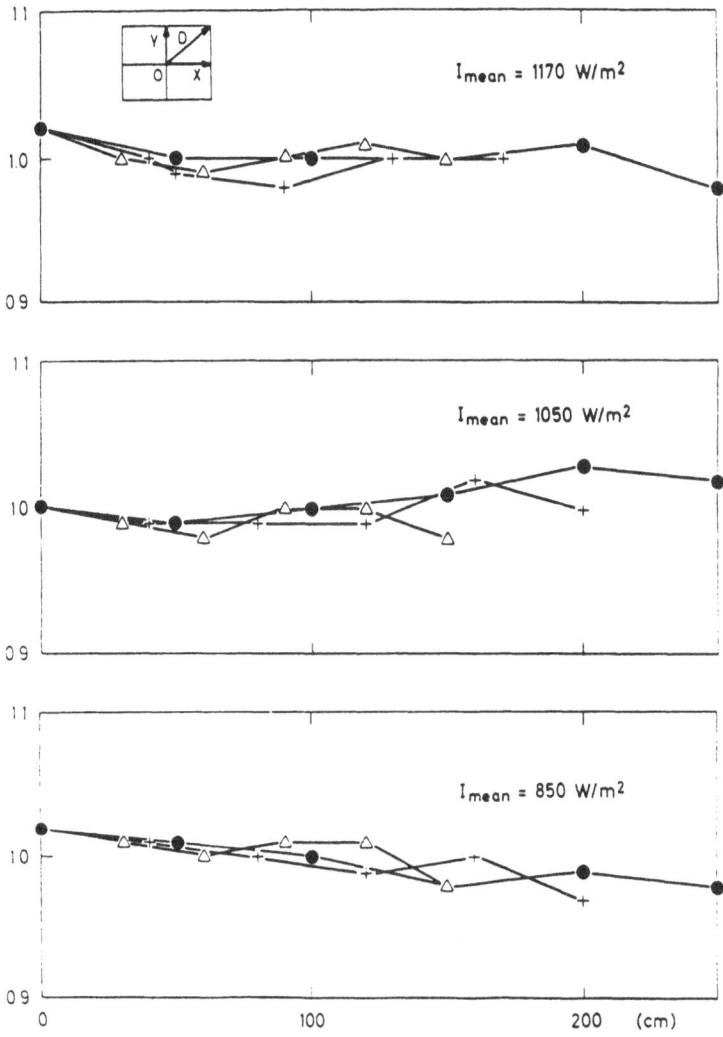

Fig.11 : RELATIVE INTENSITIES AT A DISTANCE OF 220 cm
BETWEEN SOURCE AND TEST PLANE ALONG DIREC-
TIONS X (+), D (●), Y (△)

IRRADIANCE (W/m²)	1	2	3	4	5	6	7	8	9	10	GROUP NUMBER
	24	26	12	46	6	6	94	24	35	22	
250	+	+	+	−	−	−	−	−	−	−	62
470	+	+	+	+	+	+	−	−	−	−	120
640	+	−	−	+	−	−	+	−	−	−	164
850	+	+	−	+	−	−	+	+	−	−	214
1050	+	+	+	+	+	−	+	+	+	−	267
1170	+	+	+	+	+	+	+	+	+	+	295

Fig. 12 : LAMP CONFIGURATIONS FOR LS-1 (+ : ON, − : OFF).

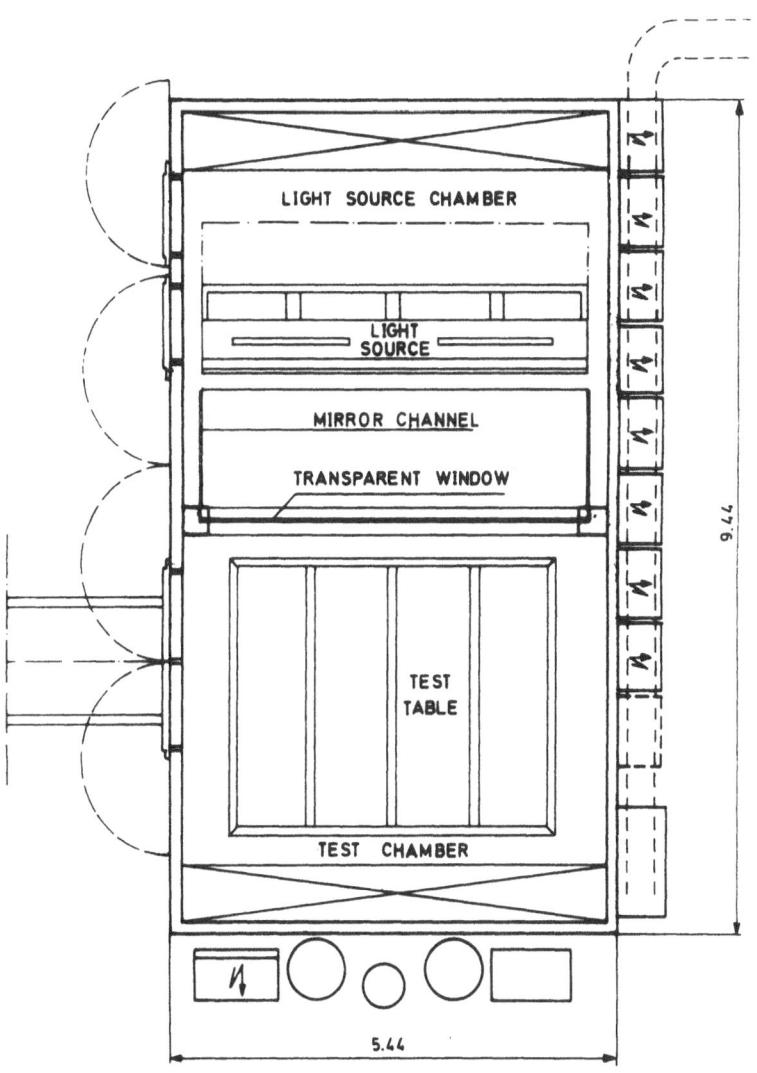

Fig.13a   LIGHT SIMULATOR LS1

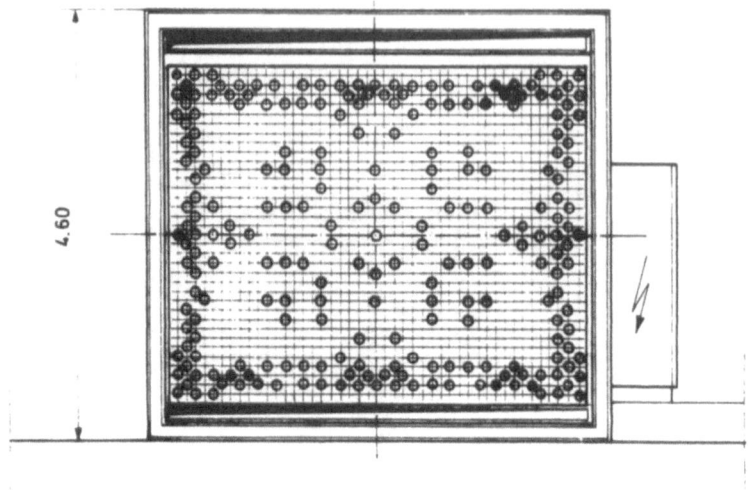

Fig.13b

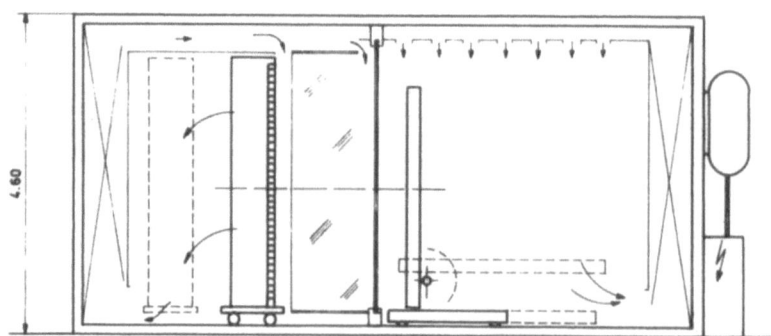

Fig.13c

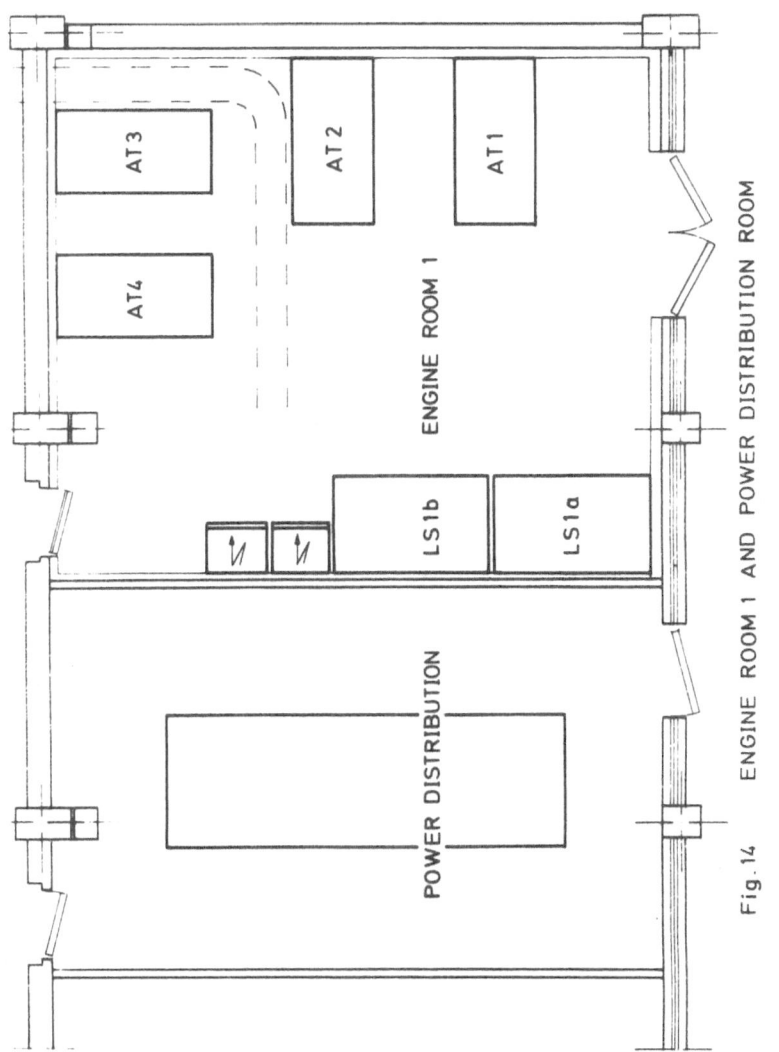

Fig. 14   ENGINE ROOM 1 AND POWER DISTRIBUTION ROOM

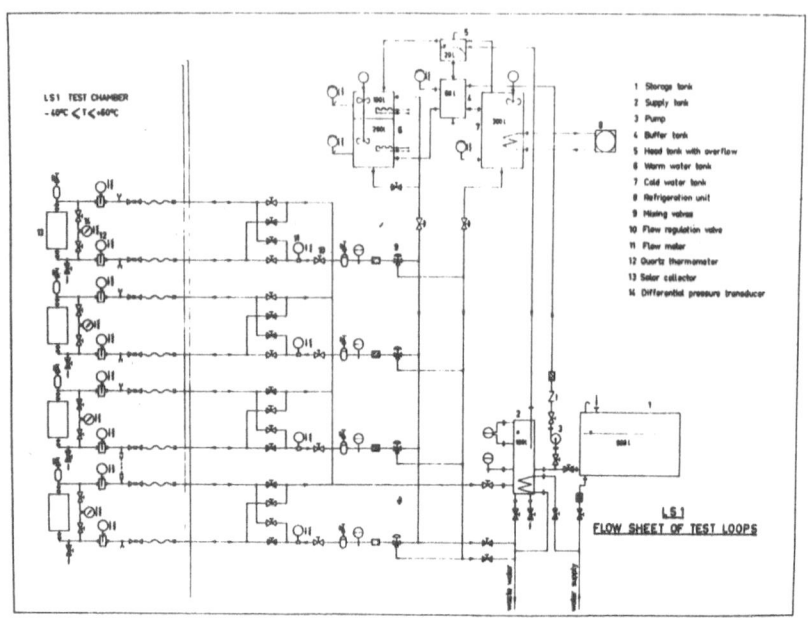

Fig. 15

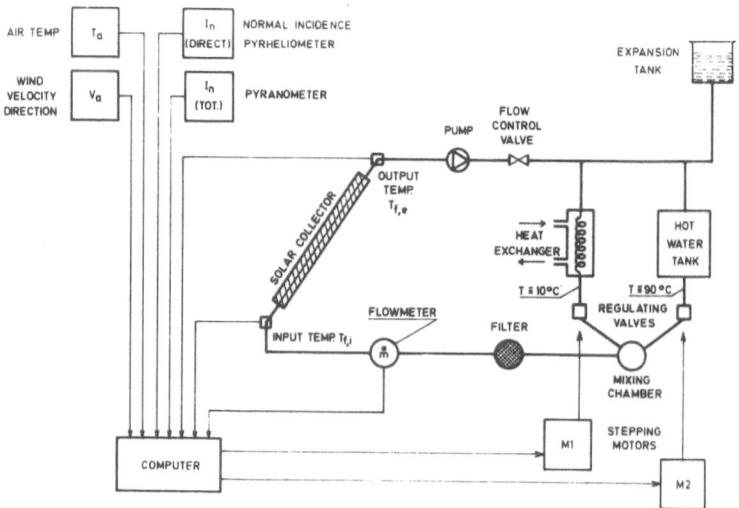

Fig. 16

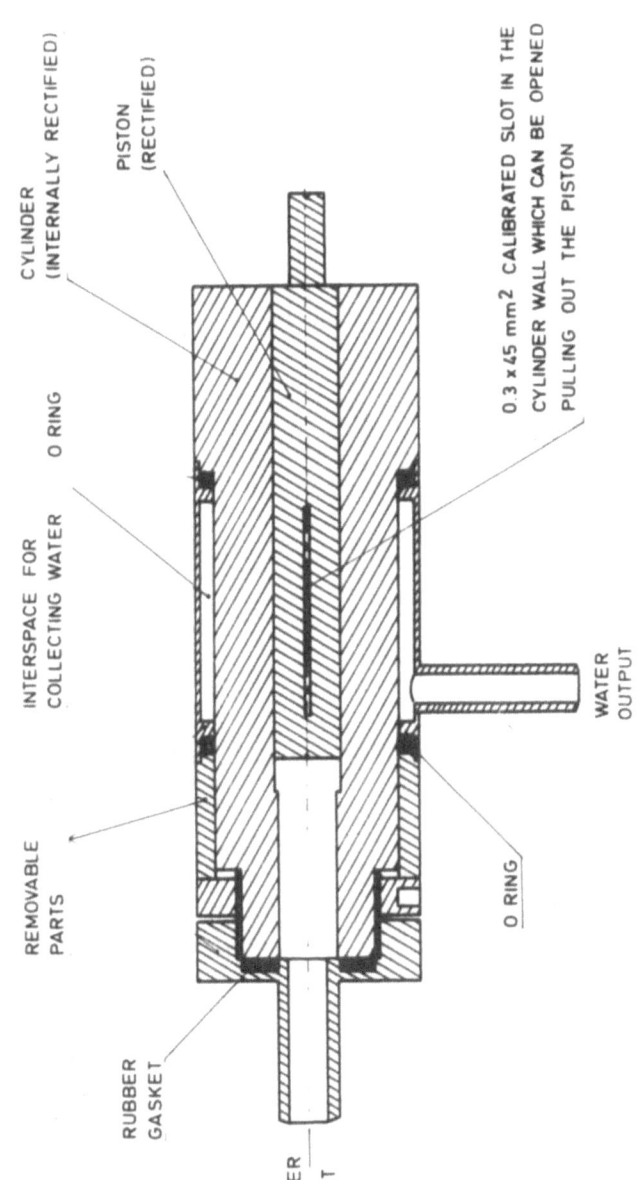

CYLINDER
(INTERNALLY RECTIFIED)

PISTON
(RECTIFIED)

O RING

INTERSPACE FOR
COLLECTING WATER

REMOVABLE
PARTS

RUBBER
GASKET

WATER
INPUT

0.3 x 45 mm² CALIBRATED SLOT IN THE
CYLINDER WALL WHICH CAN BE OPENED
PULLING OUT THE PISTON

WATER
OUTPUT

O RING

Fig.17

- 293 -

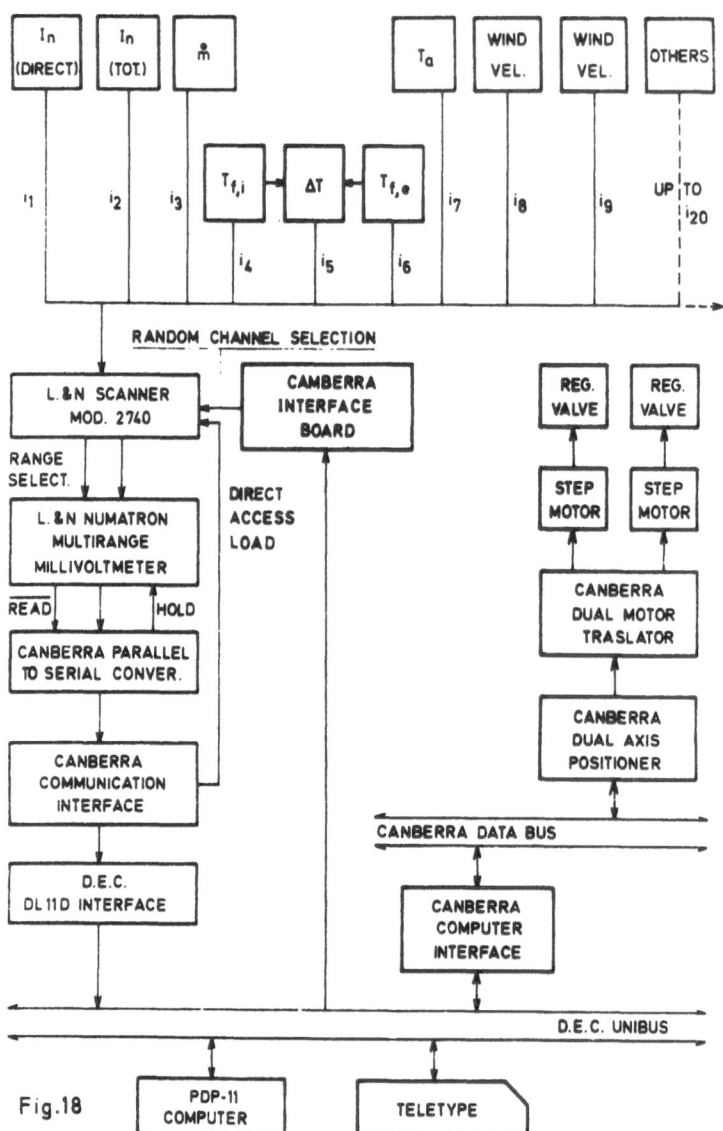

Fig.18

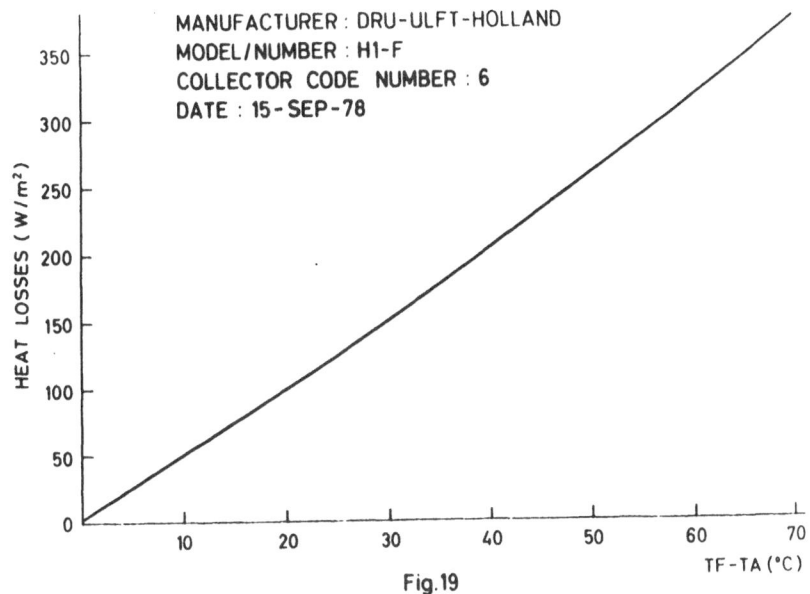

MANUFACTURER : DRU-ULFT-HOLLAND
MODEL/NUMBER : H1-F
COLLECTOR CODE NUMBER : 6
DATE : 15-SEP-78

Fig.19

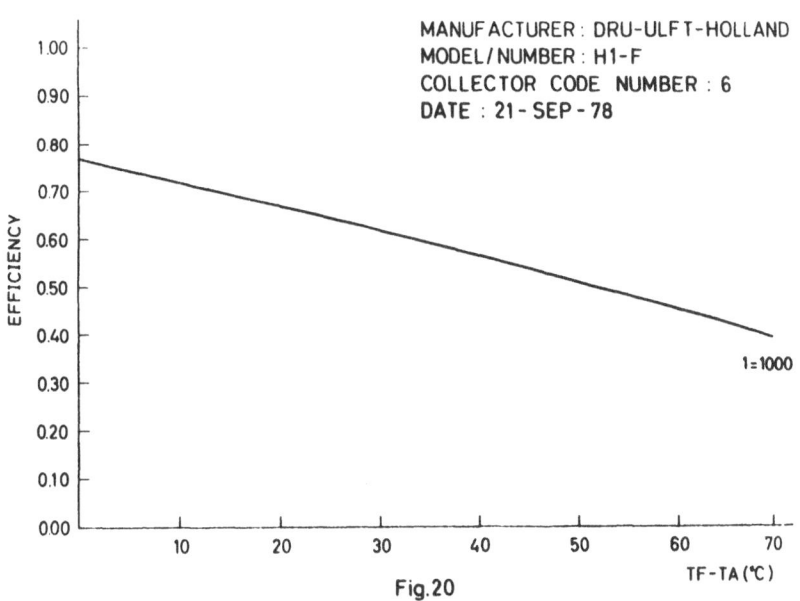

MANUFACTURER : DRU-ULFT-HOLLAND
MODEL/NUMBER : H1-F
COLLECTOR CODE NUMBER : 6
DATE : 21-SEP-78

I = 1000

Fig.20

QUALIFICATION AND DURABILITY TESTS: APPLICATIONS FOR THERMAL COLLECTORS AND PHOTOVOLTAIC MODULES

G. Riesch

Commission of the European Communities
Joint Research Centre - Ispra Establishment
I-21020 Ispra (Va), Italy

1. DEFINITION OF QUALIFICATION AND DURABILITY TESTS

1.1 Qualification tests

A qualification test on solar energy converters (thermal collector or photovoltaic module) verifies that at the moment of testing the converters can withstand certain demands, i.e. that they meet the requirements established in the test standard. Examples are:

- resistance of the thermal collector to a specified internal pressure, or of the photovoltaic module to high voltage tension;
- resistance of the converters to the impact of hailstones;
- resistance of the whole collector to a thermal shock by cold water spray;
- resistance of the converters to a uniform life load (e.g. wind-load);
- etc.

These types of test do not necessarily say anything about the expected lifetime of the converter because these tests do not have any long-term effect on the converter. They just decide whether or not for instance the collector withstands a pressure of 10 bars or the photovoltaic module resists a voltage of 1000 V applied between the metallic frame and the output terminals. They "qualify" the specimen. They determine the "endurance" of

G. Beghi (ed.), Performance of Solar Energy Converters: Thermal Collectors and Photovoltaic Cells, 297–325.
Copyright © 1983 ECSC, EEC, EAEC, Brussels and Luxembourg

the specimen. These tests do however have some relation to the lifetime of the converter, because the events, simulated in the tests, could occur in reality during the life of the converter and if it then fails its useful life would be over.

For instance an overpressure in the absorber might occur as a consequence of some failure in the pressure regulating system; or a hail shower could occasionally hit the roof-mounted collectors, or the photovoltaic array field, or an overtension might build up in a larger array as a consequence of some line interruption.

1.2 Durability tests

A durability test on the other hand implies that the test specimen (the collector or the module) is exposed for some time to the influence of (potentially) damaging agents, e.g. high temperature, moisture, ultraviolet irradiation and so on. For these tests it is supposed that the influencing agents change the properties of the specimen, as a function of the duration and of the intensity of the influence of the agent. The purpose of the durability test is of course to provoke in a shorter time the change in certain properties which would occur in real operation over many years of service so that lifetime predictions become possible within a short testing time.

1.2.1 Accelerated tests. We call a durability test an "accelerated" test if the intensity of the damaging agent is increased. It is supposed that in doing this, the same degradation as in real life exposure can be obtained.

Some example: Expose plastics (e.g. polyethylene foils) to ultraviolet irradiation. In principle one could say that if the UV-intensity is 10 times as high as the natural UV-intensity (which is of the order of 60 W/m^2 in bright sunlight) the same effect (e.g. yellowing) is produced in one tenth of the time. Such a simple relationship might be true in some special, isolated cases, however in general the correlation between effects of "real conditions" aging and "accelerated" aging is much more complicated.

It is not simply the dose (the product of dose rate - UV-intensity - and time) which determines the effect but both factors act in a separate way. There are very small intensities (dose rates)

where even with an infinite irradiation time nothing happens
because certain processes need a certain threshold intensity
(one might also say that the healing rate is higher than the damaging rate). There are also high intensities where the threshold
for another type of processes is exceeded, and then quite different effects can occur. Let us refer at this point to the well-known
example of the egg which, if kept at about 39°C, after three
weeks gives a chicken and which, if kept at about 100°C, after
some minutes gives a boiled egg.

To make things yet more complicated, there are other parameters which can strongly influence the aging behaviour: for instance in UV-irradiation some coloured tissues will hardly fade.
If the samples are sprayed periodically with water the bleaching
effect will be much stronger. So humidity is an important parameter for colour testing under UV-irradiation. The temperature
of the sample during UV-irradiation is also very important.
There are certain threshold temperatures where new photo-
chemical reactions appear in certain plastics.

1.2.2 Abbreviated test. Besides the just described "accelerated"
aging procedures where the intensity of the main damaging agent
is increased, there are the "abbreviated" aging procedures. An
example of this is thermal cycling. If, for instance, a collector
on a cloudless sunny day makes one thermal cycle per day (between -10°C to +25°C (ambient temperature) and the normal
operating conditions) and if the associated thermal expansion
has some degrading effect on the adherance of the absorber
coating (for collectors) or on the lamination between the glass
and the silicon cells (for modules), then the same degrading
effect can be produced by making a thermal cycling between the
same temperature extremes, for instance about 150 times per
day (10 min. cycles). In such a way the testing time is "abbreviated" by a factor of 150. Most mechanical fatigue tests are
based on this principle: close a contact one million times in a
few days, compress a spring ten thousand times and so on.

There are also combinations of accelerated and abbreviated
tests where both the intensity of the damaging agent and the frequency of its action are increased. For instance: thermal cycling
at higher temperature extremes than occur in reality.

1.3 Remark on durability tests

At this point it is important to emphasise once more that durability tests do not necessarily give an indication of the behaviour
of the test specimen in real life conditions, and they do not allow
an easy extrapolation to an expected lifetime of the specimen.
Often these tests are simply "conventions", test standards,
which have been proposed on rather weak theoretical grounds.
Despite this fact they are often used for relative tests, that is
for comparison of the behaviour of similar, slightly different
test samples, or for the comparison of the products of different
manufacturers.

As there is no really convincing reason why a given test has to
be carried out exactly as laid down in a "standard test procedure" there is always ample room for criticism, scepticism
and proposals for improvement. An "accepted" standard test
procedure can only result from a proposal made by someone,
ample tests executed by different people according to this proposal, evidence that this durability test gives "reasonable" results, and then a consensus of all concerned experts in a standard setting committee to endorse the test.
This is of course a long procedure and in a new field of technology such as solar energy no generally accepted tests exist as
yet (1981). We are in the phase of proposals which are being
tried by different people. Reference to existing proposals is
given in the next paragraph. Many of these proposals are based
on accepted test standards used in other fields, e. g. environmental tests for electrical equipment; testing of materials,
structural components and equipment, etc.

1.4 Literature review of tests on thermal collectors

The proposed tests are normally composed of different test procedures (qualification tests and durability tests) which are already "accepted" for tests on certain materials and components,
and these tests are completed by some new test proposals, specifically adapted to solar collectors. As these proposals are
normally unofficial and just serve as a working basis for people
who are interested, they are not widely known; hence the literature references given here are certainly far from being exhaus-
tive: they do not refer to officially released documents and might
already be obsolete.

In Germany the Bundesverband Solarenergie, BSE, proposes ref. (1). It is an outdoor exposure (with certain temperature restrictions) lasting one year. There is also ref. (2), specifying certain requirements for absorber, containment, coating and corrosion resistance and fixing a procedure for the measurement of the maximum dry exposure temperature.

In France the Centre Scientifique et Technique du Bâtiment, CSTB, has proposed ref. (3), which provides a one-year outdoor exposure, a falling steel ball test, an overpressure resistance test of the absorber and a test for resistance to wind suction on the glazing.

In another working document the following tests are envisaged: resistance of mounting brackets, resistance of manifolds to torsion, rain tightness, salt spray test for enclosure, and artificial aging of absorber coating under ultra-violet irradiation cycles.

Another organization, CETIAT (Centre Technique des Industries Aéroliques et Thermiques) has recently proposed a series of tests (4) which will be discussed elsewhere in this course.

In Italy, Phoebus, research for solar energy, uses a series of tests which are described in ref. (5). These tests are similar to the ones proposed by the JRC, ESTI, and we will therefore speak about them later.

In the United Kingdom three "codes of practice" for the solar heating industry exist, but none of these seems to emphasise tests for qualification or durability. It appears that there are activities at the Solar Energy Unit of University College, Cardiff (measurements of thermal efficiency before and after natural outdoor aging of box-type collectors), (measurements of changes of the optical properties of the absorber surface and the transmittance of covers due to natural or accelerated aging) (6). Another organization active in this field in the U.K. is the Building Research Establishment, BRS.

In Denmark the Thermal Insulation Laboratory of Copenhagen University executes tests with dry exposure (with inner and outer thermal shocks), tests for rain penetration and tests for wind loads (7).

In the USA the best known test proposals are given in ref. (8) and ref. (9). The NBS proposal lists 12 qualification and durability tests. It seems that there has been criticism from industry that some of these tests are too severe. However, an extensive experimental programme based on this proposal is underway. The Florida Solar Energy Center proposal, centred around a thirty-day outdoor exposure with spray test, seems to be more usable in practice.

In Canada a set of tests is proposed in ref. (10). Proposed qualification tests are: overpressure resistance of absorber, thermal cycling and uniform positive and negative loads on the cover. Proposed durability tests are: thirty-day outdoor no-flow exposure (with dry boiling). The tests are hence similar to the ones proposed in refs. (9) and (8).

After this short review of the tests proposed by others, we shall now describe the qualification and durability tests proposed and being investigated experimentally at ESTI.

1.5 Literature references for tests on solar modules

Only a few proposals are known for qualification and durability tests on photovoltaic modules. In ref. (11), SERI, (USA) has fixed the proposals for test methods as part of a DOE Photovoltaic Performance Criteria and Test Standards Project.

An earlier proposal was made by JPL in ref. (12). According to these test specifications 53 modules were tested in 1979/1980.

A European proposal for tests is made in ref. (13). These tests will now be discussed in detail in the following, first the qualification tests, then the durability test. The sequence in which the tests will be performed is slightly different from the enumeration in the next paragraph.

2. QUALIFICATION TESTS ON THERMAL COLLECTORS

2.1 Materials and construction evaluation

First and very important elements for the evaluation of a collector are given by a complete and detailed description of the materials used and of the construction. Required are:

- drawings of all parts of the collector and an assembly drawing,
- material descriptions for all parts of the collector (cover, absorber, coating, enclosure and its components, insulation, sealants or gaskets or caulkings) and of the accessories (fasteners, mounting brackets, hoses),
- specifications: maximum tolerable temperatures, pressures, flow rates (and possibly, especially for plastic covers, maximum wind, snow and ice-loads).

An inspection programme of installed solar equipment recently started by the IEA has revealed that many of the failures observed are due to design errors. Most of such failures can certainly be predicted by an experienced expert just by examining the construction and the materials used for the construction of the collector. Hence the availability of a detailed description of the collector is an important part of the test procedure.

2.2 Overpressure resistance of the absorber

A first simple, yet important test is the determination of the resistance of the absorber to overpressure. Normally, the applied overpressure is 1.5 times the normal operating pressure specified by the manufacturer, e.g. if the manufacturer specifies a 6 bar operating pressure, the absorber should withstand 6 x 1.5 = 9 bars overpressure. If a collector is specified for operation at atmospheric pressure, it should withstand at least 1.5 bars. If it is not specified at all, but intended for use in pressurized systems the absorber should withstand 12 bars.

In Ispra we started these tests with a water-pressurizer which is cheap and safe. Filling the absorber with water, however, spoils the inside of the absorber. We are therefore now considering carrying out the overpressure test with an air-pressurizer. The safety aspects of a possible explosion of the pressurized absorber have yet to be studied.

2.2 Leak test

We believe that it is important to perform a leak test of the absorber after the overpressure test. Small leaks are not detectable by the overpressure test, yet they might rapidly cause failure of the collector because small amounts of liquid, entering the collector enclosure during operation, can cause the thermal insulation to deteriorate and cause corrosion and condensation inside of the collector.

To detect leaks we use air, pressurized to the nominal operating pressure of the collector. We then watch for 15 min. to see whether the pressure of the collector remains constant after it has been isolated from the pressure source. The leak test with pressurized air is much more sensitive than a leak test with water for two reasons: temperature effects during the 15 min. observation time are much smaller with air than with water and air escapes through small leaks more easily than water.

Presently we are also studying the possibilities offered by a helium leak test.

The facility used for the pressure tests is called AT-5. A short specification of the different test facilities at ESTI is given in Appendix 1.

2.3 Rain penetration

Another test, which we feel is very important, checks the rain water tightness of collectors. The already mentioned inspection programme of the IEA revealed that rain water leakage into collectors is a frequent and serious problem.

Our test facility consists of a vertical mounting of the collector behind a grid of spray nozzles. The water is at ambient temperature and is used as (originally) demineralized water in a closed circuit driven by a water pump. (A coarse water filter will now be added to the circuit in order to avoid blockage of the spray nozzles.)

Normally full-cone nozzles are used giving a uniform spray on the collector surface of the order of 2 l/min per m^2 for a duration of 15 to 20 min. The water jets have no prescribed pressure, however, it is possible to apply a differential air pressure of a few hundred Pascals between the inside and the outside of the collector during the spray cycle. In our installation the pressure difference is obtained by connecting the inside of the collector

to the atmosphere and by pressurizing the outside of the collector (the test chamber in which the sprays are applied) to the
desired value.

The possible water penetration is measured by accurately weighing the collector before and after the spray treatment.

Much care must be paid in these tests to the openings in the collector enclosure provided for ventilation. There are collectors
with supposedly "tight" enclosures where the "breathing" occurs
at unknown places, probably at the passages of the nipples of the
absorber through the case or at the joints between case and cover.
In these collectors all parts, which in real operation are exposed
to rain, will be exposed to the water spray.

For collectors with ventilation openings it should be decided
case by case whether these openings have to be closed before-
hand or protected during the spray test. (This is the case if in
real use they are protected from rain and also from wind-driven
rain.) The openings can also remain as they are (if they are probably the points where rain penetrates in real use). It has also
to be decided whether in the case with ventilation openings it is
sufficient to apply a differential pressure between the inside and
the outside of the collector. As rain is often combined with wind,
which creates a pressure difference in real use, normally a collector should remain rain-tight under a slight underpressure of
about 100 Pa.

For the rain-tightness test we use our facility AT-3.

2.4 Resistance to life loads

This test verifies whether the glazing and its fixings to the collector case withstand certain life loads, i.e. wind, snow and ice
pressure on the cover or wind suction on the cover.

The uniform positive or negative loads are generated in our test
facility by air pressure in the test chamber AT-3.

The test chamber AT-3 can be charged by a reversible blower
with air pressures up to 6500 Pa or with negative air pressures
up to 4000 Pa (against atmospheric pressure). The air used to
pressurize or depressurize the chamber is taken not from the

surroundings, but from another chamber of about equal volume. This is necessary because the tests can be made between room temperature and -15°C and hence a reservoir of cold air is necessary.

The collector is mounted and secured on the rear wall of the chamber in such a way that its inside remains at atmospheric pressure (e. g. by holes in the side- or back walls of the enclosure or by a hose connected between the inside of the collector and the atmosphere). The front cover of the collector is exposed to the varying positive or negative pressures. Instead of applying a negative pressure on the outside of the collector, it is also possible to inflate the inside of the collector.

Three types of pressure tests can be made: (see Figure 1)
a) Measurement of a load-deflection curve.
   Pressure steps of 300, 600, 900 and 1200 Pa positive and then negative are applied in steps. The deflection of the cover plate can be measured by a dial position indicator or a gauge head. A 600 Pa pressure is applied for initial positioning before the measurement.
b) Resistance to maximum load.
   A single pressure pulse of 2400 Pa is applied to the outside of the collector. This simulates the pressure of 130 km/h wind with a safety factor of 3 for gusty wind.
   If snow and ice pressure are also to be simulated, then 5400 Pa of pressure are used instead of 2400 Pa. For suction a maximum value of 2400 Pa is used. (It should be noted that the quoted pressure values are rather high. They are however smaller than those proposed in ref. (8) and the snow loads correspond to the values given in DIN 1055, part 5, for roofs with less than 30° inclination).
c) Dynamic loads.
   Alternating pressure and suction of 600 Pa can be applied in order to simulate dynamic wind effects which might provoke fatigue damage, especially to plastic covers. (This is, however, no longer a qualification test but a durability test.)

2. 5 Resistance to hailstone impact

For plastic covers especially it is worthwhile testing the resistance of the cover to hailstones. There are standards which use steel ball dropping for this kind of test (DIN 52306/7). It has been

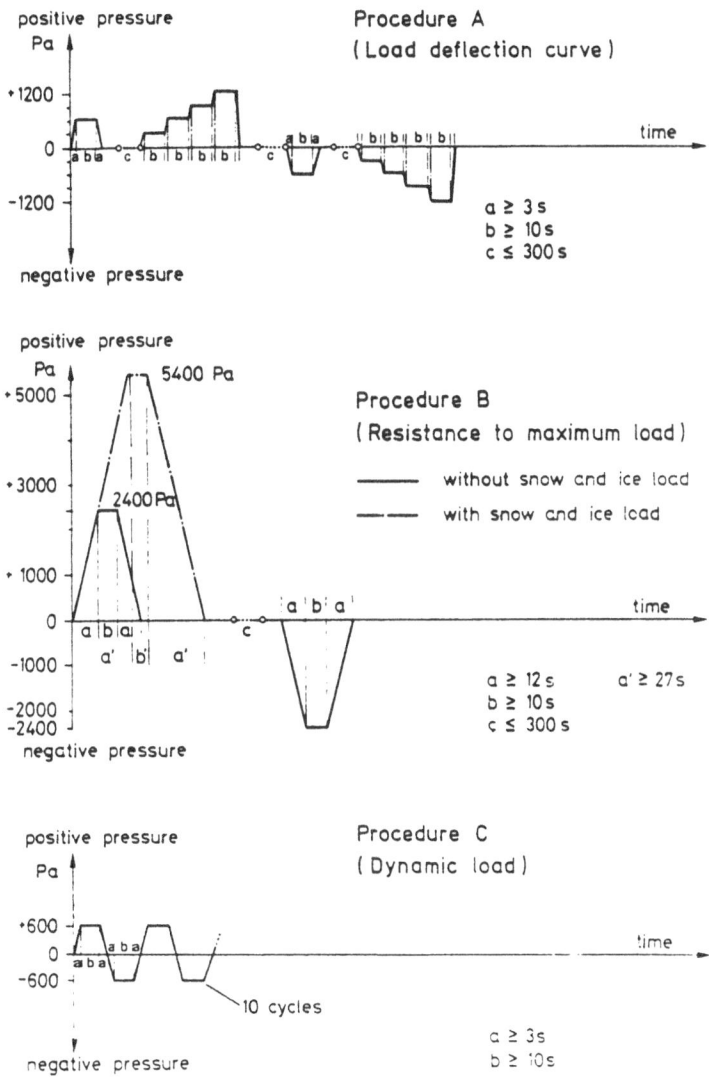

Figure 1. Pressure test cycles.

noted, however, that this test cannot be reasonably correlated to hailstone damage, as the elastic behaviour of a steel ball is quite different from that of an ice ball.

Our facility (AT-4) allows us to shoot ice balls of different diameters (from 1/2 inch to 1 1/2 inch) horizontally by air pressure at velocities between 16 m/s to 30 m/s onto the test specimen. The air pressure is applied by a quick opening valve and the ice ball velocity is measured by a light barrier.
After the shot a visual inspection of the target is made. A collector should withstand without damage the impact of hailstones of 2.5 cm diameter and of 23 m/s velocity.

2.6 Thermal shock

2.6.1 By water spray. This test simulates to a certain extent the events which may occur to a collector in normal use, when a rain shower falls on the hot collector. This test will be performed at ESTI in the AT-6 facility which is not yet available (1981), but which is under construction. The collector is mounted with an inclination of 45° under an array of 88 lamps (Osram Power Star HQI-R 250 W). The irradiance falling on the collector is adjustable up to 1100 W/m^2. This "aging simulator" AT-6 will be equipped with a water spray facility and with a water fill facility. The water spray facility allows us to spray demineralized water at a rate of up to 4 l/min per m^2 onto the hot collector. Before the spray cycle, which lasts 15 min., the collector is brought to equilibrium temperature by a one hour irradiation at 950 W/m^2. During the spray cycle the lamps are shut off.

Two inspections are made after the spray cycle: a visual inspection for any mechanical damage to the collector and a weight comparison of the collector before and after treatment in order to detect any water which has entered. This water tightness test can be skipped if the collector has already proved to be watertight at room temperature (as checked in test 3, already mentioned).

2.6.2 By water fill. Another thermal shock test is made by filling a hot, empty absorber with the coolant fluid prescribed for the collector. In that case the collector is again irradiated at 950 W/m^2 for at least one hour before filling of the absorber from the bottom starts at a rate of about 1 l/min per m^2 of absorber area. In the first few moments the fluid will evaporate, so the absorber must remain open in order to avoid pressure build-up. Then circulation of the coolant liquid will be maintained for 5 min.

A visual inspection is made after the thermal shock.

3. DURABILITY TESTS ON THERMAL COLLECTORS

3.1 Dry exposure

The essential durability test proposed by many experts (1), (8), (9), (10) is an outdoor weathering during longer times (from 30 days up to 1 year) with certain minimum irradiation values (normally 17 MJ/m^2 per reckonable day of exposure. The 30-day test can only be executed in a few months calender time with favourable weather conditions. In these tests it is further prescribed that during a 4 hour period an irradiance of minimum 950 W/m^2 must exist. It is not easy to observe this prescription.

At ESTI we are therefore trying to perform a test which should give equivalent results in an indoor simulator.

This facility, called AT-6, has already been mentioned because it will also be used for the thermal shock tests. We have no experience yet whether the possible damage to a collector exposed to the 30-day outdoor test can also be generated in a short indoor test. However, if it is not the duration of the exposure, but the temperatures reached during the 950 W/m^2 irradiance period, or the thermal cycling which are important, then we can expect to obtain a valid durability test with our aging simulator. The irradiance in this facility can be varied up to 1100 W/m^2 and the temperature of the air surrounding the collector will be high enough so that the conditions prescribed in the 30-day outdoor tests can easily be exceeded. Different irradiation conditions, lengths and thermal cycles will be applied to the collectors in order to define a valid durability test. A correlation to what happens to a collector on an outdoor dry exposure test will be obtained by the test described hereafter.

3.2 Dry exposure with temperature analysis

This test has two aims: in the first place it is a long outdoor exposure of a collector in dry conditions and gives correlation data for the indoor exposure test described above.

In the second place the collector is equipped with a temperature measurement detector on the absorber and furthermore the ambient temperature and the incoming solar irradiance are measured continuously. The temperature of the absorber (and of the whole collector) is equal to the ambient air temperature when no irradiation falls on the collector and it increases during the day and returns to the air temperature during the night. If an energy balance is made for time intervals for which the initial and the final collector temperatures are equal then the integral over the absorbed radiant energy during this integration interval and the integral over the heat losses of the collector are equal.

The integral over the absorbed radiant energy is equal to the incident solar energy multiplied by the effective $\tau\alpha$-value (τ = transmission of cover, α = absorption of absorber). The integral over the heat losses is equal to the integral over the temperature difference $(T_{collector} - T_{air})$ multiplied by an effective overall heat transfer coefficient U.

$G(t)$ (global irradiance), $T(t)$ (collector absorber temperature) and $T_a(t)$ (ambient air temperature) are measured continuously. The ratio of the two calculated integrals ($\int (T-T_a)dt$ and $\int G\,dt$) is equal to $\tau\alpha / U$.

If the collector degrades during the outdoor exposure $\tau\alpha$ will become smaller and U will become larger and hence the ratio of the integrals cited is a measure of the quality of the collector.

The test installation for this analysis at ESTI is called STA (stagnant temperature analysis). It is mounted on the roof of building 45.

3.3 Exposure to UV-irradiation

At ESTI we have a provisional UV-irradiation facility composed of three lamp columns containing 6 lamps each. The lamps are commercially available Uvistra lamps, 400 W electric each, mounted in a housing with filter. At 1 m distance from the lamps a rather homogeneous irradiation of 60 W/m^2 and 300 W/m^2 total is obtained on a test area of 2 x 2 m. The UV-irradiation is counted between 300 and 400 nm).

The temperature of the ambient air and the humidity cannot be adjusted in this proof-of-concept set-up. This facility is called AT-2.

Another UV-irradiation facility composed of four 2000 kW UV-lamps has recently been put into operation. With lateral reflecting walls this facility gives about 110 W/m^2 UV (from 300 to 400 nm) and a total of about 1000 W/m^2 at 1 m distance from the lamps on a test area of 1.5 x 1 m.

3.4 Exposure to ozone-containing atmosphere

Our largest test chamber with inside dimensions of 2 m width, 3.7 m depth and 2.2 m height, called AT-1, allows three different operating modes.

One of these operates with air at a temperature between 20°C and 40°C, humidity between 10% and 95%, and an ozone concentration up to 4 volume per mille. With this chamber we can hence produce the test atmospheres as prescribed in DIN 53509 (25°C, 55% relative humidity, 0.5 ppm O_3) or as prescribed in ASTM D 3395-79 (40°C, 0.5 ppm O_3).

3.5 Exposure to sulphur dioxide

In the same test chamber AT-1, sulphur dioxide gas can be applied instead of ozone. The temperature and humidity levels are the same as already mentioned. The SO_2 content can be as high as 1000 ppm. This value has been chosen, because DIN 50018 envisages in cyclic tests initial SO_2 concentrations of 670 ppm (volumes). It seems that this SO_2 concentration is rather high. A newer test procedure (IEC 68-2-42) prescribes only 25 ppm SO_2. Such a low SO_2 concentration is about the minimum level which can be detected and controlled by the SO_2-measuring instrument of the AT-1 facility. It might also be interesting to add CO_2-gas to the SO_2-gas (fume chamber).

3.6 Exposure to cyclic damp heat

The third operational mode possible with the AT-1 chamber allows a thermal cycling of the specimen between 20°C and 40°C at 95% relative humidity. These conditions are prescribed in DIN 40046, part 31 and in IEC 68-2-30. The main purpose of this test is to generate on the outside and on the inside of "breathing" equipment condensations which might damage the equipment. The outside condensation of solar collectors occurs during the temperature raise phase, when the specimen is colder

than the surrounding air and the inside condensation occurs
during the temperature fall phase when the humid inside air is
warmer than the collector walls.

3.7 Outdoor correlation test field

As already explained the damage caused by accelerated aging
tests must be correlated to the damage which occurs in real
outdoor exposure. In order to have reliable data for outdoor degradation under well-defined conditions an outdoor test field is
in preparation at ESTI. In this test field three specimens of five
different collector types will be exposed. On a data logger collector temperatures, air temperature, air humidity, solar irradiation and SO_2-concentration will be recorded.

3.8 Salt mist test

For these tests a salt spray (fog) chamber will be available to
execute tests on parts of collectors according to DIN 50021 or
ASTM B 117-73.

4. QUALIFICATION TESTS ON PHOTOVOLTAIC MODULES

4.1 Visual inspection and module description

Each module to be tested shall be fully described, including data
on materials used, especially sealants and frame materials.
By visual inspection broken or cracked windows, cell breakages
which are observable, faulty connections, failures of adhesive
bonds and so on will be identified. Suspect features will be examined with magnification (up to 40 times) under a microscope.
The modules will also be weighed and measured.

4.2 Robustness of terminations

There are modules with wire or flying lead terminations and
modules with tags, threaded studs, screws or other terminations.
Wire connections will be tested for tensile strength (e. g. 4 kp
for wires with more than 1.2 mm^2 of cross section) and they
will also be subjected to a bending test.

Other terminations will be tested in a similar way. If they are protected in a connection box a cable will be fixed to this box and tensile and bending tests will be performed on this cable.

For terminations with threaded studs or screws a torque test will be made.

4.3 Mounting twist test

Another mechanical test checks whether the module can be slightly twisted on being mounted: three corners are fixed and the fourth corner will be lifted by 20 mm per metre of diagonal length.

4.4 Insulation test

An electrical qualification test prescribes the application of a voltage of 1000 V between the shorted output terminals and the metal frame or (in the absence of such a frame) a structure simulating the array-supporting structure. The insulation resistance must not be less than 100 $M\Omega$. (This test is not required for modules to be used in low voltage (< 50 V) arrays.)

4.5 Hail resistance test

Because of the thinness of the cover glasses it seems that photovoltaic modules are rather vulnerable to hailstones. A test procedure has therefore been developed which allows us to examine the resistance of photovoltaic modules to hailstones.

There exist standards which use steel ball dropping for this kind of test (DIN 52306 and 52307). It has been noted, however, that this test cannot be reasonably correlated to hailstone damage as the elastic behaviour of a steel ball is quite different from that of an ice ball. The damage caused by a steel ball will generally be greater than that caused by an ice ball with equal impact energy, because the impact are is smaller.

In photovoltaic modules, especially, it might occur that the glass cover is not damaged, whereas the silicon cell underneath the target point is broken. Hence, careful visual inspection with a microscope will be necessary after the shot.

Our facility (AT-4) allows us to shoot by air pressure an ice ball of a diameter to be chosen from 12.5 mm to 40 mm horizontally at a velocity to be chosen between 16 m/s and 30 m/s on the solar module. The target point can be chosen deliberately and easily because the whole gun is mounted on a table which is adjustable in height and in lateral position.

The shooting distance can also be chosen at will; (normally this distance is around 1 m from the barrel end to the module). One ice ball is shot at a time by loading it into the barrel from behind and applying an air pressure of around 0.5 bar by a quick opening valve. The velocity of the ice ball is measured by an electro-optical device. (In the Appendix the main specifications of the ESTI test facilities are summarized.)

4.6 Resistance to life loads

This test verifies whether the module glass itself, its attachments in the frame and the frame itself, when mounted as prescribed by the manufacturer, can withstand certain life loads. Such life loads are wind pressure or suction on the glass or snow and ice pressure on the glass.

The uniform positive and negative loads on the module glass are generated in our test facility by air pressure in the test chamber AT-3. This test chamber can be charged by a reversible blower with air pressures up to 6500 Pa or with negative air pressures up to 4000 Pa (against atmospheric pressure). The air used to pressurize or depressurize the chamber is taken not from the surroundings but from another chamber of about equal volume. This is necessary because the tests can be performed between room temperature and $-15°C$ and hence a reservoir of cold air is necessary.
The module is mounted on a thick wooden frame which simulates the array-supporting structure in a rigid-non-flexible form.
In this frame the module is mounted closely but without touching the frame, in an almost air-tight way. If the air-tightness is not sufficient to allow a pressure difference between the front and the back of the module by the existing blower, tightness is improved by thin masking tape around the slits. The module thus forms a diaphragm between the front chamber and the back chamber.
Three types of pressure tests can be performed (see Figure 1).

a) Measurement of a load-deflection curve.
Pressure steps of 300, 600, 900 and 1200 Pa positive and then negative are applied in steps. The deflection of the whole module can be measured by a dial position indicator or a gauge head. A 600 Pa pressure is applied for initial positioning before the measurement.

b) Resistance to maximum load.
A single pressure pulse of 2400 Pa is applied to the upper side of the module. This simulates the pressure of 130 km/h wind with a safety factor of 3 for gusty wind.
If snow and ice pressure have to be simulated too, then 5400 Pa of pressure are used instead of 2400 Pa.
For suction a maximum value of 2400 Pa is used. The snow loads correspond to the values given in DIN 1055, part 5, for roofs with less than 30° inclination.

c) Dynamic loads.
Alternating pressure and suction of 600 Pa can be applied in order to simulate dynamic wind effects which might provoke fatigue damage. (This should, however, already be considered as a durability test.)

4.7 Ice formation test

In this test the module is mounted vertically or slightly inclined in a climatic chamber and then the air temperature (and hence the module temperature) is decreased to $-10^{\circ}C$ at a rate of about $5^{\circ}C$ per hour until a temperature of $+2^{\circ}C$ is reached. Water is sprayed continuously onto the front surface of the module so that it is drip wet and water penetrates into any existing slits. At temperatures below $0^{\circ}C$ water is sprayed only intermittently until an ice layer of at least 2 mm is formed. The iced module is then kept for 1 hour at $-10^{\circ}C$ and is then allowed to warm up at a rate of less than $5^{\circ}C$ per hour. This freeze test is repeated once.

With this test possible damage due to the expansion of freezing water in small slits is provoked; this test is hence quite close to reality and simulates conditions which do occur in winter when melting snow covers the modules.
At ESTI, this test is executed in chamber AT-3.

5. DURABILITY TESTS ON PHOTOVOLTAIC MODULES

5.1 Humidity freeze test

This test, although it simulates certain damaging effects which
might occur in real operation, should be considered as a durability test. The module is first kept for 48 h at +40°C in a humidity chamber at about 95% relative humidity.
It is then transferred into a climatic chamber and its temperature is decreased to -40°C in about 20 min. or more (decrease
rate smaller than 3° per min.). After 1 h at -40°C the module
is brought back to room temperature at a rate of not faster than
1° per min. The temperature cycle is then repeated once.

Clearly, this test stresses the module more than would occur
in real use, but on the other hand freezing of a humid module is
quite possible in real use and might happen quite often in winter.
The test hence has the characteristics of an accelerated aging
test.

It should be noted that the panel is not wet on the outside, when
it is brought to the freeze chamber; there is hence no ice formation on the outside of the module; only the humidity diffused into
the module will freeze in this test. For this reason, the test is
quite different from the one described in 2.7.

At ESTI this test is executed in chamber AT-8 (for humidity
pre-treatment) and in LS-1 (for temperature treatment).

5.2 Temperature cycling

The module will be subjected to 50 temperature cycles between
-40° and about +80°C at a rate of less than 2°C per min. At the
extreme temperature limits (of -40°C and NOCT +40°C) the
temperature of the module will remain stable for at least 10 min.
(Temperature cycling is a well-known accelerated test method
and is described in many test standards, e.g. (14)).

At ESTI this test is performed in the climatic chamber LS-1
(or AT-0 for small modules).

5.3 Damp heat, long storage

For long exposure to damp atmosphere we have at ESTI the climatic chamber AT-8. This chamber, without elaborate automatic controls, allows us to maintain a temperature of about 40°C and a humidity of about 95%. As the temperature is not very accurately regulated and as also some parts of the chamber walls are colder, where condensation does occur, the relative humidity is not defined very precisely inside the chamber. With three ventilators the air is continuously moved and passes over an open bath of water at 40°C. Hence a high relative humidity of about 95% will be maintained in the chamber.

Humidity storage tests are widely known in the literature (15).

5.4 High temperature long exposure

Dry-heat long-exposure tests are specified for many different conditions (heat dissipating or non-dissipating specimen, sudden or gradual change of temperature, different severities (temperature and duration)) in the literature (16).

For photovoltaic modules it has been found that for comparative tests different modules should be tested at an equal temperature difference above the respective NOCT (normal operating cell temperature). In real use a module with a higher NOCT will be at higher operating temperatures than a module with lower NOCT and hence the damage due to temperature effects will also be higher in the first case.

The test temperature envisaged at ESTI is therefore not a fixed one but NOCT +50°C i.e. about 90° to 105°C depending on module characteristics. The selected temperature should be constant in time and in local distribution to $\pm$ 5°C.

After a first test period of 30 days there will be an inspection and an electrical performance measurement and then another test period of 90 days will follow.

The necessary temperature chamber for these tests is in preparation at ESTI (1981).

5.5 Ultraviolet irradiation

Ultraviolet irradiation for large specimens is a costly procedure; commercially available UV-irradiation apparatus has irradiation areas of the order of some square decimetres to 0.2 m^2
maximum. For the irradiation of whole photovoltaic modules a
new, large area irradiation facility was designed. The AT-7
facility of ESTI gives on a horizontal test area of 1 x 1.5 m^2
with lateral reflectors a uniformity of about $\pm$ 10% in irradiance.
The UV-irradiance (between 300 nm and 400 nm) is 115 W/m^2
and the total irradiance (between 300 and 2500 nm) is 1000 W/m^2.
In the UV-region the spectral distribution corresponds rather
well to the typical UV-spectral distribution in daylight. Several
filters are available to cut the intensity of the lower wavelength
(i.e. below 300 nm or below 330 nm). The temperature of the
irradiated modules can be adjusted between certain levels (35°
to 100°C) by choosing an appropriate cooling air speed.

The lamps are specially constructed for long lifetime (1500 h).

In order to establish, on a relative basis, the behaviour of different modules and in order to detect possible major defects,
the UV-dose is limited for the first tests to 40 MJ/m^2, which
corresponds to 4 days irradiation time.

5.6 Ozone test

In our climatic chamber AT-1 the atmospheric conditions as
prescribed in ref. (17) can be obtained: they are 0.5 ± 0.1 vpm
ozone, 25 ± 2^oC and $55 \pm 5\%$ relative humidity. For the tests
on photovoltaic modules a higher test temperature of 40 ± 2^oC,
as prescribed in ref.(18), has been selected which gives an
acceleration in the rate of ozone attack. The test duration is 5
days. No dynamic strain conditions are applied.

5.7 SO$_2$ test

In the same test chamber AT-1, sulphur dioxide gas can be
applied as well as ozone. The temperature is 25°C and the humidity is 75% relative humidity. The SO$_2$ content could be as high
as 1000 ppm. (This value has been chosen because DIN 50018
envisages in cyclic tests initial SO$_2$ concentrations of 670 ppm
(volume).) It appears that this SO$_2$ concentration is rather high.

TABLE 1

Test Number and Description	Module Number							
	1	2	3	4	5	6	7	8
2.1 Visual inspection, weight, dim.	x	x	x	x	x	x	x	x
2.2 Robustness of terminations					x			
2.3 Mounting twist					x			
2.4 Insulation	x	x	x	x	x	x	x	x
- Performance (STC)	x	x	x	x	x	x	x	x
2.5 Hail resistance					x			
2.6 Life loads (mechanical)					x			
2.7 Ice formation					x			
- Determination NOCT				x				
- Performance NOCT				x				
- Hot spots				x				
3.2 Temperature cycling	x	x						
- Performance (STC)	x	x						
3.1 Humidity freeze	x	x						
3.3 Damp heat						x		
3.4 High temperature							x	
3.5 UV-irradiation + perf. (STC)								x
3.6 Ozone + perf. (STC)								x
3.7 SO$_2$ + perf. (STC)								x
3.8 Salt mist + perf. (STC)								?
2.4 Insulation	x	x	x		x	x	x	x
- Performance (STC)	x	x	x		x	x	x	x
2.1 Visual inspection	x	x	x		x	x	x	x

A newer test procedure (IEC 68-2-42) prescribes only 25 ppm SO$_2$ for tests on electric contacts. For the tests of photovoltaic modules a SO$_2$ concentration of 50 $\pm$ 20 ppm will be used. The test duration is 5 days.

5.8 Salt mist

For these tests ESTI will be completed by a salt spray (fog) chamber, which will allow us to execute tests on photovoltaic modules according to DIN 50021 or ASTM B 117-73.

REFERENCES

1. BSE Richtlinien und Hinweise für die Bestimmung der Gebrauchstauglichkeit von Solarkollektoren. B. Langzeit-tests von Solarkollektoren, January 1979.
2. DIN 4757 Teil 3: Sonnenheizungsanlagen, Sonnenkollektoren, Begriffe, Anforderungen, Prüfungen. February 1979.
3. Groupe spécialisé No. 14: Guide technique spécialisé pour la constitution d'un dossier de demandes d'avis technique, procédés solaires", March 1979, CSTB.
4. CETIAT (+ COMES): Projet de cahier des charges pour la normalisation.
5. PHOEBUS: Programma di prove di breve durata per collet-tori solari piani. 80-PBD-001, May 1980.
6. Hayward, K., University College, Cardiff: 1981, Methods of obtaining information on the durability of solar collectors. Helios (newsletter) ISSN 0141-2965, No. 12 July '81.
7. Svendsen, S.: 1981, private communication, January.
8. NBSIR 78 - 1305 A: 1978, Provisional flat plate solar collec-tor testing procedures, National Bureau of Standards. June '78.
9. FSEC-77-5: 1979, Test methods and minimum standards for solar collectors. Florida Solar Energy Center, January '79.
10. CSA Preliminary Standard F 378-M: 1979, Canadian Stan-dards Association, August '79.
11. SERI, Solar Energy Research Institute, Golden, Colorado: 1980, Interim Performance Criteria for Photovoltaic Energy Systems. SERI/Tr-742-654, April '80.
12. JPL, Jet Propulsion Laboratory, Pasadena, California: 1978, Photovoltaic Module Design, Qualification and Testing Specification, DOE/JPL-1012-78/7, March '78.
13. CEC, Joint Research Centre, Ispra Establishment of the Commission of the European Communities: 1981, "Photovol-taic Module Control Test Specifications", Specification No. 501 (1981), Report EUR 7545 EN.
14. IEC, International Electrochemical Commission: 1974, Test N: Change of temperature, 68-2-14; or similar: DIN 40046, Teil 14.
15. IEC, International Electrochemical Commission: 1969, Test Ca: Damp heat, steady heat, 68-2-3; or similar: DIN 40046, Teil 5; or more like our test 3.3: DIN 50017 (type SK).

16. IEC, International Electrochemical Commission: 1974, 68-2-2, Tests B: Dry heat; or similar: DIN 40046, Teil 4.
17. DIN 53509, Blatt 1, Beschleunigte Alterung von Gummi unter der Einwirkung von Ozon.
18. ASTM D 3395-75, Standard Method of Test for Accelerated Ozone Cracking of Rubber under Dynamic Strain Conditions.

APPENDIX

SPECIFICATION OF ESTI DURABILITY AND QUALIFICATION
TEST FACILITIES

(AT-X = indoor facilities)

AT-0   Climatic Chamber

Inside dimensions
(width x height x depth):   80 x 70 x 70 cm
Temperature range: -40°C to $+90^{\circ}$C
Humidity range: up to 100% relative humidity
Special feature: 17 cm diam. window on top
                                   for irradiation

AT-1   Climatic Chamber

Inside dimensions
(width x height x depth):   200 x 320 x 370 cm
Temperature range: 20° to 40°C
Humidity range: 10 to 100% relative humidity
Special features:
    Ozone gas: up to 4 ppm
    Sulphur dioxide gas: from 25 to 1000 ppm
    Damp cyclic heat: according to DIN 40046 part
                                   31 or IEC 68-2-30
    Sample holders: two carriages for 6 thermal
                                   collectors each
    Penetrations: 6 x 2 ports for lamps or for
                                   service

AT-2   UV-irradiation lamps

Lamps:	3 columns of 6 lamps each - lamp housing 50 x 200 cm; 18 x 400 W electric
Irradiation area:	200 x 200 cm vertical (at 1 m distance from lamps)
Irradiance:	60 W/m^2 of UV-irradiance (between 300 and 400 nm) 300 W/m^2 of total irradiance
Special feature:	Operated in laboratory, without temperature or humidity adjustment

AT-3   Lifeload Chamber

Inside dimensions:	220 x 300 x 70 cm
Temperature range:	ambient to -15°C
Pressure range	+ 6500 Pa to - 4000 Pa
Special features:	double chamber (for cold air reservoir) for dynamic pressure changes; spray facility for rain tightness tests; possibility for freezing water

AT-4   Hail Gun

Hailstone diameter:	from .5 to 1.5 inches
Hailstone speed:	from 10 to 50 m/s
Operation mode:	air pressure driven hailstones; quick-opening valve; light barrier speed measurement

AT-5   Overpressure test

Pressure source:	Hydraulic (hand pump), up to 20 bars; pneumatic (compressor), up to 12 bars
Special feature:	leak test with air pressure by precision manometers or with helium leak test

AT-6   Aging simulator

Inside dimensions (width x height x depth):	340 x 390 x 430 cm
Irradiation area:	120 x 340 cm (45° inclined)
Irradiance:	up to 1100 W/m^2 (adjustable)
Lamps:	88 Power Star lamps 250 W electric each, 5000 h lifetime
Special features:	Facility for water spray on hot specimen. Facility for filling of collectors with coolant fluid

AT-7   UV-irradiation

Lamps:	4 units with 1 2000 W electric lamp each
Irradiation area:	lateral reflectors, 100 x 150 cm (horizontal) at 1 m distance from lamps
Irradiance:	115 W/m^2 UV-irradiance (between 300 and 400 nm); 1000 W/m^2 total irradiance
Special features:	UV-spectrum well adapted to natural UV-spectrum; Different filters available; Operated in laboratory without temperature or humidity adjustment

AT-8   Humidity chamber

Inside dimensions (width x height x depth):	150 x 150 x 100 cm
Temperature range:	up to 50°C
Humidity range:	100% relative humidity

(xxx = outdoor facilities)
(mounted on the roof of building 45 or on ground level)

CTF 0   Correlation test field zero

Prototype test circuit for checking of obtainable irradiation conditions

CTF 1   Correlation test field 1

Test field for 5 x 3 collectors for outdoor exposure under controlled conditions

STA   Stagnant temperature analysis

Rack for two collectors, equipped with temperature and irradiance measurement devices for analysis of degradation

FACILITIES FOR THE STUDY OF BEHAVIOUR IN CORROSIVE ATMOSPHERES AND THEIR APPLICATION TO THERMAL AND PHOTOVOLTAIC CONVERTERS

P. Weisgerber

Commission of the European Communities
Joint Research Centre - Ispra Establishment
I-21020 Ispra (Va), Italy

1. MATERIALS USED IN SOLAR HEAT COLLECTOR CONSTRUCTION

In order to obtain an overview of the materials to be involved in future corrosion tests, some statistical data will be outlined. These data were elaborated from the "Katalog der in den neun Ländern der Europäischen Gemeinschaften produzierten Sonnenkollektoren" (Catalogue of Solar Collectors Manufactured in the Nine Countries of the European Community) [1], which was edited in 1978.
Our data, therefore, refer to the number of 137 collector models commercially available in the EEC at that time, and not to the collectors in use, as no data about production quantities were available.

These data refer to flat collectors for heating of liquids, being generally composed of the following construction elements: absorber plate, transparent cover, frame, thermal insulation, bottom plate.
The materials applied were given in percent of the total number of models listed in Tables 1-5.

To this variety of materials, which can be combined, various auxiliar materials should be added, such as the selective layers on the absorber surface (45.6% black paint, 4% pigmented plastics, the rest other, or not specified) and also materials for

G. Beghi (ed.), Performance of Solar Energy Converters: Thermal Collectors and Photovoltaic Cells, 327–336.
Copyright © 1983 ECSC, EEC, EAEC, Brussels and Luxembourg

seals and caulkings.

As to temperature and humidity, the climatic conditions in the
EEC will not be explained in this chapter.

2. MATERIALS USED IN PHOTOVOLTAIC MODULE CON-
   STRUCTION

The actually manufactured and commercialized photovoltaic
modules use silicon photovoltaic cells. These are composed of
the so-called "wafer", made of ultra pure and in most cases
monocrystalline, or, at this moment by one manufacturer only,
rather large grain polycrystalline silicon.
The upper surface, i.e. the one exposed to solar irradiation,
has undergone the known treatments for the formation of the p-n
junction and of the antireflex layer.

Metallic electric contacts are applied to both surfaces of the
wafer; uniformly on the lower one and in a grid pattern, in order
to permit the access of light, on the upper surface. Widely used
are metal contacts which are composed of three layers, i.e.
titanium, palladium and silver, being deposited by electrodeless
plating one after the other.

The lower surface is in many cases entirely coated with an additional layer of tin base soldering alloy. The upper grid is provided with a small nickel strip for connecting purposes, which is
fixed with a tin base soldering alloy. Given the rather low voltage
of a single cell, these are connected in series by soldered nickel
strips in order to form arrays. The array is embedded between
two layers of glass fibre scrim and then impregnated by a transparent adhesive, which is applied by coating or melting during
lamination. The adhesive is also bonded to the insulating, transparent cover plate, made of mineral or plastic glass and to the
insulating bottom plate usually also made of plastic.

The whole sandwich is hermetically sealed round the edges and
is inserted into a frame of metal profile for mechanical reinfor-
cement. The metal used for the frame, in many cases aluminium,
gives support also for the mounting of an electric connections
box.

As long as the module remains hermetically sealed, which depends on the stability against degradation of cover and seal materials, only the outer parts, i.e. the metal frame and the electric connection devices can be attacked by the atmospheric corrosion of metals.

If, however, degradation of the plastic covers and seals occurs, the "breathing" from the day-night cycle favours penetrating of the atmosphere into the sandwich and the condensation of humidity. Deleterious effects are then to be expected, which will be enhanced by the presence of corrosive air pollutants. Even without solar irradiation the variety of metals combined in the array of photovoltaic cells offers sufficient different components for the formation of electrochemical corrosion elements. The deleterious effects of the latter will be increased by the conditions of temperature and electrical potential, which occur under solar irradiation.

3. CORROSIVE ENVIRONMENTS

It is generally known that apart from air pollutants such as salts and certain combustion products, high humidity favours metallic corrosion. As corrosion processes are chemical reactions directed to the formation of the most stable compounds in systems being chemically not in equilibrium, the reaction rates are increased with rising temperature. (Thumb rule: every 10 K increase rates by a factor of 2, but care is to be taken for many exceptions).

During temperature fluctuations of the day-night cycle, if constant absolute air humidity is given, the relative humidity increases during the night. These fluctuations, of temperature and relative humidity, are also reflected at the inside of a collector case because more than half of the models is provided with a construction which permits "breathing", and for the remaining models, which are hermetically sealed during manufacturing, in many of them breathing may occur anyhow due to leaks.

As air pollutants enhancing metallic corrosion, salt mists are to be mentioned which are found usually near the sea and in coastal districts, and the combustion products of numerous fuels (coal,

heavy and light fuel oil, certain natural gases). The latter are CO_2, SO_2 and last but not least H_2O which becomes also effective increasing the relative humidity of the diluting air.
Carbon dioxide and sulphur dioxide form acids in the presence of water: H_2CO_3 and H_2SO_3, particularly with the water which is kept by adsorption on surfaces. The acids are weak and rather volatile. Unfortunately, H_2SO_3 oxidizes, preferentially by surface catalysis, with the oxygen of the air forming sulphuric acid: H_2SO_4, which is a strong and very little volatile acid and which remains at the place where it is formed, with deleterious effects on nearly all usual metals.

In industrial and urban atmospheres the SO_2 content of the air, in extreme cases, can superate in transients the limit of 150 vpb (0.15 vpm) (with noticeable immediate biological effects). Medium concentrations of somewhat less than 100 vpb (0.1 vpm) for longer periods are not unusual in certain regions, especially during the cold seasons.

The resistance of the usual metals to corrosive atmospheres of the above kind and also suitable protection measures are known from practical experience (painting, plastifying, plating, galvanizing, etc.). This is not true under conditions of relatively high temperatures to which construction materials of solar components are exposed. Moreover, the fact that a rather large variety of metals can be used together in the same device, can complicate the problem due to the formation of electrolytic couples.

Very high SO_2 concentrations can occur in the micro-climate on the roof of a house, where the chimney releases combustion gases. By far the most collectors actually installed will be exposed to such micro-climates, because collectors are intended to support but not to replace the existing combustion heating system. Undiluted combustion gases can contain up to 0.5 vol% or 5000 vpm SO_2. Some cases of collector failures by external corrosion within one or two years of service, were probably due to combustion gas release [2].

By far the most photovoltaic modules actually installed, will not be exposed to such micro-climates, because they are rarely provided for residential service. But with increasing use they will meet more frequently the neighbourhood of the existing combustion heating systems.

Ozone (O_3) is an air pollutant which is responsible for the degradation of a great part of organic materials, such as paints
and plastics used for surface protection, putties and mastics for
seals and caulkings and even resins and plastics used as construction materials or for thermal insulation.
Its concentration in the low atmosphere is usually very low, and
somewhat related to the presence of other pollutants and UV-
irradiation. It seems that organics which resist to UV-irradiation
also resist to ozone.

4. THE AT-1 TEST FACILITY

Our largest test chamber with inside dimensions of 2 m width,
3,7 m depth and 2.2 m height, called AT-1, permits three different operating modes.

4.1 Exposure to Cyclic Damp Heat

This operation mode makes possible a thermal cycling of the
specimen between 20°C and 40°C at 95% relative humidity.
These conditions are prescribed in DIN 40046, part 31 /3/ and
in IEC 68-2-30 /4/. The main purpose of this test is to generate
on the outside as well as on the inside of "breathing" equipment
condensations that might damage the equipment. The outside
condensation on solar collectors occurs during the temperature
raise phase, when the specimen is colder than the surrounding
air and the inside condensation occurs during the temperature
decrease phase when the humid inside air is warmer than the
collector walls.

4.2 Exposure to Sulphur Dioxide

In the same test chamber AT-1 alternatively corrosive gases
such as ozone or sulphur dioxide can be applied. The temperature and relative humidity levels are 20° to 40°C and 10 to 95%
respectively. The SO_2 content can be as high as 1000 ppm. This
value has been chosen because DIN 50018 /5/ foresees in cyclic
tests initial SO_2 concentrations of 670 ppm (volumes). It seems
that this SO_2 concentration is rather high. A more recent test
procedure (IEC 68-2-42) /6/ prescribes only 25 ppm SO_2. Such
a low SO_2 concentration is about the minimum level that can be
detected and controlled by the SO_2 measuring instrument of the

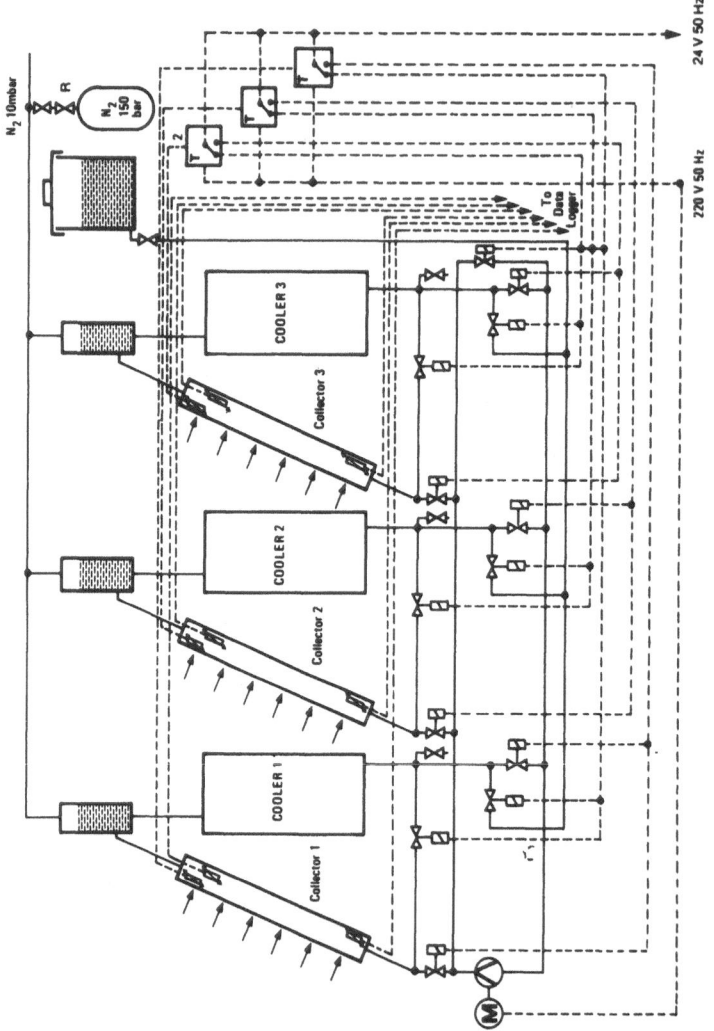

Figure 1

AT-1 facility. It might also be interesting to add CO_2-gas to the SO_2-gas (fume chamber).

4.3 Exposure to Ozone Containing Atmosphere

The third mode possible with the AT-1 is operating with air of a temperature between 20°C and 40°C, a relative humidity of between 10 and 95%, and an ozone concentration of up to 4 volume per mille. With this chamber we can therefore produce the test atmospheres as prescribed in DIN 53509 (25°C, 55% relative humidity, 0.5 ppm O_3) [7] or as prescribed in ASTM D 3395-79 (40°C, 0.5 ppm O_3) [8].

5. OUTDOOR CORRELATION TEST FIELD

The above mentioned standard test procedures were developed time ago for other purposes as solar technologies were not yet in use. It must be found out by trial and error whether they are useful for testing solar equipment, or whether they have to be modified in order to obtain suitable accelerated aging tests.

As already has been explained, the effects caused by accelerated aging tests must be correlated to the effects that occur in real outdoor exposure. In order to have reliable data for outdoor degradation under well-defined and known conditions, outdoor test fields are in preparation at ESTI. In this test field three specimens of five different collector types will be exposed. The collector types were chosen on the basis of the materials used in their construction, reflecting fairly the above mentioned statistical considerations. To make the exposure conditions realistic, the collectors are connected to simple cooling circuits, filled with the heat transport fluid, recommended by the respective manufacturers. A simple temperature regulation system makes it possible to keep all collectors at the same maximum temperature as soon as solar irradiation permits (Fig. 1). Collector temperatures, air temperature, air humidity, solar irradiation and SO_2-concentration will be recorded.

The system has the following advantages compared to dry heating or stagnant boiling:

- All collectors are at the same internal fluid temperature which enables comparison under equal conditions corresponding to real service.

- The fluid recommended by the manufacturer is used, which avoids damage by freezing and absence of the corrosion inhibitors indispensable for the protection of the inner surface of the absorber plates.

Prototype circuits, CTF-0, are in service without interruption for more than half a year.

One test field, CTF-1, will be installed this season, as soon as the Ispra outdoor facility is ready to receive it.
Another test field, CTF-2, equipped in exactly the same way, with the same type and number of collectors, will be installed in the vicinity of a thermal power plant in order to make tests under conditions of known SO_2-fallout.

In addition to the thermal collectors also photovoltaic modules will be installed in the same correlation test fields.

6. SALT MIST TEST

A salt spray (fog) chamber will be available for ESTI. This chamber would make it possible to carry out tests on collector parts according to DIN 50021 /9/ or ASTM B 117-73 /10/.

This facility may possibly be supported by an outdoor correlation test field, CTF-3, in a coastal district.

7. TABLES

TABLE 1 - Absorber Plates

Aluminium	34.5
Carbon Steel	23.5
Copper	14.5
Plastics	9.0
Stainless Steel	5.5

(rest: 13% other, or not
  specified)

TABLE 2 - Transparent Cover

Window Glass	39.2
Special Mineral Glass (such as mirror glass or tempered glass)	22.8
Plastic glasses	25.3

(rest: 12.7% other, or not
  specified)

TABLE 3 - Frame

Aluminium	44.7
Galvanized Steel	19.2
Fibre-reinforced resins	9.2
Stainless Steel	2.8

(rest: 24.1% other, or not
  specified)

TABLE 4 - Thermal Insulation

Foamed Polyuretane resin	40.6
Glass wool or mineral wool	32.2
Glass wool and foamed polyuretane resin	5.6

(rest:21.6% other, or not
  specified)

TABLE 5 - Bottom Plate

Aluminium	27.5
Galvanized Steel	15.2
Fibre-reinforced resins	10.2
Stainless Steel	2.2

(rest: 44.9% other, or not
  specified)

8. BIBLIOGRAPHY

/1/ "Katalog der in den 9 Ländern der Europäischen Gemeinschaften produzierten Sonnenkollektoren", JRC-Ispra and
Franklin Institute GmbH-München (1978).

/2/ "Indagine sul comportamento nel tempo di alcuni tipi di
collettori solari"; CIAB-Bologna, Study Contract No. 1356-
80-08 ISP-I, JRC-Ispra/CIAB-Bologna.

/3/ DIN 40046, Teil 31; "Umweltprüfungen für die Elektrotech-
nik", Prüfgruppe D, Prüfung Db: Feuchte Wärme zyklisch
(12 + 12 Stunden).

/4/ IEC-Recommendation 68-2-30 (1968), Part: "Cyclic Damp
Heat".

/5/ DIN 50018,"Beanspruchung in Kondenswasser Wechselklima
mit schwefeldioxidhaltiger Atmosphäre" (1978).

/6/ IEC 68-2-42 ; Basic Environmental Testing Procedures,
Part 2: Tests - Test KC: "Sulphur Dioxide Test for Contacts
and Connections" (1968).

/7/ DIN 53509, "Beschleunigte Alterung von Gummi under der
Einwirkung von Ozon" (1964).

/8/ ASTM Designation D 3395-75, "Standard Methods of Test
for Accelerated Ozone Cracking of Rubber under Dynamic
Strain Conditions", (1975).

/9/ DIN 50021, "Sprühnebelprüfungen mit verschiedenen Natrium-
Chloridlösungen" (1975).

/10/ ASTM Designation B 117-73, "Standard Method of Salt Spray
Testing", (1973).

PERFORMANCE OF A HYBRID SOLAR HEATING SYSTEM OF THE SOLAR LABORATORY AT THE JRC-ISPRA

D. van Hattem, E. Aranovitch and P. Actis-Dato

Commission of the European Communities
Joint Research Centre - Ispra Establishment
I-21020 Ispra (Va), Italy

1. INTRODUCTION

The average European winter weather conditions are ill-suited to cost-effective solar space heating systems. Conventional active solar systems cannot be expected to deliver over 150 kWh (thermal) per square metre of solar collector annually, thus leading to excessively long pay-back periods.

For this reason a great deal of attention has been paid in recent years to solar systems which will be operated both over longer periods of time per year and at lower temperature levels. In this respect, the so-called hybrid systems, combining a solar system with a heat pump and a long-term energy storage system, look particularly promising.

The introduction of a heat pump in large solar heating systems with a long-term heat storage, can significantly improve the performance of the system, while the cost of the heat pump is only a small fraction of the total system costs.

At the Joint Research Centre of the European Commission, located at Ispra, in Northern Italy, such a system has been used to heat the solar laboratory and has intensively been monitored over a period of three years. In order to study the influence of the main parameters, such as collectors area, storage volume, heat pump characteristics, scale effect, etc., a simplified ma-

G. Beghi (ed.), Performance of Solar Energy Converters: Thermal Collectors and Photovoltaic Cells, 337–354.
Copyright © 1983 ECSC, EEC, EAEC, Brussels and Luxembourg

thematical model has been developed. As could be expected, the influence of scale is particularly felt as far as the storage system and the costs are concerned.

2. DESCRIPTION OF THE SURROUNDING ENVIRONMENT

The project location is Ispra, Italy, at about 220 m above sea level. Ispra, at the South-side of the Alps (45°48′N, 8°37′E), has a humid climate. The yearly number of degree-days for heating is of the order of 2700 (basis 20°C). The yearly global solar irradiation at an inclination of 60° is approximately 1300 kWh/m^2. For the period from the end of September to the end of March it is 525 kWh/m^2. The average cloud cover is 4.1 octas. More information is given in Table 1.

Table 1 - Principal Climatic Data for Ispra
(average over 20 years)

Average temperature of the year	11.5 °C
Average temperature of June	18.8 °C
Average temperature of December	2.5 °C
Yearly solar radiation on horizontal m^2	4363 MJ
Solar radiation on horizontal m^2 June	598 MJ
Solar radiation on horizontal m^2 December	129 MJ
Yearly average wind speed	1.7 m/s

3. DESCRIPTION OF THE OVERALL SYSTEM

The hybrid solar heating system, which is being dealt with in this report, is exclusively used for space heating. The system is schematically shown in Figure 1.
It essentially consists of:

- an array of various types of collectors, selective and non-selective, mounted on the inclined wall at 60°. Total surface: 41 m^2;

- an insulated concrete water vessel of 50 m^3, built under the solar laboratory. In order to reduce heat losses by evaporation-condensation processes, the water surface is covered with a thin oil film;

- a water-to-water heat pump; max. heating power 17 kW;

- a floor heating system operating in the temperature range of
  25 to 35°C;

- a small heat storage of 2 m³ of water used as daily storage
  for the collectors and as buffer for the heat pump.

The system is built in the solar laboratory of the Joint Research
Centre (Figure 2).
The main characteristics of this building are:

- ground surface 160 m² (20 x 8 m);

- walls are made of a sandwich structure with two brick layers
  and 12 cm of polyurethane in between. Windows are double-
  glazed;

- 64 m² of various types of solar collectors can be accommodated on the wall facing South, inclined at 60°, for space heating
  experiments;

- 80 m² of collectors can be put at variable inclinations on the
  flat roof, essentially for solar cooling studies and collector
  testing purposes;

- the storage systems consist of a 2 m³ insulated stainless steel
  tank and a 50 m³ insulated concrete basement vessel both with
  inside heat exchangers;

- the auxiliary heating systems are an electrical resistance
  system and a 17 kW water-to-water heat pump;

- the internal heat distribution systems consist of a floor heating system, air convectors and a thermal ventilation system
  with recuperation;

- a 4.5 kW lithium-bromide absorption machine for cooling experiments;

- instrumentation and a data acquisition system for the monitoring of the various experiments.

During the year various heating and cooling experiments are
carried out in this laboratory which is characterized by its flex-
ibility.
The heating requirements of the building are between 200 and
300 W/°C, depending on the occupancy and ventilation rate.
The peak demand for space heating is 9 kW. The design yearly
heating load is about 55 GJ. About one third of this load is met

by passive solar contribution, lights and instruments and occupants.

4. SYSTEM OPERATION

The operation of the system is schematically shown in Figure 3, it can be divided into four main phases:

  i) A period of heat storage from May to end September by the solar collectors into the 50 m^3 water vessel;

 ii) From mid-September to mid-December the space heating requirements are met by the solar system, with the heat accumulated during the preceding period being used as back-up;

iii) Only when, around mid-December, the temperature in the water vessel falls below 30°C, does the heat pump enter into operation, "pumping" heat from the 50 m^3 basin into the small daily reservoir of 2 m^3 at the required temperature level of 35°C, before entering the heat distribution system. During this phase the solar collectors are connected to the heat exchanger in the large storage. They recharge the storage. Due to the low temperature of the storage ($\sim$ 10°C), the collectors operate at very high efficiencies. They can keep the large storage at a temperature level which enables the heat pump to operate with a high C.O.P.;

iv) At the end of February when the supply of solar energy increases, the collectors are again connected to the heating system of the building. The heat pump is used in the same manner as in phase iii), but now as back-up only. The large storage is not recharged anymore and its temperature therefore will further decrease. In this way the minimum storage temperature occurs at the end of the heating season and the use of the heat pump is minimum.

4.1 Control Strategy

a) <u>Collector Control.</u> The collector pump is controlled by means of a simple differential controller. During the second phase a second differential controller checks whether the collectors charge the daily storage or the large storage. When the temperature in the small storage is 10°C higher than that in the large storage, the collectors are automatically connected to the large storage.

b) <u>Heat Pump Control.</u> The heat pump on/off control is done
with a temperature switch, which has its sensor in the 2 m^3
buffer storage. The heat pump keeps this storage at the minimum temperature required for the floor heating system. Due
to the buffer the heat pump operates in periods of at least
2.5 h.

c) <u>Floor Heating Control.</u> The water temperature of the floor
heating system is related to the outdoor temperature, in such
a way that the room temperature is maintained at 20°C. If
the room temperature, due to passive contributions, exceeds
20.5^{o}C, the circulation pump is shut-off.

5. THERMAL PERFORMANCE

In this chapter the system performance for two heating seasons
is presented and discussed.

The system has been monitored over a period of three years (the
results of the first year have already been published (Aranovitch and co-workers, 1979)), during which, from one year to
another, various modifications have been applied in order to improve the cost-effectiveness. The collector surface was reduced
from 53 to 41 m^2, the heat pump was changed for a more adapted
one and the system operation has been improved.

The scope of the monitoring was not only to determine the performance of the system but also to obtain all the elements necessary for the validation of mathematical models, describing the
behaviour of the various components of the system.

From the continuous recording of the essential parameters (climatic data, system´s and building´s temperatures, mass flows,
electricity consumption, etc.) which were then reduced to hourly,
daily and weekly averages, the experimental data were processed
with a computer-based data acquisition system for two purposes:
first for the characterization of the various sub-systems (building, large storage, collector array, heat pump), secondly for
the determination of the overall performance of the system.

The system operated without major problems for three years.
The heat pump turned out to be very reliable and did not require

any maintenance.
The floor heating system performed very well as far as room
comfort and operating temperatures are concerned.

6. SYSTEM PERFORMANCE

The results presented concern the periods during which the
house heating system is operating (phases 2, 3 and 4). Since the
large storage is used during the summer months for the storage
of chilled water, only one month is left to heat up the storage
for the heating in winter. Therefore the storage temperature is
relatively low at the start of the heating season; 47.7°C on
22 - 10 - 1979 and 41.6°C on 23 - 10 - 1980. The data concerning the complete heating season are given in Table 2 and in
Figures 4 and 5.
The weekly performance is illustrated in Tables 3 and 4.
Figure 6 shows the weekly average of the large storage temperature and the C.O.P. of the heat pump.
The system performance indicators are given in Table 5.
For the season 1980 - 1981 the principal system parameters
are shown on a daily basis in Appendix I.
The useful amount of solar energy per unit collector area is
about 740 MJ/m^2 .

Considering these results it is realized that the effect of seasonal storage, that is to use heat captured in summer for house
heating in winter, was not very strong in these experiments.
To obtain a correct idea of the potential yearly energy output per
unit area, one should add about 450 MJ/m^2, which brings the
total to 1190 MJ/m^2. This is 2.5 times more energy than one
can expect from a conventional active solar heating system.

7. CONCLUSIONS

Results obtained after several years of monitoring this hybrid
solar heating system have shown that such systems will considerably increase the yearly output of a solar system, compared
to the performance of a conventional active solar heating system.

Obviously, the size of the system presented here is too small to
be economic. There are important effects of scale, especially
the performance of the large storage and the specific costs of
the heat pump change rapidly when the system size increases.

Therefore, this type of system is not to be recommended for single family dwellings, but could be very interesting in appartment buildings, schools, offices or groups of houses.

As far as the energy saving potential of the system is concerned, it can be said that the savings of primary energy are in the order of 50% to 75%, depending on the storage and on the C.O.P. of the heat pump.

8. BIBLIOGRAPHY

Aranovitch, E., M. Ledet, C. Roumengous, and D. van Asselt (1977). Description et performance d'un système de chauffage solaire fonctionnant à basse température. Conf. Int. "Chauffage Solaire dans le Batiment", Liège, 1977.

Mustacchi, C., V. Cena, and M. Rocchi (1978). Stockage saisonnier de la chaleur solaire dans le sol. Contract No. 872-78-05 SISPI-ADES.

Aranovitch, E., M. Hardy, M. Ledet, C. Roumengous, D. van Hattem, and P. Actis-Dato (1979). Performance of a solar heating system combined with a heat pump and a long-term heat storage device. American Ins itute of Mechanical Engineers, Annual Meeting on Solar Energy, San Francisco, Nov. 1979.

Aranovitch, E., M. Hardy, and D. van Hattem (1981). Results of three years of operation of a solar hybrid system with heat pump in Northern Italy. ISES, Brighton, August 1981.

9. NOMENCLATURE

CE   = Collector efficiency defined as $CE = (Q\ 137 + Q\ 136 + Q\ 132)/Q\ 120$,
Q 120 = Global solar radiation,
Q 132 = Controlled energy flow from collector subsystem directly to the ground-coupled subsystem,
Q 136 = Controlled energy flow from collector subsystem directly to the heat pump subsystem,
Q 137 = Controlled energy flow from collector subsystem directly to the building load,

Q 237 = Controlled energy flow from ground-coupled subsystem directly to the building load,

Q 603 = Controlled operating energy to the heat pump subsystem,

Q 637 = Controlled energy flow from heat pump subsystem to the building load,

SPF = Total seasonal performance of the heat pump defined as: SPF = Q 637 / Q 603,

T_{st} = Temperature of the large storage.

Table 2 - Integrated Energy Flows for Two Heating Seasons

Energy Flow [Units: GJ]	1979/ 1980	1980/ 1981
Total operating energy for pumps etc.	1.2*	1.4*
Total non-usable heat losses from other than collector and large storage subsystems	0	0
Controlled energy flow from collector subsystem directly to the large storage subsystem	19.0	25.1
Controlled energy flow from collector subsystem directly to the building load	6.1	3.5
Controlled energy flow from collector subsystem directly to the heat pump subsystem	0	0
Controlled energy flow into large storage subsystem from the surrounding environment	0	0
Controlled energy flow out of large storage subsystem into surrounding environment	0	0
Controlled energy flow from large storage subsystem directly to the building load	3.1	2.5
Controlled energy flow from large storage subsystems to the heat pump subsystem	20.2	26.0
Controlled energy flow from heat pump subsystem to the large storage subsystem	0	0
Controlled operating energy to the heat pump subsystem	6.8	9.0
Controlled energy flow from heat pump subsystem to the building load	26.6	34.5
Controlled cooling energy from the building to the heat pump subsystem	0	0
Total auxiliary energy for direct heating	0	0
Global solar radiation	67.4	69.7

* estimation

Table 3 - Energy Flows of Hybrid Solar Heating System. Period
22-10-1979 to 30-3-1980. Daily Averages over One-Week
Periods /Unit: MJ/.

Week	Q 120 [per m²]	$\frac{Q\,137 + Q\,136}{Q\,132}$	CE [%]	Q 637	Q 603	SPF	$\frac{Q\,137 + Q\,637}{Q\,237}$	Q 237	Q 132	Q 137
1	2.4	2.1	2	-	-	-	64.1	62.0	0	2.1
2	13.2	201.0	37	-	-	-	119.2	15.4	97.3	103.8
3	8.7	114.5	32	-	-	-	135.7	73.2	52.0	62.5
4	7.2	129.2	44	-	-	-	220.2	160.0	75.1	54.1
5	11.1	154.4	34	-	-	-	179.8	119.6	94.1	60.2
6	11.3	170.8	37	197.4	47.8	4.1	225.6	18.4	161.0	9.8
7	11.1	177.7	39	225.1	49.4	4.6	225.1	-	177.7	-
8	8.0	127.1	39	229.7	51.9	4.4	229.7	-	127.1	-
9	7.7	124.5	39	301.1	69.4	4.3	301.1	-	124.5	-
10	10.6	189.2	44	241.4	57.5	4.2	241.4	-	189.2	-
11	13.5	241.4	44	286.9	67.6	4.3	286.9	-	241.4	-
12	7.3	105.9	33	349.2	87.0	4.1	349.2	-	105.9	-
13	3.6	42.7	29	299.6	85.6	3.5	299.6	-	42.7	-
14	9.2	202.3	54	271.5	85.5	3.3	271.5	-	202.3	-
15	12.6	271.4	52	272.9	78.0	3.5	272.9	-	271.4	-
16	14.0	294.6	51	206.4	52.4	4.0	206.4	-	294.6	-
17	10.7	192.4	44	246.4	61.2	4.1	246.4	-	192.4	-
18	14.3	275.9	47	226.2	55.6	4.2	226.2	-	275.9	-
19	17.1	188.7	27	33.0	7.9	4.3	221.7	-	-	188.7
20	11.4	91.4	20	147.9	37.2	4.0	239.3	-	-	91.4
21	9.4	85.0	22	107.5	29.6	3.7	192.5	-	-	85.0
22	8.7	80.3	23	80.3	23.2	3.5	160.6	-	-	80.3
23	12.6	142.6	28	76.0	23.1	3.4	218.6	-	-	142.6

Table 4 - Energy Flows of Hybrid Solar Heating System. Period
23-10-1980 to 2-4-1981. Daily Averages over One-Week
Periods [Units: MJ].

Week	Q 120 [per m²]	Q 137+ Q 136+ / Q 132+	CE [%]	Q 637	Q 603	SPF	Q 137+ Q 637+ / Q 237+	Q 237	Q 132	Q 137
1	13.4	214.7	25.4	-	-	-	76.1	-	138.6	76.1
2	6.3	64.5	11.4	-	-	-	143.7	125.0	45.7	18.7
3	5.6	65.0	13.1	-	-	-	180.9	130.8	14.9	50.1
4	8.4	134.9	18.9	27.5	6.6	4.2	203.6	105.8	64.5	70.3
5	2.3	16.9	5.1	203.4	47.7	4.3	203.4	-	16.9	-
6	7.0	109.2	20.2	260.1	59.8	4.4	260.1	-	109.2	-
7	13.2	228.7	37.9	344.2	79.1	4.4	344.2	-	228.7	-
8	4.7	54.2	18.3	332.8	82.9	4.0	332.8	-	54.2	-
9	8.8	165.1	36.9	342.5	98.6	3.5	342.5	-	165.1	-
10	13.4	276.8	50.3	347.5	97.6	3.6	347.5	-	276.8	-
11	11.9	236.4	42.3	303.4	81.9	3.7	303.4	-	236.4	-
12	9.9	182.4	33.8	387.6	108.6	3.6	387.6	-	182.4	-
13	12.0	253.1	46.8	320.8	91.9	3.5	320.8	-	253.1	-
14	15.2	319.5	51.0	369.3	101.5	3.6	369.3	-	319.5	-
15	15.9	328.6	48.8	303.2	77.9	3.9	303.2	-	328.6	-
16	12.1	231.9	38.0	242.4	58.2	4.2	242.4	-	231.9	-
17	9.1	144.2	33.9	316.1	77.4	4.1	316.1	-	144.2	-
18	12.5	233.5	41.9	276.0	73.5	3.8	276.0	-	233.5	-
19	11.2	200.6	38.6	263.8	66.1	4.0	763.8	-	200.6	-
20	11.9	237.2	45.1	160.9	40.2	4.0	160.9	-	237.2	-
21	14.9	225.5	33.4	73.7	17.1	4.3	189.1	-	110.1	115.4
22	15.8	123.0	14.8	-	-	-	123.0	-	-	123.0
23	7.5	49.2	7.1	58.4	14.0	4.2	107.6	-	-	49.2

Table 5 - System Performance Indicators for Two
Heating Seasons

Indicator	1979/ 1980	1980/ 1981
Large storage effectiveness	0.85	0.93
Total seasonal performance heat pump	3.9	3.8
Collector efficiency	0.37	0.41
Total solar fraction	0.78	0.74

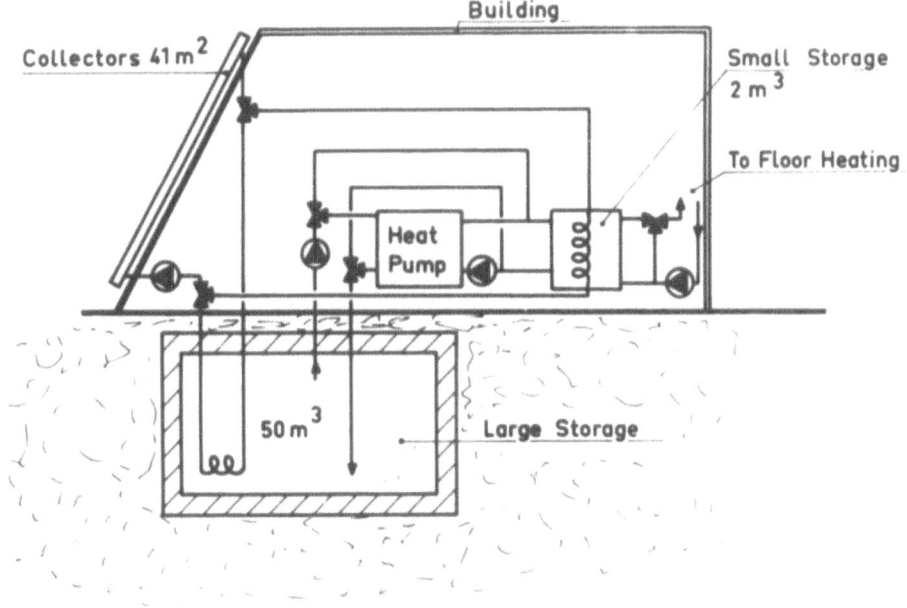

Fig. 1 SCHEMATIC VIEW OF HYBRID SOLAR HEATING SYSTEM OF THE SOLAR LABORATORY OF THE J.R.C.

Fig. 2 - The Solar Laboratory of JRC

1° phase "Summer"

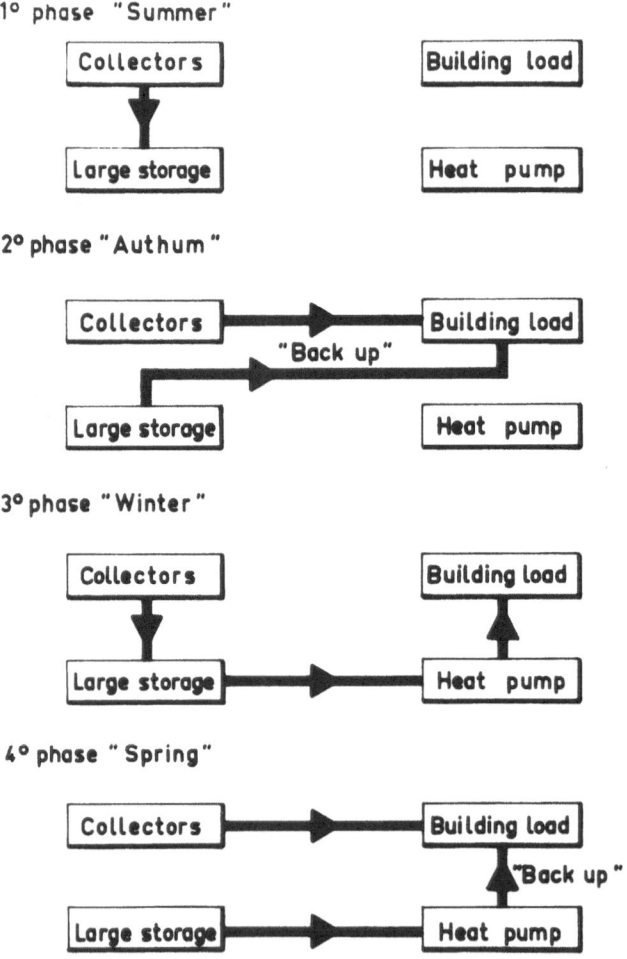

Fig. 3 SCHEMATIC REPRESENTATION OF THE OPERATION
MODES OF THE HYBRID SOLAR HEATING SYSTEM

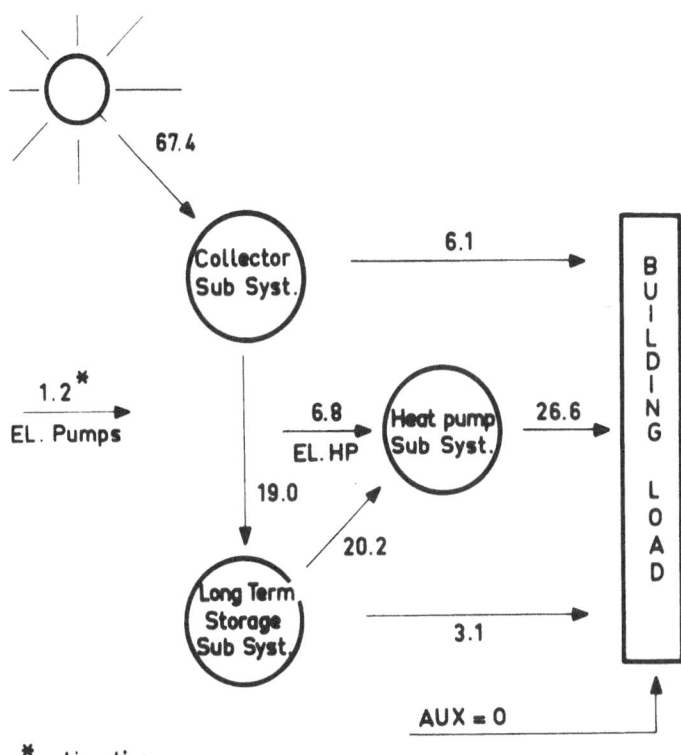

67.4

Collector
Sub Syst.

6.1

1.2*

EL. Pumps

6.8

EL. HP

Heat pump
Sub Syst.

26.6

19.0

20.2

Long Term
Storage
Sub Syst.

3.1

AUX = 0

B
U
I
L
D
I
N
G

L
O
A
D

*estimation

Fig. 4   ENERGY FLOW DIAGRAM FOR HYBRID SOLAR
HEATING SYSTEM.
PERIOD 22·10·1979 to 30·3·1980.
(units : GJ)

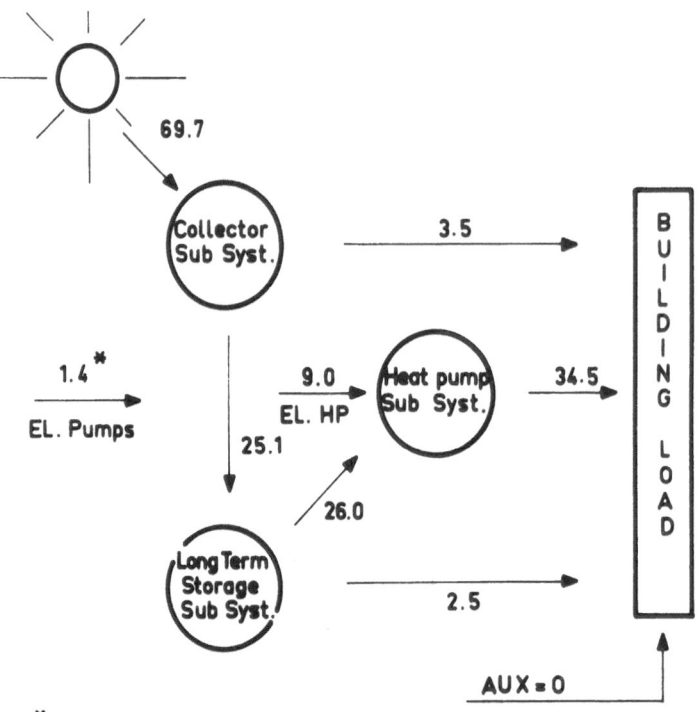

* estimation

Fig. 5 ENERGY FLOW DIAGRAM OF HYBRID SOLAR
HEATING SYSTEM.
PERIOD 23·10·1980 to 2·4·1981

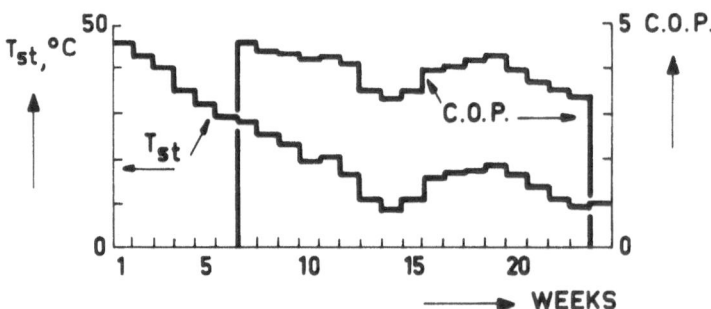

Fig. 6 AVERAGES OVER ONE WEEK PERIODS OF
THE LARGE STORAGE TEMPERATURE AND
THE C.O.P. OF THE HEAT PUMP.
(heating season 1979–1980)

SOLAR HEATING 1980-1981 (23/10-6/4). DAILY RESULTS

SOLAR IRRADIATION, 60 DEGR.

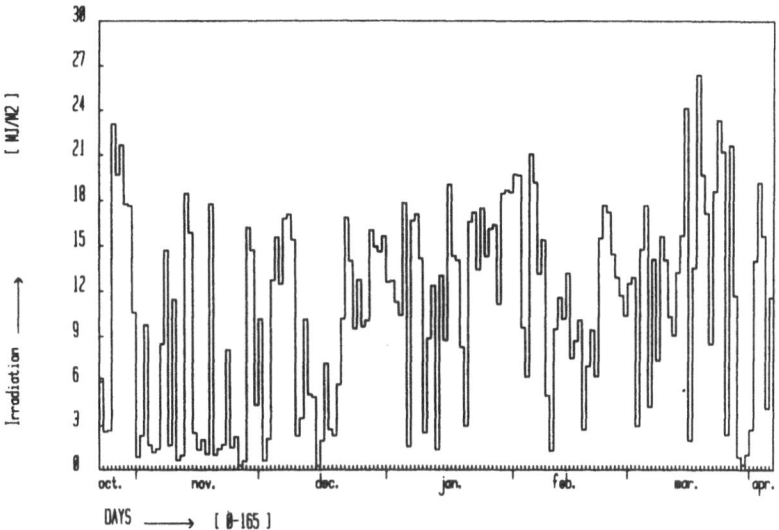

SOLAR HEATING 1980-1981 (23/10-6/4). DAILY RESULTS

COLLECTOR EFFICIENCY

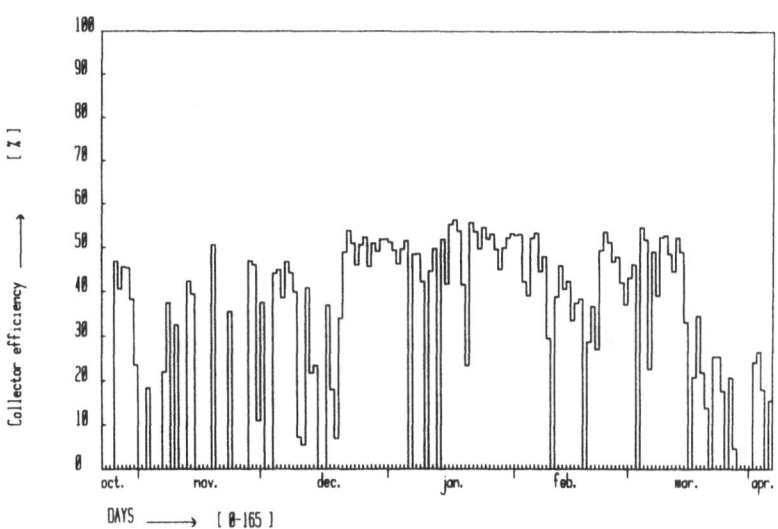

SOLAR HEATING 1980-1981 (23/10-6/4). DAILY RESULTS

OUTDOOR TEMPERATURE

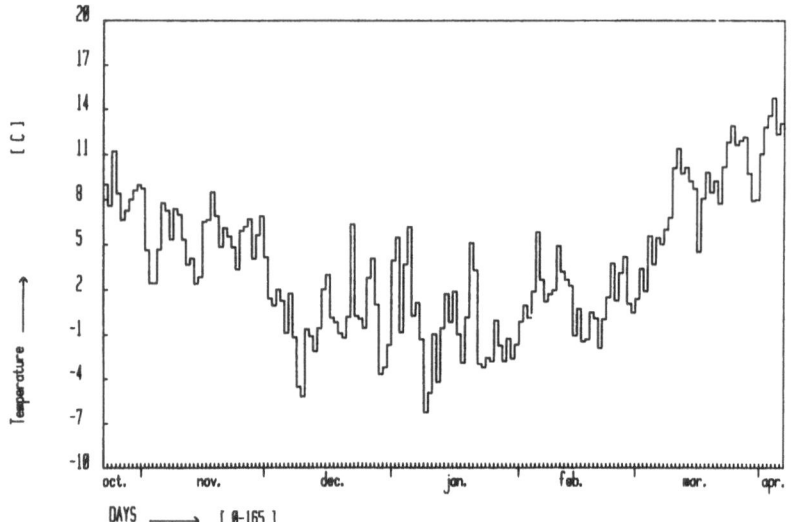

SOLAR HEATING 1980-1981 (23/10-6/4). DAILY RESULTS

ENERGY USED VIA FLOORHEATING SYSTEM

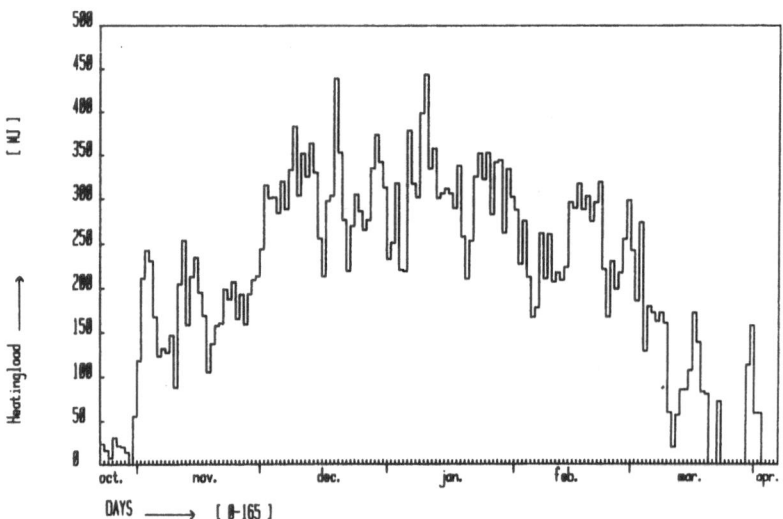

SOLAR HEATING 1980-1981 (23/10-6/4). DAILY RESULTS

ELECTRICITY USED BY THE HEATPUMP

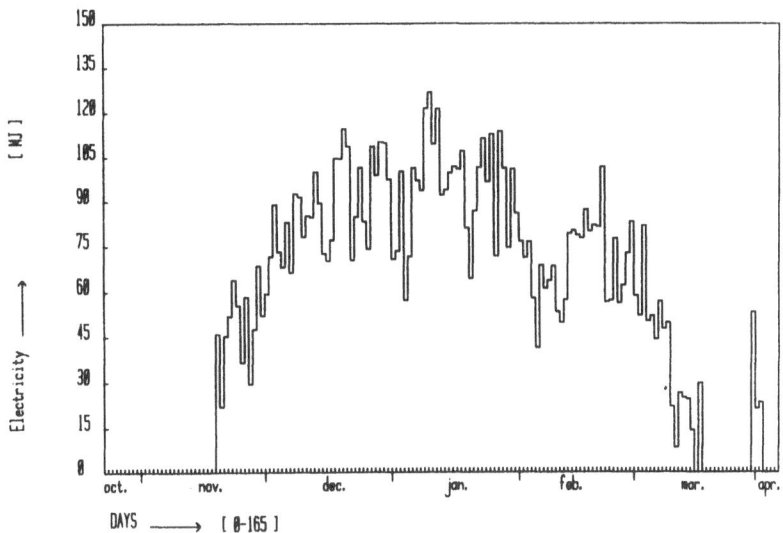

SOLAR HEATING 1980-1981 (23/10-6/4). DAILY RESULTS

TOTAL HEAT OUTPUT HEATPUMP

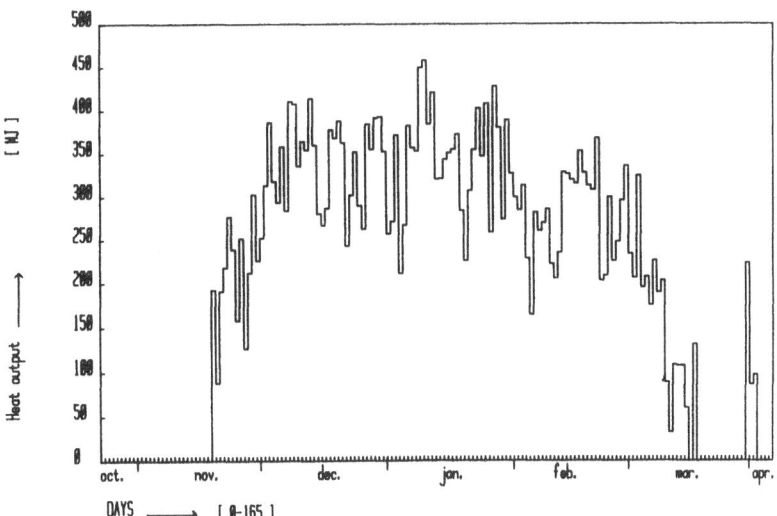

SOLAR HEATING 1980-1981 (23/10-6/4). DAILY RESULTS

ENERGY OUTPUT SOLAR COLLECTORS

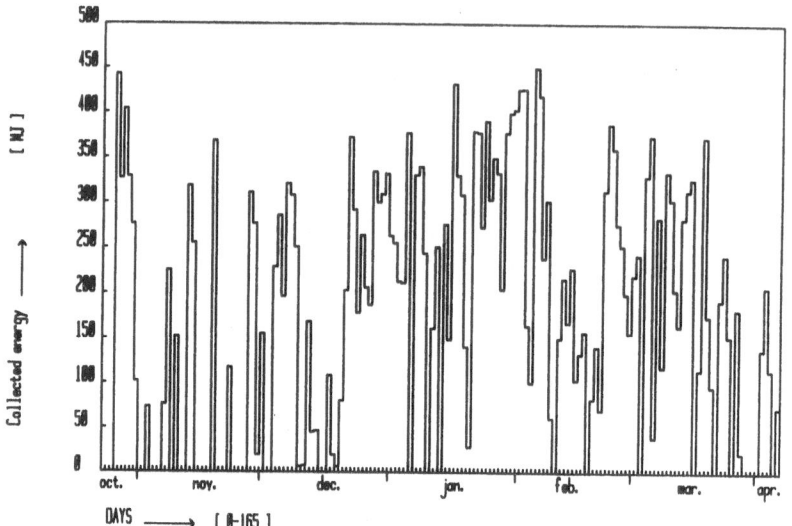

SOLAR HEATING 1980-1981 (23/10-6/4). DAILY RESULTS

LARGE STORAGE TEMP.

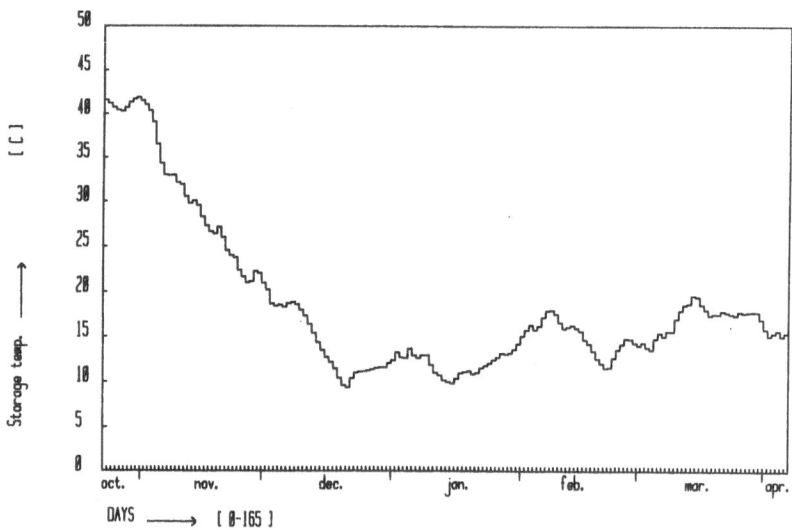

- 354 -

Introduction to the EC Solar Heating Programme

T.C. Steemers

Commission of the European Communities
Directorate-General XII
Science, Research and Development
Energy R and D

Up to now the EC Solar Heating Programme concentrates mainly
on solar energy applications in dwellings, although it is realised
that very soon the attention should widen towards buildings in
general, including office blocks, schools, hospitals, etc., as
there is a wide scope for energy conservation by means of solar
applications in this latter area.

When writing an R and D programme one should have a clear
picture of the final goals, e.g. for a solar heating programme,
one should be able to imagine clearly what an architect or
building engineer, further called the designer, in the future
needs in order to be able to design and build a passive or active
solar heating system for a dwelling.

The designer should first of all be well trained in the subject.
This will in the future undoubtedly happen during his professional
education, where he will become familiar with design tools,
components, systems, system optimisation from both a technical as well as economic point of view, etc. The inclusion of
courses on solar heating design and technology in the curricula
at the universities is however a long term goal, which is pre-
ceeded by a learning phase of the teaching staff by means of
research, workshops, conferences and study of the available
literature.

In the meantime mid-career designers should have the opportunity to become conversant with solar heating design and tech-

G. Beghi (ed.), Performance of Solar Energy Converters: Thermal Collectors and Photovoltaic Cells, 355–359.
Copyright © 1983 ECSC, EEC, EAEC, Brussels and Luxembourg

nology. Consequently the short term goal should be the provision of handbooks and manuals on design and construction of solar heating systems.

The technologies used in the thermal application of solar energy are in principle conventional, but careful studies of first generation systems have shown that there is more to it than constructing a system from available components. The disappointing efficiencies and economies of these first generation systems, and in particular of the space heating systems, have several causes, but is mainly due to lack of technical and economical system optimisation (dimensioning, tuning, control, etc.). The proper tools for such an optimisation were not available at this time.

This leads us to the following programme requirements:
. component and system development;
. system optimisation.
The past has learned us that component and system development should go hand in hand; e.g. an excellent collector will not function well in a bad system.

The current EC Solar Heating Programme is based on the above mentioned considerations and has the following overall goals:
. the production of design tools for both passive and active solar heating systems;
. component and system development;
. technical and economical system optimisation and simplification.
The production of test procedures for components and simple systems is seen as a task particular fitting for a European programme.

In order to introduce the next paper on Solar Pilot Test Facilities, it may be useful to explain the key-role these facilities play in the programme.

The production of design tools for active solar heating systems is one of several major objectives in thermal solar energy research. Although this has already been recognised years ago, the absence of reliable and useful data prevented progress in the validation of simulation models necessary for the development and testing of simplified design tools.

The influence of the occupants of solar houses on the system performance is such that a relative great sample size, typically in the order of 25 well monitored identical solar houses, would have been necessary to average out the influence of that occupancy. Such projects are extremely costly and, moreover, a variety of systems has to be looked at in order to perform a reliable validation of the mathematical models.

The solution adopted was to use a purpose built Laboratory installation called a Solar Pilot Test Facility (SPTF) to obtain data for validation and to carry out system development studies. A SPTF consists of a real solar system with collectors, storage, controls and associated piping but with the dwelling thermal distribution system replaced by a physical load simulator. The simulator being capable of producing a typical thermal load for a house, interactive with the actual weather and taking account of the effect of the occupants.

Figures 1, 2 and 3 give the schematics of the reference system identical in all eight participating countries.

This solution has, apart from the exclusion of the influence of the occupants, a number of advantages over employing a real installation. Not only can all the parameters affecting the solar system performance be economically monitored and controlled, it is also possible to carry out tests of a short duration to investigate alternative components, system arrangements and controls. Additionally it is possible to operate two different systems simultaneously using the same simulated house load.

Parallel to the SPTF-project, a collaborative and concerted action in the field of modelling by means of the European Modelling Group for Solar Heating Systems and Domestic Hot Water assures the direct transfer of the SPTF-data for the purposes they are needed: the validation of a sophisticated research model (EMGP-2) as a means to produce simplified design tools.

This information sheet not only describes the reference solar system identical in all eight participating countries, but also other parts of the installation which vary little from one country to another : the interface and cooling system.

1. SOLAR SYSTEM ONE

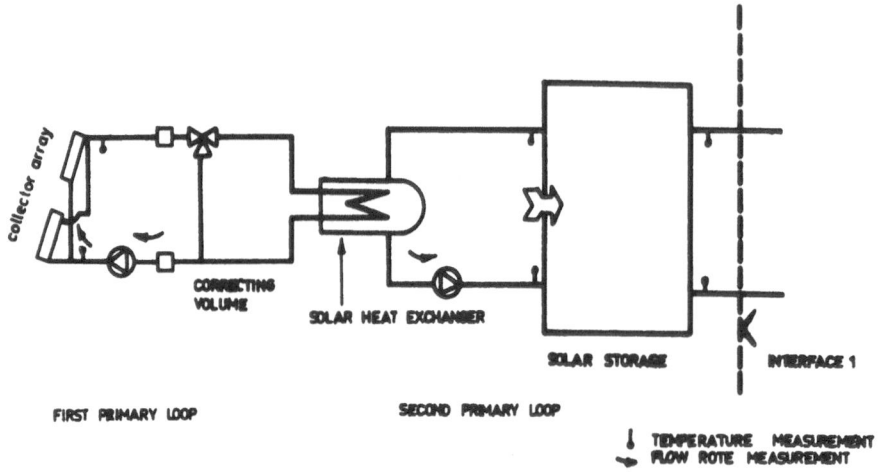

COLLECTORS :

— Flat plate, single glazed, selective surface
— Tilt : latitude
— Orientation : South
— Arrangement : 46 m² in 40 collectors
— Back of collectors sheltered

SOLAR HEAT EXCHANGER :

Type : shell-and-tube

SOLAR STORAGE :

— Medium : water
— Volume : 3 m³
— Heat loss : less than 5 W K⁻¹

PIPING AND FLUIDS :

— Heat loss : less than 15 W K⁻¹
— Antifreeze in the first primary loop : 30 % or 40 % solution of monoethylene glycol
— Correcting volume to ensure a given total volume of the first primary loop
— Flow rates in first and second primary loops adjustable from 0.625 m³ h⁻¹ to 2.5 m³ h⁻¹. Pumps always on.

CONTROL DEVICES

— Three-way valve with thermo-motor
— Differential thermostat : ranges 5K (on), 1K (off)

2. INTERFACE ONE

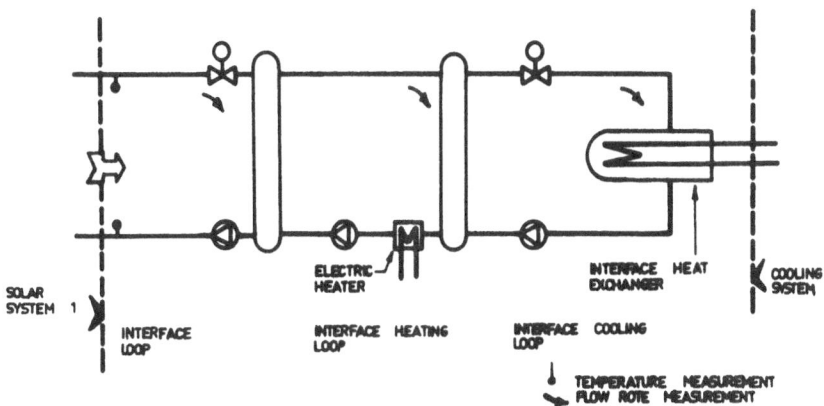

— Electric heater power : 3 KW
— Interface heat exchanger power : 10 KW
— Two motorized two-way valves of industrial quality
— Flow rates : interface loop : 0 to 0.43 $m^3 h^{-1}$
 interface heating loop : 3 $m^3 h^{-1}$
 interface cooling loop : 1 $m^3 h^{-1}$

3. COOLING SYSTEM

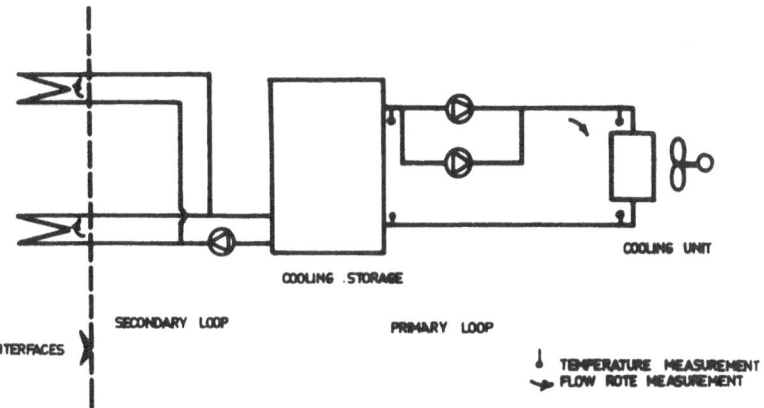

— Cooling unit power : 10 KW for a corrected logarithmic mean temperature difference of 10 K.
— Cooling storage volume : 0.5 m^3
— Fluid : use of antifreeze solution
 flow rates : primary loop 8 $m^3 h^{-1}$, secondary loop 2 m h^{-1} to 4 $m^3 h^{-1}$

— Cooling unit fan and primary loop pump are stopped when the cooling storage temperature is less than 0°C or less than air temperature plus 3°C.

Validation of Simulation Models of Solar Heating Systems with
Data from the Solar Pilot Test Facilities

Ir. P. Deceuninck

Katholieke Universiteit Leuven - Belgium

INTRODUCTION

Since solar energy has gained a lot of interest, it was felt that
there was a need for a computer program, capable of predicting
the performance of any solar system with a good accuracy and
a well known uncertainty interval for every system and for every
European climat. The Solar Pilot Test Facility Group has been
created in order to provide detailed measurements of the behaviour of solar systems, and the European Modelling Group
has been created to develop a reliable computer simulation model, calibrated against the PTF-data.

1. THE SOLAR PILOT TEST FACILITY GROUP (SPTF)

A Solar Pilot Test Facility consists of a real solar system complete with its collectors, storage, controls and piping; with
solar collectors subject to actual weather conditions, and the
system load provided by a simulator capable of producing a
typical house load interactive with the weather conditions /1/.
Each PTF is composed of a reference solar system (SS1), as
far as possible identical in each country for comparison purposes, and another system (SS2) reflecting the current national
design of solar systems.

1.a Solar System 1

Figure 1 shows a simplified schematic of the reference system

G. Beghi (ed.), Performance of Solar Energy Converters: Thermal Collectors and Photovoltaic Cells, 361–380.
Copyright © 1983 ECSC, EEC, EAEC, Brussels and Luxembourg

SS1, where:
OCT1   = outlet collector temperature
ICT1   = inlet collector temperature
FPF1   = collector loop flow rate
SPF1   = second primary loop flow rate
OPT1   = outlet temperature from the storage tank to the
           second primary loop
IPT1   = inlet temperature from the second primary loop to
           the storage tank
SST1   = solar storage temperature
IIT1   = inlet interface temperature
OIT1   = outlet interface temperature
OIF1   = interface flow rate
RDT1   = requested distribution temperature
DF1   = flow rate of the distribution system
AT   = ambient temperature

The collector loop, called the "first primary loop" mainly consists of the solar collector panels, the circulation pump, a
three-way valve and the primary side of the heat exchanger.
The three-way valve is positioned by a servomotor and an
ON-OFF temperature difference controller having one temperature sensor in the collector array outlet pipe and the second
sensor in the storage tank. The heat transfer fluid in the collector loop is a waterglycol mixture protecting the system from
freezing, requiring the use of a heat exchanger. In order to
allow for a correct collector outlet temperature measurement,
even when the three-way valve is in the bypass position, it
was decided that the collector loop circulation pump should run
continuously.

The second primary loop or storage loop consists of the secondary side of the heat exchanger and a 3 m^3 storage tank, using
water to store the heat extracted from the collector loop. Since
it was decided that SS1 should be a system without stratification
in the storage tank, the circulation pump in this loop also runs
continuously, and the fluid enters the storage tank close to the
bottom and leaves the storage tank at the top. These measures
together with a high flow rate in the storage loop make the tank
to be well stirred and hence prevents stratification to build up.

The heat from the storage tank is removed by means of a physical simulator according to a mathematical model of the
dwelling being considered. This physical simulator is called
interface. The heating system of the house is directly connected
to the storage tank. The requested distribution temperature
(RDT1) depends linearly on the outdoor air temperature (AT).

The mass flow rate through the house heat distribution system (DF1) is assumed to be constant. When the temperature at the interface inlet (IIT1) exceeds the requested distribution temperature, the modulated three-way valve adjusts the mass flow rate (OIF1) through the storage tank such that the resulting mixing temperature equals the requested distribution temperature. When the storage temperature decreases below the requested distribution temperature, the full flow rate of the heating system passes through the storage tank and the auxiliary heating system supplies the energy needed to obtain the requested distribution temperature. The return temperature (OIT1) results from the distribution temperature and the requested heating power.

Every participant has made a detailed study (measurements or calculations) of the values to be assigned to the installation parameters in order to have a well known uncertainty interval for these parameters (f. i.: heat loss coefficients, heat capacities, etc.). Accurate measurements are performed of at least all the temperatures and flows mentioned on Fig. 1, and of the climatic conditions. The data is recorded by a computer every five minutes and this five minute data, or hourly averages of it, are stored on magnetic cassettes.

After running his SS1 for a certain period in its basic configuration, every participant applied some minor changes to the system, in order to investigate one (or more) of the following subjects:
. on-off control: value of the temperature difference between the outlet of the collector and the storage tank at which the collector loop flow goes through the heat exchanger or in bypass;
. value of the flow rates in the collector loop (FPF1) and in the storage loop (SPF1);
. pump control: instead of setting the three-way valve in the collector loop in bypass position, the pumps in the collector loop and in the storage loop are shut off;
. ratio of storage volume/collector area;
. stratification of the storage tank;
. value of the requested distribution temperature;
. value of the heating demand;
. components and technical arrangements.

1. b Solar System 2

Each PTF has a different SS2. Fig. 2 shows a simplified schematic of SS2 of Belgium /2/ (the meaning of the symbols is the

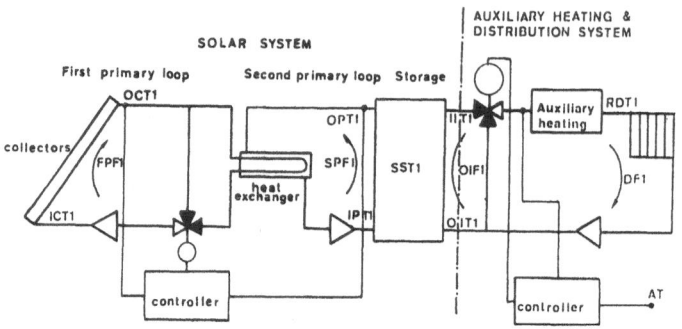

Figure 1 - Simplified schematic of SS1

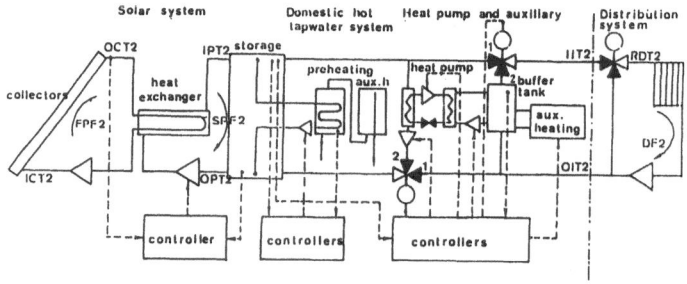

Figure 2 - Simplified schematic of SS2

same as for Fig. 1). To allow stratification in the storage tank of SS2, an ON-OFF control of the circulation pump in the storage loop is used instead of the ON-OFF control of a three-way valve in the collector loop as in SS1. Therefore, the SS2-collector loop is a single loop, without three-way valve and bypass, such that the collector fluid always passes through the heat exchanger. The differential thermostat for ON-OFF control of the circulation pump in the storage loop has one temperature sensor in the collector outlet pipe and a second sensor in the bottom section of the storage. The circulation pump in the collector loop is started and stopped by a clock commanded switch.

Heat is withdrawn from the solar storage in three different modes, one to supply heat for preheating of domestic tapwater and two different modes of operation for space heating. The solar storage can always supply the full heat demand whenever the temperature in the top part of the storage tank equals or exceeds 40°C. When this condition is satisfied, the three-way valves, which are simultaneously actuated by servomotors, are in position 1 and the storage is connected directly to the interface. When the temperature at the top of the storage tank decreases below 40°C, the three-way valves switches over to position 2. The storage is then connected to the heat pump evaporator and can be used as a low grade heat source for the heat pump. Because a sufficiently high mass flow rate has to be used for the heat pump to have a satisfactory performance, on the evaporator as well as on the condensor side, it follows that stratification in the storage will now be destroyed more rapidly and also that the possibly low interface mass flow rate cannot be used directly to extract the heat from the heat pump condensor. Therefore an intermediate loop, extracting the heat from the heat pump condensor is used together with a small buffer tank of about 250 liters. The house heat distribution system is coupled to this buffer tank when the three-way valves are in position 2. The heat pump is now ON-OFF controlled by a thermostat in the 250-liter buffer tank. The use of the buffer tank also reduces the intermittance of the heat pump operation. The storage temperature decreases as heat is extracted by the heat pump, but meanwhile, heat can be supplied to the storage by the solar system with increasing collector efficiency. The use of the storage as a low grade heat source for the heat pump is however limited because the water temperature at the heat pump evaporator outlet may not decrease below 2°C. Whenever this condition occurs, the heat pump is switched off and an auxiliary system has to take over. The auxiliary heat source has been installed as a 12 kW-heating element in the buffer tank.

The auxiliary heating system is ON-OFF controlled by a thermostat in the buffer tank. The domestic hot water system is coupled to the 5000 liter storage tank. It consists of a preheating tank with an internal heat exchanger and an electric boiler in series with the preheating tank. The preheating loop, taking water from the top of the solar storage and returning it to the bottom of the storage, is ON-OFF controlled by a differential thermostat with one sensor at the top of the storage tank and a second sensor at the bottom of the domestic water preheating tank. The hot water draw-off is realised by opening and closing of electromagnetic valves at the outlet of the electrical boiler. These valves are actuated by the central control system.

At least one of the SS2 has implemented one of the following components:
. a heat pump;
. a stratified storage vessel;
. a domestic hot water system;
. air collectors;
. a rock bed storage;
. an absorption cooling machine.

2. THE EUROPEAN MODELLING GROUP (EMG)

2.a Introduction

The European Modelling Group for Solar Systems is mainly concerned with the development of computer simulation models, detailed as well as simplified and the knowledge of the accuracy and validity range of these models /3/. Twelve participants, spread all over the European Community, are working to achieve this ambitious objective.
Various models have been used during the previous phase of the programme and the experience which has been gained from comparison of models should culminate in the development of European models which have been validated in detail against experimental data of high quality. Solar System 1 of the SPTF-programme, being a common reference system in each of the eight Pilot Test Facilities, offers the possibility of validating and improving the computer simulation model for a large variety of climatic conditions while the eight Solar Systems 2, being all different, enable the Modelling Group to validate computer models for a variety of different systems.
Based on the SPTF-experimental data, solar system simulation models for nine different systems are to be developed and to be validated and improved such that an acceptable level of accuracy

can be associated with each of these system simulations in the different climatological zones of the European Community. The ultimate goal of the modelling activity consists in the development of one European simulation programme of large general applicability, i. e. capable of simulating a large variety of solar systems with a satisfactory and known accuracy in the climatic zones of Europe.

2. b Detailed Simulation Model Development

During the first phase of the modelling activity, the participants have gained ample experience in the development of models and the subroutines for the simulation of the solar system components. Each participant calculated the performances of well-defined systems, using the same climatological data. It appeared to be very difficult to explain the differences in the results obtained with the different models. On Fig. 3, the solar percentage for every month of a yearly calculation is presented for eight different models /4/. In fact, these differences are of a very complex structure since as well the physical equations of various components as the mathematical algorithms to solve the equations differ from one model to another. Differences in the interpretation or definition of input and output quantities made such a comparison of models even more difficult. From the validation calculations with PTF data it could be concluded that several models can reproduce the measured behaviour of SS1 more or less satisfactory, resulting in a growing confidence in the capability of simulation models. It was however not possible to conclude that one of the models is superior in accuracy than any other of the simulation models used for validation calculations. Finally a completely new program has been constructed, using all the advantages of the different models. The principles of the program and the development and usage of this simulation model will be treated in the following paragraphs.

2. b. 1 Principles of the Program

- Integration method /5/

The differential equations are solved by a central integration routine considering the whole set of differential equations simultaneously. In matrix notations the set of differential equations can be written as an initial value problem of the following type:

$$\dot{Y} = F(X, Y, Y_a, VD(X)) \qquad (1)$$

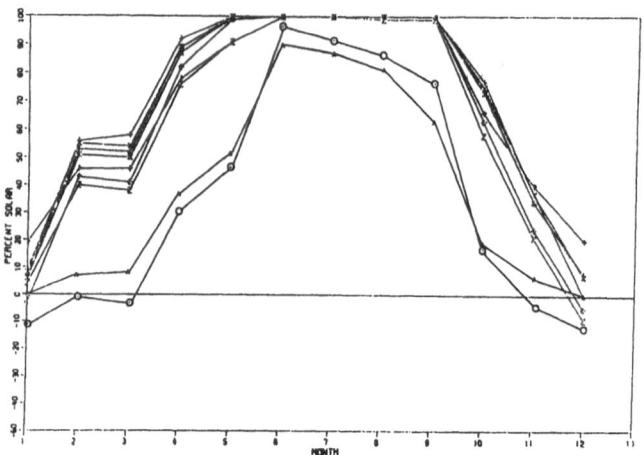

Figure 3

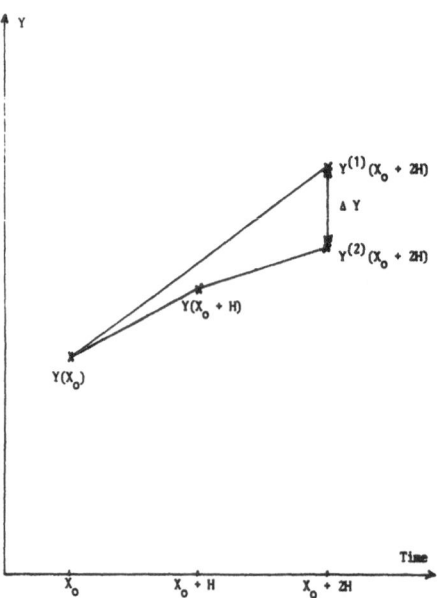

Figure 4

where:

Y   = the state vector

Y_a = a vector of additional variables resulting from the so-
       lution of the algebraic equations to be considered

X   = the time

VD(X)= the time dependent input variables such as weather
       data and heating power.

The mathematical algorithm used to solve this equation has to
be such that no numerical stability problems arise, no significant time step dependence occurs and with a sufficient accuracy.

In the European model, a fourth-order Runge-Kutta integration
procedure is used, which is stable and self-starting, i. e. only
the functional values at one previous point are needed to obtain
the functional values at the next point. This is particularly
interesting because it makes it easy to change the time step H
at any step in the calculation whenever this would be required
for reasons of accuracy.
In fact, the time rate of change of the input variables as well as
the effective time constants of some components are time
dependent such that during the calculation it may be required to
continue for some time with a smaller time step, which may then
again be increased when the time rate of change of the system
state slows down again. The Runge-Kutta method however requires the evaluation of the right-hand side of the system equations four times every time step, such that the required computer time may be higher than with other methods. Control of
accuracy and time step is done by comparison of the results obtained with a double and two single time steps. The algorithm
being used is Gill's modification of the classical Runge-Kutta
formulas with compensation of accumulated roundoff-errors.

In the integration procedure, the control of accuracy and the
correspondingly required time step selection proceeds as follows
(Fig. 4). At any instant of time, two successive time step intervals are considered, thus considering the process in the interval
$(x_o, x_o + 2H)$. $Y(x_o + 2H)$ is calculated in two different ways.
The first value, denoted by $Y^{(1)}(x_o + 2H)$, is obtained from a
calculation as given above, using the double time step 2H (for
VD(X) the average of the input data over the 2H interval is taken).
A second calculation evaluates $Y(x_o + H)$ and a third calculation
evaluates $Y(x_o + 2H)$ in a single step of length H, starting from
$Y(x_o + H)$. This last value of $Y(x_o + 2H)$, denoted as $Y^{(2)}(x_o + 2H)$
is of course more accurate than $Y^{(1)}(x_o + 2H)$ since it has been
obtained with two single steps. In the $Y(x_o + H)$ calculation VD(X)

is the average input over the $(x_o, x_o + H)$ interval and in the calculation of $Y^{(2)}(x_o + 2H)$ from $Y(x_o + H)$, $VD(X)$ is the average input over the $(x_o + H, x_o + 2H)$ interval. The values of each of the corresponding components of $Y^{(1)}(x_o + 2H)$ and $Y^{(2)}(x_o + 2H)$ should not differ from each other by more than the requested accuracy for the time step being used to be acceptable.

When this accuracy criterion turns out to be satisfied, the values of $Y(x_o + H)$ and $Y^{(2)}(x_o + 2H)$ are accepted and the calculation proceeds with the same procedure from $(x_o + 2H)$ onwards. When at least one of the Y-components does not satisfy the accuracy test, the calculation goes back to x_o and starts again with a halved time step.
If the discrepancy between $Y^{(2)}(x_o + 2H)$ and $Y^{(1)}(x_o + 2H)$ is less than 1/50 of the requested precision, the time step is doubled. This procedure allows to obtain the requested precision during every step of the calculation.

- Example of the Physical Equation of a Component

An essential component in the system and perhaps the most difficult to deal with in modelling work is the collector. In the first version of the European simulation model, the flat plate collector is described by the following equation, applicable to every collector segment characterised by one temperature:

$$\text{C.A.} \ \frac{dT}{dt} \ = \ AI\eta + \delta_W \dot{V} \rho \ c_p (T_{in} - T) \tag{2}$$

where:

T   = element temperature
C   = thermal capacity per unit surface area
A   = surface area
I   = incident collector global radiation
      = collector efficiency
$\dot{V}$ = volumetric flow rate
ρ = fluid density
c_p = fluid specific heat

δ_W = 0 when flow is zero, 1 with fluid flow

T_{in} = inlet temperature

The collector efficiency is given by:

$$\eta \ = \ (\alpha\tau)F' - \frac{F'U_t}{I}(T - T_{at}) - \frac{F'U_b}{I}(T - T_{ab}) \tag{3}$$

CHARACTERISTIC	COLLECTOR CEC 4
Name	DRU Hi F
Manufacturer	Koninklijke Fabrieken Diepenbrock en Reigers, Ulft, Holland.
Gross dimensions (m) Aperture Area (m^2)	1.62 x 0.79 x 0.115 1.185
<u>Aperture Cover Assembly</u> Number of cover plates Material Thickness (m) Solar transmittance Thermal emittance	 1 Glass 0.004 0.88 0.84
<u>Absorber Plate Assembly</u> Material Manufacturing process Coating Solar Absorptance (Nominal) Thermal Emittance (Nominal)	 Stainless Steel Spot welded S_nO_2 Enamel 0.91 0.20
<u>Air Space (m)</u> Between cover and absorber	 0.025
<u>Insulation</u> Material Thermal conductivity ($Wm^{-1}K^{-1}$) Case Material	 Mineral wool 0.036 Aluminium

Figure 5 : The collector characteristics

where:

$(\alpha\tau)$ = effective transmittance-absorptance product

F' = collector efficiency factor

U_t = top heat loss coefficient

U_b = back side heat loss coefficient

T_{at} = ambient temperature at the top side

T_{ab} = ambient temperature at the back side

The top heat loss coefficient is calculated from Klein's formula, with temperatures in K:

$$U_t = \left[\frac{N}{\frac{C}{T}\left(\frac{T-T_{at}}{N+f}\right)^e} + \frac{1}{h_w} \right]^{-1}$$

(4)

$$+ \frac{\sigma\,(T+T_{at})\,(T^2+T_{at}^2)}{(\epsilon_p + 0.00591\,N h_w)^{-1} + \dfrac{2N+f-1+0.133\,\epsilon_p}{\epsilon_g} - N}$$

where:

N = number of glass covers

h_w = ambient heat transfer coefficient

ϵ_p = absorber plate emittance

ϵ_g = glass cover emittance

f = $(1 + 0.089 h_w - 0.1166 h_w \epsilon_p)(1 + 0.07866\,N)$

C = $520\,(1 - 0.000051\,\beta^2)$ for β 70°

     = 390.052 for β 70°

β = collector tilt angle

e = $0.43\,(1 - \frac{100}{T})$

h_w = 5.7 + 3.8 V

V = wind speed

Measurements have been made on each PTF to determine the back and side heat loss coefficient and the heat capacity of the whole collector battery, including the connections between the collectors. In fact, every installation parameter, for which it was possible, has been measured by the participants of the PTF group. This allows the modelling group to give good values to the parameters of the physical equations of the simulation model.

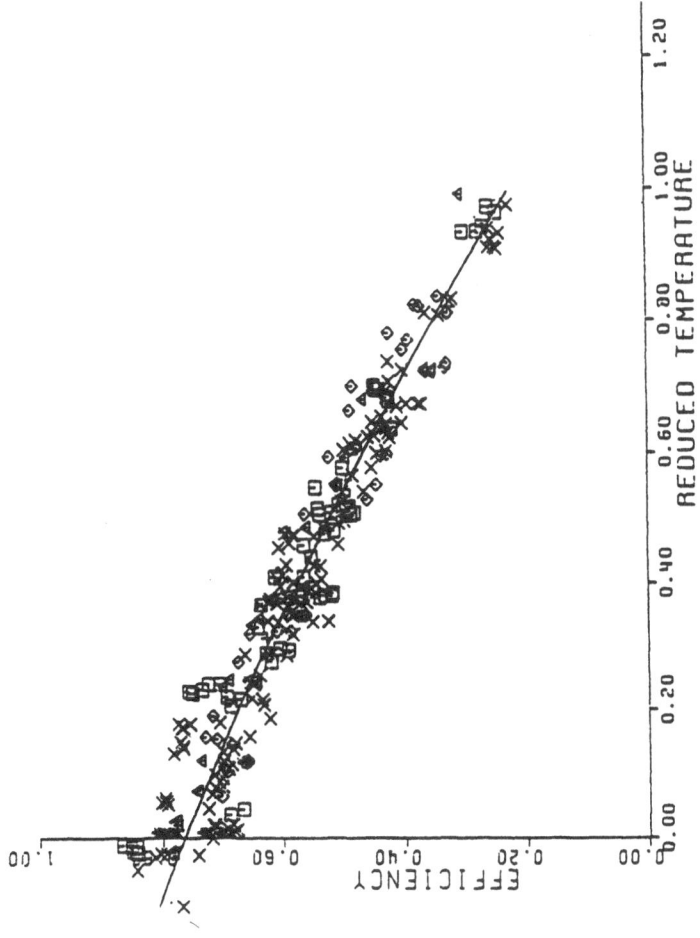

Figure 6 : Efficiency data from 16 facilities

The results of calculations using the first collector model will be compared to the results obtained with an alternative model, based on measured efficiency curves instead of using equations (3) and (4). The SS1 collectors, which are identical in all the Pilot Test Facilities have been tested by the Collector Testing Group of the European Community. The characteristics of the collector, and the efficiency data from 16 facilities are given on Figs. 5 and 6 /6/. A least square fit has been given to the data points in the form:

$$\eta = \eta_o = a_1 T^* - a_2 (T^*)^2 \tag{5}$$

where:

η = efficiency

T^* = reduced temperature

$\quad = \dfrac{U_o (T_m - T_a)}{I}$

U_o = normalized coefficient

T_m = collector temperature

T_a = ambient temperature

I = irradiance

η_o, a_1, a_2: coefficients to be determined by the least square fit

The alternative collector model will be based on this efficiency curve.

Since the SS2 of the English PTF participant contains air collectors, a model for air collectors is being developed. The absorber plate and the air itself will be considered as two different elements, each with its own temperature and calculated with its own equation.

2.b.2 Development and Usage of the Detailed Simulation Model (Fig. 7)

Short term validation calculations

A method has been developed to provide statistical criteria to judge and interpret the discrepancy between theoretical and experimental results. The method accounts for the uncertainty associated with measured input variables, as well as the standard deviations to be taken into account for each of the input variables and parameters have to be known. The estimation of

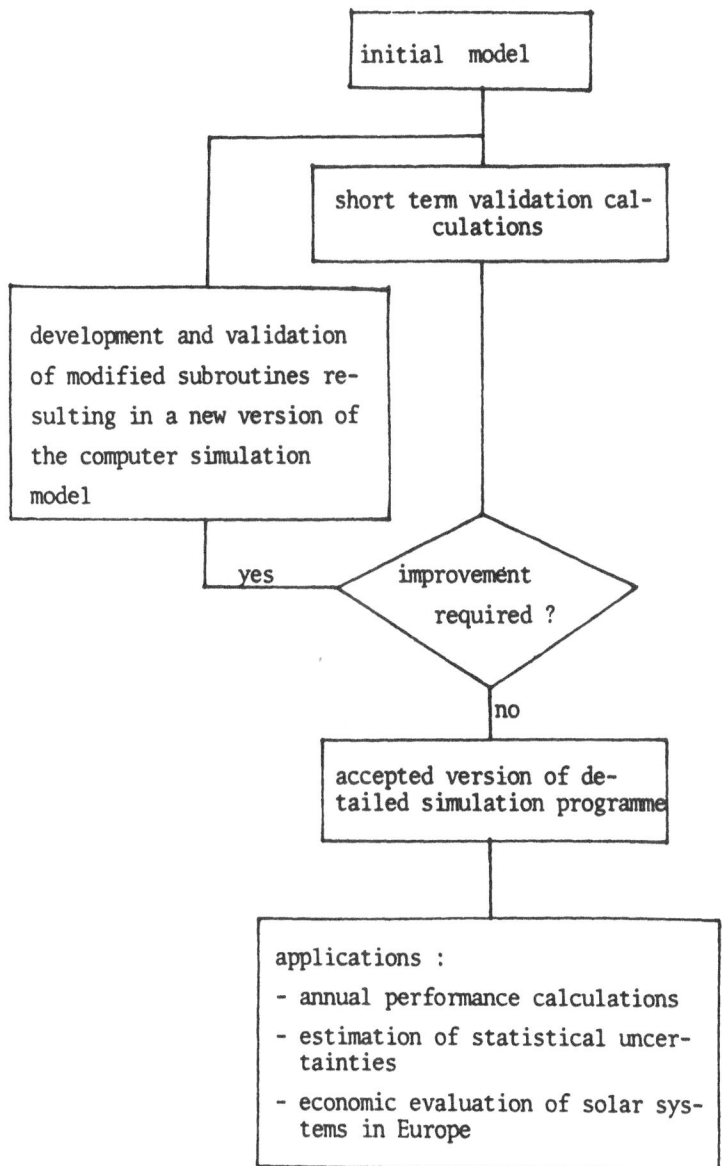

Figure 7

these standard deviations as well as the parameters to be taken
into account is part of the work of the PTF group.
Figures 8 and 9 /7/ dhow the results of a statistical validation
exercise on a period of 24 days. The curves are cumulative:
at the end of each day, the totals from the beginning of the
period until the end of the considered day are evaluated. The
meaning of the symbols used is the following:

SSG1 $= E_2$: integrated solar system gains (= input to the
     storage tank)

E^m_2 : measured solar system gains

E^c_2 : calculated solar system gains

σ^a : absolute standard deviation

IL1/HP $= \eta_4$: ratio of integrated interface losses and inte-
     grated house heating demand (= solar percentage)

η^m_4 : measured solar percentage

η^c_4 : calculated solar percentage

The uncertainty on the measured quantities is caused by measurement errors of temperatures and flows. The uncertainty on
the calculated quantities is caused by measurement errors on
variable input data (solar irradiance, ambient temperature, etc.)
and by inaccurate values of the installation parameters. The
application of this method for various systems and under different climatic conditions will allow to predict the accuracy of the
model.

Component Validation and Model Improvement

When model improvements are required, the main origin of the
faulty behaviour of the model has first to be located. This may
be clear already from the detailed output of the system validation calculations. Because errors propagate from one quantity
to another and back to first, due to the physical feedback in the
system, the identification of the subroutine(s) to be improved
may however also require validation calculations on separate
components.
An improved physical model of the questionable component(s)
may then be developed. An improvement can be based f.i. on
a physically more sophisticated component model, accounting
for effects which were not included in the initial model, such as
expansion tanks, heat exchanger heat capacity, distributed pump
power dissipation, etc., or it may be obtained by using another
mathematical algorithm or even be obtained from a model sim-

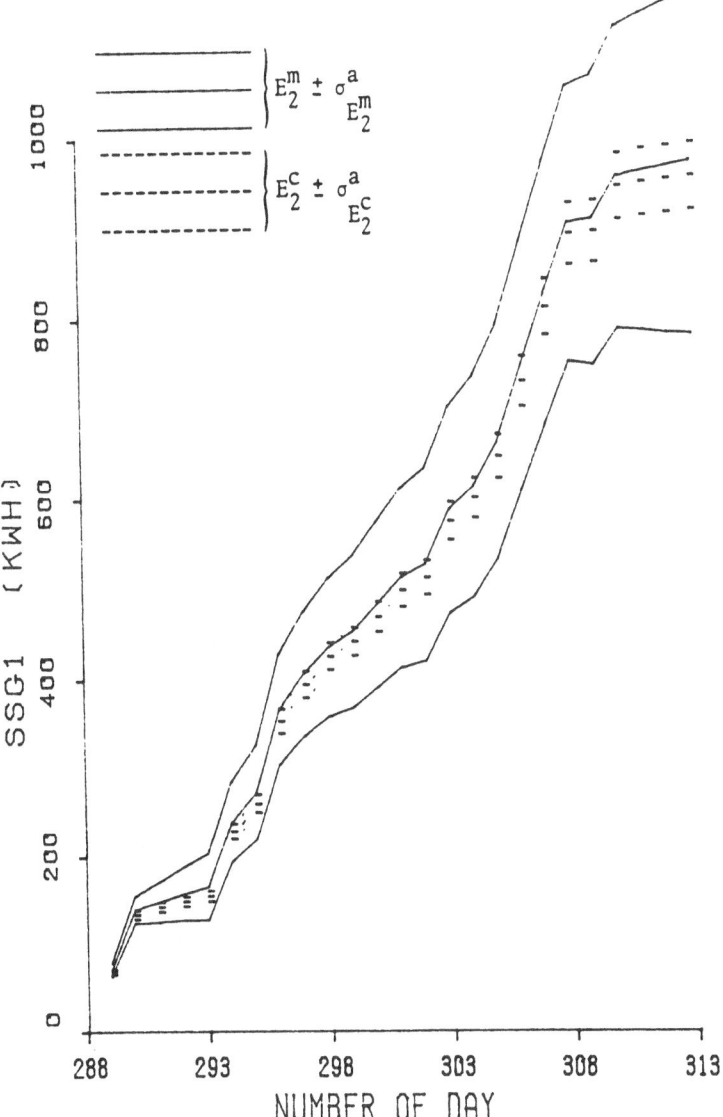

Figure 8

plification.

The modified subroutines are then to be tested in new validation calculations of the considered components, not only for the periods which gave difficulties, but also for the periods for which the original model was found to be satisfactory already. The modified version should indeed be proven to be more generally valid than the original version.

The improved and by the European Modelling Group accepted new version of some subroutines is then to be implemented, yielding a new version of the complete system simulation programme.

Calculations with the Accepted Model

The models (for the different systems being considered) finally obtained from the validation procedure, may be rather different from the model that was used initially and can be considered as validated models with a known accuracy, to be used for long term calculations with different climatic reference years as occurring in the EEC. For this long term calculation too, a statistical error analysis is to be performed for a few typical cases in order to be able to specify the accuracy of the long term performance prediction and to account for the uncertainty of installation parameters and variability of the climatic conditions and heating load.

Based on the results of these annual performance calculations, an economic evaluation study of the considered class of solar systems in Europe can be performed.

2. c Simplified Method Development

For the further development of simplified models, a procedure which is very similar to the detailed model development procedure can in principle be applied, accounting however for the specific nature of the simplified method.

Many simplified methods for computing long-term performance of space heating and domestic hot water solar systems have been published in the recent years. However, no systematic comparison has been made, and designers generally have some difficulties to choose the "best" method. Sixteen published methods have been analysed and they all fit into one of the following three categories /8/:

Analytical models

Their common principle is to compute a simplified thermal balance of the collection loop, based on the notion of critical radiation intensity (radiation is only useful above a certain level,

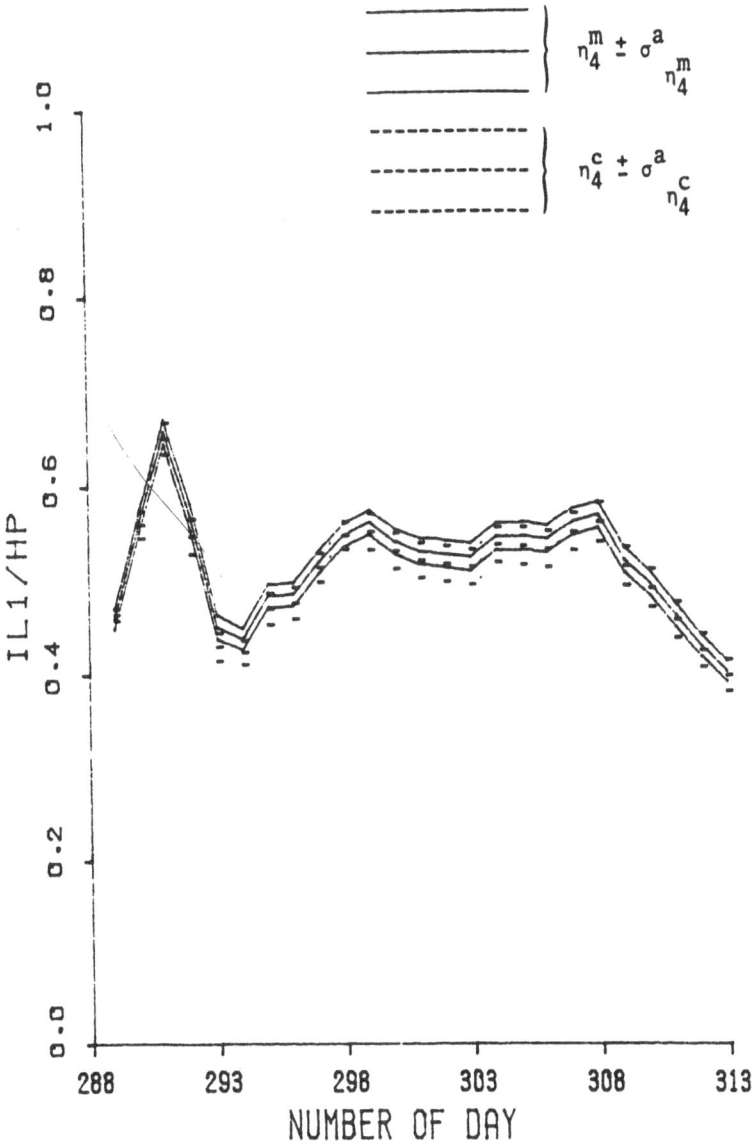

Figure 9

depending on the base temperature, the ambient temperature and the collector characteristics).

Empirical Methods

A computer simulation model is run for a lot of system parameters and climatic hourly data files. These monthly results are then fitted as a function of a few groups of characteristical parameters. The fitting is presented on easy to use charts.

Hybrid Methods

These methods are a synthesis of the two preceeding approaches. They are established in three steps:
. an estimate of "available energy" at a minimum temperature level is made;
. a computer simulation of the system is run;
. a statistical corrective factor is fitted according to the simulation results and applied to the "available energy".

There are seven methods which may be applied to a large geographical zone and with a minimum range of parameters. These methods are retained for further investigation and development according to a procedure almost identical to the one used for detailed simulation models. The calibration of the simplified method is however not performed against detailed measurements from the SPTF group, but against the (already calibrated) detailed model, or against measurement data from the Performance Monitoring Group (which monitors solar houses all over Europe).

REFERENCES

1. Strategy paper for the Solar Pilot Test Facilities. G. Olive.
2. Design, Construction and First Operation of the Solar Pilot Test Facility in Belgium. W. L. Dutré, A. Debosscher and P. Deceuninck.
3. Research Programme of the European Modelling Group for SHS and DHW, July 1981 - June 1983. W. L. Dutré.
4. European Modelling Group for SHS and DHW Coordination Meeting of Contractors, January 1981. O. Jörgenson.
5. Description of EMGP1. W. L. Dutré.
6. Results and Analysis of the Round Robin Testing of the 4th Solar Collector in the European Community Programme. A. Derrick.
7. Statistical Analysis of PTF-validation Calculations. W. L. Dutré, A. Debosscher and P. Deceuninck.

THE PHOTOVOLTAIC PILOT PROJECTS OF THE EUROPEAN COMMUNITY

SCHNELL Werner

Commission of the European Communities

Abstract

The Commission of the European Communities has started in 1980 a programme for the design and construction of a series of photovoltaic pilot projects in the range of 30-300 kWp. Virtually all important industries and other development organisations in Europe working on photovoltaic cells and systems are involved in this programme.

The different technologies which are being developed concern the modules, the cabling of the array, structure design, storage strategy and power conditioning. The various applications include powering of an island, villages, recreation centres, water desalination and disinfection, powering of radio transmitters, emergency power plants, dairy farm, training school, cooling, water pumping, powering of a solar heated swimming pool and last but not least, hydrogen production.

1. PROGRAMME DESCRIPTION

In February 1980, the Commission of the European Communities has, as part of its second solar energy R+D programme, launched a call for tenders for the design and construction of a series of photovoltaic pilot projects in the European Community. Contracts for the construction of 15 projects with a total power of approximately 1 MW were signed in 1981. The total cost of the programme is about 30 mio ECU, one-third of which is borne by the Commission, the rest being contributed by national programmes, regional funds, utilities and the industry itself.

G. Beghi (ed.), Performance of Solar Energy Converters: Thermal Collectors and Photovoltaic Cells, 381–390.
Copyright © 1983 ECSC, EEC, EAEC, Brussels and Luxembourg

The programme was started with the aim of :

- giving European industry the opportunity to develop the necessary expertise on complete pv systems in the range of intermediate loads;

- comparing various technologies under field conditions;

- starting a process of innovation where the technology of system components is reviewed and brought to a high level;

- strengthening cooperation within European industry.

Until now, system experience in Europe was concentrated in the power range up to 1 kW. Above that level, only a few photovoltaic plants were built by European industry, for instance, 4 photovoltaic systems in the 5 kW range as part of the Commission's R+D programme, a 26 kW irrigation pump in France and a 10 kW system for a hospital in Mali, Africa. All these systems were built at the end of the 1970's.

The pilot projects range from 30 to 300 kW and cover a wide variety of applications.

Table 1 : Technical data of EEC photovoltaic pilot projects

Peak power in kW	Site	Application
300	Island of Pellworm (Germany)	Power supply for a vacation centre
100	Kythnos Island (Greece)	Power supply to the village in connection with diesel generators and 5 aero-generators
80	Alicudi Island (Italy)	Electrification of an island with 120 inhabitants
30	Marchwood near Southampton, later Scottish island (UK)	Power supply to the grid
65	Tremiti Islands (Italy)	Water desalination (reversed osmosis)
63	Chevetogne, Province of Namur (Belgium)	Powering solar heated swimming pool and sports centre
50	Montbouquet (France)	Power supply to radio and television transmitter
50	Nice (France)	Nice airport power management and control

Peak power in kW	Site	Application
50	Fota Island, near Cork (Ireland)	Electricity for a dairy farm
50	Terschelling Island, (Netherlands)	Power supply to a marine training school
50	Aghia Roumeli, Crete Island (Greece)	Electrification of an island village
45	Giglio Island (Italy)	Water disinfection, ice-making for agricultural coldstore
44	Rondulinu, Corsica (France)	Power supply to dwellings, a dairy and a workshop, water pumping
35	Kaw, French Guyana	Village electrification
30	Hoboken, Province of Antwerpen (Belgium)	Hydrogen production by electrolysis for Hoboken semi-conductor factory

A specific objective of the programme is the comparison of various technologies under field conditions and the acquisition of particular information on:
- cost
- performance
- operation
- reliability
- maintenance

For this purpose, the Commission has contractually engaged industry to operate the projects for a period of at least five years. Discussions are held for the installation of standardized monitoring systems (see 3.2).

The projects will be constructed by international Consortia in which several utilities and almost all European photovoltaic companies participate. The design phase is closely monitored by the Commission which regularly organizes coordination meetings where all Consortia present their concepts and discuss their technical options. The cooperation between the participating companies is excellent. Design studies will not only focus on the integration of components but all possibilities for increasing the efficiency of each subsystem will be evaluated. Final design review took place in December 1981 (1).

To avoid cost overruns in the programme and to stimulate industry to optimize their cost calculations, contract provisions were made for a fixed financial contribution from the Commission per kW installed with payment of a 20% bonus if all technical specifications defined in the contract were met on schedule.

2. SUBSYSTEMS

2.1 Modules

All cells for the pilot projects will be of mono- or poly-crystalline silicon and manufactured by European companies. Modules will be tested on performance and lifetime at the ESTI laboratory of the Joint Research Centre, Ispra, Italy, according to specification EUR 501 (2). Tests of 2 months' duration are performed on 4 prototype modules or, when series production has started, on 8 series modules and include i.a.

Table 2 : EEC Photovoltaic Pilot Projects - Module Tests

Visual Inspection
Dimensions and Weight
Robustness of Terminations
Mounting Twist Test
Insulation Test
Electrical Performance at Standard Test Conditions (STC)
Determination of Nominal Operating Cell Temperature (NOCT)
Electrical Performance at Nominal Operating Cell Temperature
Hot-Spot Heating Test
Temperature Cycling Test
Humidity-Freezing Test
Hail Resistance Test
Mechanical Loading Test
Damp Heat Long Exposure Test
High Temperature Long Exposure Test
UV Test
O_3 Test
SO_2 Test
Salt Mist Test
Ice Formation Test

Manufacturers have given a 5 year degradation guarantee which states that each module must maintain at least 80% of its nominal power and that loss of power may not exceed 10% for the total array.

2.2 Arrays

The Commission has requested that power losses for mismatch, cabling (ohmic losses) and diodes may not exceed 5 percent. Arrays will be subdivided into subarrays separately connected to the central control station to ease detection of faulty modules.

Table 3 : EEC Pilot Projects - Details on arrays and modules

PROJECT	ARRAY kWp (nom)	ARRAY V	MODULE Maker	MODULE Wp	MODULE n° of cells per module in series	MODULE n° of cells per module in parallel	Total n° of modules
Pellworm	300	354	AEG	19.2	20	1	15.648
Kythnos Is.	100	160	SIEMENS	120	18	8	800
Alicudi	80	220	ANSALDO	33	12	3	1.210
			SOLARIS	66	12	6	605
Tremiti	65	145	{ACE	120	24	6	378
			{ANSALDO	33	12	13	648
Chevetogne	63	220	IDE	33	36	1	1.920
Montbouquet	50	220	PHOTOWATT	66	72	1	760
Nice	50	220	PHOTOWATT	66	72	1	760
Fota Is.	50	220	AEG	19.2	20	1	2.610
Terschelling	50	262	AEG	19.2	20	1	2.610
Aghia Roumeli	50	297/253	FRANCE PHOTON	68	71	1	384
Giglio Is.	45	230	ANSALDO SOLARIS }	64	12	6	720
Rondulinu	44	144	FRANCE PHOTON	68	72	1	648
Kaw	30	100/110	"	72			492
Hoboken			IDE	33	36	1	912

All pv array structures installed previously in Europe
were heavily overdesigned, partly because existing building
regulations do not include pv arrays and are based on wind
speeds at excessive heights (10 m). Although many government-
funded studies, wind tunnel tests and field experience were
performed in the US, the effect of wind loads on large-scale
arrays (turbulences) are insufficiently known. Present experience
has shown that wind loads can be reduced, by order of effectiveness, by

- fences to protect array
- arrays in fields
- decreasing array height
- increasing array porosity
- increasing width to height ratio

Wind loads and specially-designed inexpensive wooden
structures will be studied in a test field near Florence, Italy,
in the framework of the research programme of the Commission.

For the pilot projects, the cost of structures including
the preparation of the construction site, digging works,
foundations, mounting of modules on support structures and array
wiring ranges between 45 and 190 ECU/m^2 which corresponds to
4-19% of the total project budget. Considering the rapid
decrease of module costs, the development of low cost structures
is an important task for the future.

2.3 Power conditioning

Small inverters suited for pv applications have previously
been developed for uninterrupted power supplies, wind turbines,
military applications etc. but most of them work with high
efficiencies only at full load. In the European climate where
most of the light is diffuse during wintertime, high efficiencies
even at low solar irradiance are important. For this reason,
inverters have been developed for the pilot projects by more than
half a dozen European manufacturers. Efficiency is above 90%
over the range of 10% to full load. Some designs involve DC/DC
converters mostly with an associated MPPT. Efficiency is above
90% for irradiance between maximum and 10%. Although system
performance increases with the use of a Maximum Power Point
Tracker (MPPT), its use is still very controversial. Critics
quote that :

- bypass diodes have created many localized maximums which have
  diminished effectiveness of MPPT;

- MPPT large increase of daily performance does not imply
  corresponding increase of annual performance;

- MPPT adds complexity and decreases system reliability the. eby;

- MPPT adds additional capital cost;

- controlling within 10 % voltage variation is all that is needed for an MPPT to deliver within 0.1% full range power;

- annual performance has been found to be increased only 1-4% with an MPPT.

The in-field experience and monitoring of the pilot projects will indicate the commercial future of MPPT's.

2.4 Batteries

The call for tenders of the Commission explicitly made provisions for conventional and advanced batteries, hydrogen and hydrostorage, but except for one hydrogen project, only lead acid batteries will be installed for cost reasons. Factors affecting the selection of batteries for a pv system are:

- usable depth of discharge;
- life cycle cost;
- voltage;
- efficiency;
- cycle lifetime;
- maintenance requirements;
- self-discharge rate;
- constituent material;
- reliability.

For pv applications, nominal life of batteries is expected to be between 5 and 15 years (depending on the frequency of discharges), the self-discharge rate 1% per month, the efficiency 70-80% and the usable depth of discharge 80%.

3. MAINTENANCE, DATA MONITORING

3.1 Maintenance

Only limited experience is available on maintenance operation and cost. Some of the expected maintenance tasks are shown in table 4.

Table 4 : Operation and Prevention Maintenance

Inspection tasks

- Site physical security - fencing intact, breach of security alarm test;

- Array shading - by debris, vegetation;
- Array cleanliness - dust, bird droppings;
- Cabling - damage by elements or rodents;
- Grounding paths - loose connections, corrosion;
- Battery terminals - corrosion;
- Batteries - electrolyte leakage and corrosion of support structure;
- Control equipment - cleanliness; accumulation of dirt, bird nests, rodent damage;
- Fuel/oil/water - at or above specified storage levels for backup systems.

Preventative maintenance tasks

- Clean array surface;
- Clean battery terminals and tighten connections;
- Check and refill electrolytic solution;
- Read and record all metered points;
- Perform operability tests to assure that all automatic switching and monitoring is functional and that standby backup and emergency generator units will start and operate;
- Lubrication;
- Restock stored fuel;
- Record all discrepancies observed by inspection and in performing preventive maintenance.

3.2 Data monitoring

The Commission plans to equip the pilot plants with standard monitoring and data recording systems with the aim of collecting data from all plants at one central point and exploiting them following standard procedures. The main parameters to be recorded are shown in table 5.

Table 5 : Measured and derived parameters

Measured parameter	Derived hourly parameter	Unit
Meteorological		
Global irradiance	Global insolation	$kWhm^{-2}$
Irradiance on plane of array	Insolation on array planes	$kWhm^{-2}$
Ambient air temperature (in shade)	Maximum ambient temperature	deg.C
Wind speed	Maximum wind speed	ms^{-1}

Measured parameter	Derived hourly parameter	Unit
Photovoltaic system		
Currents at sub-array output terminals Voltage at array output terminals	Array output energy	kWh
Current at load input terminals Voltage at load input terminals	Energy input to load from pv system	kWh
Battery voltage Battery charge current Battery discharge current	Energy input to battery Energy output from battery	kWh kWh
Module temperature	(Max. module temperature (Min. module temperature	deg. C deg. C
Power from back-up generator	Energy from back-up genr.	kWh
Power from or to the grid	Energy from or to grid	kWh
Start-up, close-down + switching times		

4. CONCLUSIONS

In a preliminary assessment, the Commission of the European Communities comes to the conclusion that there is ample scope for photovoltaic solar electricity production in Europe on a large scale (3).

In the meantime the Commission has taken the initiative to implement a major programme for developing the necessary expertise for photovoltaic plants within European industry. More than 40 European companies, virtually all current European industry in the field, participate in this programme as the Commission's contractors.

To the benefit of all interested European parties, a large variety of components and systems making use of photovoltaic technology are going to be developed which will result in an extensive expertise for solar electricity in Europe. Thus, it has been possible to pull together all European interests in this new and very innovative field.

References

(1) Proceedings of the Final Design Review Meeting on EC
    Photovoltaic Pilot Projects, Brussels, 30 November –
    2 December 1981, D. Reidel Publishing Co.

(2) Photovoltaic Module Control Test Specifications – Spec. 501
    August 1981, Commission of the European Communities,
    Joint Research Centre, Dr. Krebs.

(3) "Photovoltaic power for Europe, an assessment study"
    published for the Commission of the European Communities,
    by Reidel, Dordrecht, NL, within "Solar Energy R+D in the
    European Communities", Series C, Volume 2, Jan. 1983.

Annex

METHODOLOGIES

I - Recommendations for European Solar Collector Test Methods
     (Liquid Heating Collectors)
     CEC - edited by A. DERRICK and W. B. GILLETT

II - Standard Procedures for Terrestrial Photovoltaic
     Performance Measurements
     CEC Specification No. 101 - Issue 2 (1981)
     edited by K. KREBS
     (from Report EUR 7078 EN)

III - Photovoltaic Module Control Test Specifications
      Specification No. 501 (August 1981)
      edited by K. KREBS
      (from Report EUR 7545 EN)

RECOMMENDATIONS FOR EUROPEAN SOLAR COLLECTOR TEST
METHODS (Liquid Heating Collectors)

Drafted and edited by A. Derrick and W.B. Gillett
University College, Cardiff, U.K.

CONTENTS

G. Beghi (ed.), Performance of Solar Energy Converters: Thermal Collectors and Photovoltaic Cells, 393–467.
Copyright © 1983 ECSC, EEC, EAEC, Brussels and Luxembourg

1. INTRODUCTION

This document is the result of a collaborative programme directed by the European Commission which has brought together the work of twenty European laboratories in the field of Solar Collector Testing. All the recommendations are based on practical experience gained by the participants during the development of their test facilities. The participants have also compared their measurements on a series of "round robin" collectors.
The units used are those of the Systeme Internationale and the definitions of terms are consistent with the recommendations of other published codes as far as possible.
The participants have considered the direct adoption of other published test methods including the widely known American NBS and ASHRAE procedures; but have concluded that additional recommendations and test procedures are required in order to make collector testing a reasonable proposition in European weather conditions. Three methods for determining the thermal efficiency of a solar collector are recommended so that a method may be chosen to suit the local climate or the time available for testing. The methods have in common the same general experimental procedure but differ in the test conditions specified for each method.
After over three years work these recommendations are considered to be a general consensus of views, but are subject to revision and regular updating.

The work was co-ordinated by the European Commission Directorate General DG XII under an Indirect Action Programme with guidance from the Joint Research Centre, Ispra, Italy.

This document was drafted under contract by the Solar Energy Unit, University College, Newport Road, Cardiff, U.K.

2. SCOPE

This document gives recommendations for methods of testing flat plate solar collectors which employ a liquid as the heat transfer medium.
Tests for steady state and transient performance, heat loss, thermal capacity and pressure drop are included, together with instrumentation requirements and a standard format for the presentation of results.
Concentrating devices and air heaters are temporarily excluded. It is anticipated that this document will develop to include durability test procedures and recommendations for testing under simulated solar radiation.

3. UNITS AND SYMBOLS

3.1 General Recommendations

3.1.1 Units

The system of units for solar energy quantities should be the Systeme International d'Unites, (S.I.), detailed in reference 1.
S.I. units for parameters relating to solar energy are given in Tables 3-1 to 3-4.

3.1.2 Symbols

The recommendations in this document have been drafted taking into consideration the recommendations of other European scientific organisations and also
the recommendations of the Committee on Education and Standardisation of the
International Solar Energy Society which are detailed in reference 2. However it
is recognised that work is still in progress in the European Community on the
subject of units, and the recommendations given below may need to be altered
when this work has been completed.
Symbols recommended for general parameters are given in Table 3-1.
Symbols recommended for radiation quantities and related solar angles are given
in Table 3-2. The quantities given are limited to those applicable to solar collector testing.
Symbols recommended for meteorological and climatic parameters that may be
applicable to solar collector testing are given in Table 3-3.
Symbols recommended for defining the position and orientation of a solar collector or any plane are given in Table 3-4.
Recommended subscripts are given in Table 3-5.

TABLE 3-1: Recommended nomenclature - General

Quantity	Symbol	Units
Absorptance	α	dimensionless
Air Mass	M	diemensionless
Area	A	m^2
Conductivity	k	$W\,m^{-1}K^{-1}$
Density	ρ	$kg\,m^{-3}$
Efficiency	η	dimensionless
Electrical power	P	W
Emittance	ϵ	dimensionless
Energy	Q	J
Overall heat transfer coefficient	U	$W\,m^{-2}K^{-1}$
Surface heat transfer coefficient	h	$W\,m^{-2}K^{-1}$
Mass	m	kg
Mass flow rate	$\dot{m}$	$kg\,s^{-1}$
Pressure	p	Pa

.../... cont.

Quantity	Symbol	Units
Reflectance	ρ	dimensionless
Specific heat capacity	c	$J\ kg^{-1}K^{-1}$
Stefan Boltzmann constant	σ	$W\ m^{-2}K^{-4}$
Temperature	T	K or °C
Temperature difference	ΔT	K
Thermal capacity	C	$J\ K^{-1}$
Thermal power	Q	W
Transmittance	τ	dimensionless
Wavelength	λ	m

TABLE 3-2: Recommended nomenclature for radiation quantities and related solar angles

Quantity	Symbol	Convention	Units
Irradiance, total	G_T	–	$W\ m^{-2}$
Irradiance, Longwave	G_L	–	$W\ m^{-2}$
Irradiance, shortwave & solar	G	–	$W\ m^{-2}$
Irradiation, total	H_T	–	$J\ m^{-2}$
Irradiation, longwave	H_L	–	$J\ m^{-2}$
Irradiation, shortwave & solar	H	–	$J\ m^{-2}$
Radiant Energy	Q	–	J
Radiant flux	Φ	–	W
Radiant intensity	I	–	$W\ sr^{-1}$
Solar altitude angle	γ	see definition	Degrees
Solar azimuth angle	α	(clockwise from North)	Degrees
Solar hour angle	ω	(0 to 360° from solar noon)	Degrees
Solar declination	δ	(North positive)	Degrees
Angle of incidence	θ	–	Degrees
Angle of reflection	θ_r	–	Degrees
Angle of refraction	θ_R	–	Degrees

TABLE 3-3: Recommended nomenclature for meteorological parameters

Quantity	Symbol	Units
Humidity, absolute	x	dimensionless
Humidity, relative	ϕ	percent
Temperature, dry bulb	T	K or °C
Temperature, wet bulb	T˙	K or °C
Temperature, dew point	T_D	K or °C
Vapour pressure	e	Pa
Wind speed	u	ms^{-1}
Surrounding air speed	u_s	ms^{-1}

TABLE 3-4: Nomenclature for surface location and orientation

Quantity	Symbol	Unit
Altitude (above sea level)	H	m
Azimuth (of surface)	α_p	Degrees
Latitude	ϕ	Degrees
Longitude	λ	Degrees
Tilt from horizontal	β	Degrees

TABLE 3-5 : Recommended subscripts

Quantity	Subscript	Quantity	Subscript
Absorber plate	p	Inlet	i
Ambient	a	Longwave (thermal)	L
Aperture	a	Loss	ℓ
Atmospheric	A	Mean	m
Beam (direct)	b	Net exchange	X
Blackbody	b	Normal	n
Convective	c	Outlet (exit)	e
Diffuse	d	Reflected	r
Fluid	f	Shortwave (solar)	no subscript
Ground	G	Spectral	λ
Gross	δ	Total	T
Horizontal	h	Useful	u

3.2 Document Nomenclature

The following nomenclature is used in this document:

Symbol	Meaning	Units
a_0, a_1, a_2	Algebraic constants	—
A_a	Aperture area of collector	m^2
A_g	Gross area of collector	m^2
A_p	Absorber plate area of collector	m^2
c_f	Specific heat capacity of the heat transfer fluid	$J\ kg^{-1}\ K^{-1}$
C	Effective thermal capacity of collector	$J\ K^{-1}$
C_T	Effective thermal capacity of the total loop including the collector	$J\ K^{-1}$
D	Date	day-month-year

.../... cont.

...

Symbol	Meaning	Units
E	Voltage output	Volts
e	Error of curve fit	dimensionless
G	Solar and shortwave irradiance	W m^{-2}
G_d	Diffuse solar irradiance	W m^{-2}
LT	Local time	hours
$K(\theta)$	Angle of incidence modifier	dimensionless
m	Mass flow of heat transfer fluid	kg s^{-1}
N	Number of data points	dimensionless
P	Fluid heating power input	W
Q_ℓ	Power loss of the collector	W
Q_T	Total power collection of the collector	W
Q_u	Useful power extracted from the collector	W
R_0	Thermometer resistance at 273K	Ω
R_T	Thermometer resistance at temperature T	Ω
t	Time	seconds
T	Absolute temperature	K or °C
T_a	Ambient or surrounding air temperature	K or °C
T_e	Collector outlet (exit) temperature	K or °C
T_i	Collector inlet temperature	K or °C
T_m	Mean temperature of the heat transfer fluis	K or °C
T_A	Atmospheric or equivalent sky radiation temperature	K
T*	Reduced temperature	dimensionless
U_0	Normalized heat transfer coefficient of the collector	U_0=10W m^{-2} K^{-1}
U_m	Overall heat transfer coefficient of the collector	W m^{-2} K^{-1}
u_s	Surrounding air speed	m s^{-1}
V_f	Fluid content of the collector	m^3
η	Collector thermal efficiency	dimensionless
η_0	Conversion factor (η at T*=0)	dimensionless
η_0	Mean value of the measured conversion factors	dimensionless
Δp	Pressure difference of the fluid between inlet and outlet	Pa
Δt	Time interval	seconds
ΔT	Temperature difference of the fluid between outlet and inlet ($T_e - T_i$)	K

.../... cont.

Symbol	Meaning	Units
θ	Angle of incidence of direct solar radiation	Degrees
σ_e	Standard deviation of the curve fit error	—
τ	Collector time constant	seconds
ρ	Density of the heat transfer fluid	kg m^{-3}

4. DEFINITIONS

4.1 Collector Technology Concepts

Solar Collector

A solar collector is a device which absorbs solar radiation, converts it into heat and passes this heat on to a heat transfer fluid.

Flat Plate Collector

A flat-plate collector is a solar collector whose aperture area is essentially identical to the area of the absorber surface, that employs no concentration, and in which the absorbing surface is essentially planar.

Concentrating Collector

A concentrating collector is a solar collector which uses reflectors, lenses or other optical elements to concentrate the solar energy incident on the aperture onto an absorber, the subtended surface area of which is smaller than the aperture area. (Note: concentrating collectors are temporarily outside the scope of this document.)

Liquid Heating Collector

A liquid heating collector is a solar collector which employs a liquid as the heat transfer fluid.

Heat Transfer Fluid

The heat transfer fluid is the medium by which the energy retained by a collector, as heat, is removed from the collector.

Absorber

The absorber is that part of a solar collector which converts the incident solar radiation into heat and from which the heat is removed by the transfer fluid. If an absorbing liquid is used then this may constitute both the absorber and the heat transfer fluid.

Selective Surface (Absorber)

An absorber is considered to have a selective surface if it absorbs substantially all incident solar radiation while simultaneously exhibiting a low hemispherical emittance at longer wavelengths.

Aperture Cover

The aperture cover is the transparent part of a solar collector, normally positioned at the aperture which is used to reduce the heat loss from the absorber, and to provide some protection from the weather.

Aperture Area of Collector

The aperture area of a collector is the opening or projected area of a collector through which the unconcentrated solar energy is admitted.

Gross Area of Collector

The gross area of a solar collector is the overall projected area of the collector with its containing box, if present.

Aperture Factor

The aperture factor of a solar collector is the ratio of the aperture area of the collector to the gross area of the collector.

Heat Pipe

A heat pipe is a device for transferring heat by means of evaporation and condensation of a fluid in a sealed system. Heat pipes may be used as components of a solar collector.

4.2 Radiation and Solar Angle Concepts

Radiation

Radiation is the emission or transfer of energy in the form of electromagnetic waves or particles.

Radiant Flux

Radiant flux is power emitted, transferred or received in the form of radiation.

Irradiance

The irradiance at a surface is the ratio of the radiant flux incident on the surface to the area of that surface. (Note: solar irradiance is often termed "incident solar radiation intensity", "instantaneous insolation", or "incident radiant flux density".)

Irradiation

The irradiation of a surface is the time integral of the irradiance at that surface. Irradiation is often termed radiant exposure.

Solar Radiation

Solar radiation is the radiation emitted by the sun. (Approximately all of the incident solar energy is at wavelengths less than 4.0 μm and is often termed short-wave radiation.)

Direct Solar Radiation

Direct solar radiation is the solar radiation coming from the solid angle of the sun's disk.

Diffuse Solar Radiation

Diffuse solar radiation is the solar radiation as received on a surface from a solid angle of 2π with the exception of the solid angle subtended by the sun's disk.

Solar Constant

The solar constant is the solar radiation incident on the outer edge of the earth's atmosphere when the earth-sun distance is at the average value of 150×10^6 km. The solar constant value so defined is 1.353 kW m^{-2} on a surface perpendicular to the radiation beam.

Longwave Radiation

Radiation which is at wavelengths greater than approximately 4.0 μm and which is emitted by the ground and other terrestrial objects is normally termed longwave radiation.

Global Solar Radiation

Global solar radiation is the sum of the direct and diffuse solar radiation incident upon a surface from a solid angle of 2π.

Atmospheric Longwave Radiation (Sky Radiation)

Atmospheric longwave radiation is the radiation emitted by gases and particles in the atmosphere.

Sky Temperature

The atmospheric radiation received at a surface may be expressed in terms of an equivalent black body radiation temperature, i.e. the sky temperature.

Visible Radiation

Visible radiation is radiation with wavelengths that stimulate the optic nerves.

Visible radiation lies approximately within a wavelength band of 0.38 μm to 0.76 μm.

Angle of Incidence of Direct Solar Radiation

The angle of incidence of direct solar radiation is the angle between the direct solar radiation beam and the outward drawn normal from the plane of the collector aperture.

Solar Altitude

The solar altitude is the angle between a straight line from the sun to the point of observation and the horizontal plane through that point of observation.

Solar Azimuth

The solar azimuth is the projected angle between a straight line from the sun to the point of observation and due North. The angle is measured clockwise using the projections on the local horizontal plane.

Solar Declination

The solar declination is the angular position of the sun at solar noon with respect to the plane of the equator (North positive).

Solar Hour Angle

The solar hour angle is an equivalent angle ($0°$ to $360°$) for the time of day, with each hour equalling $15°$ of longitude and solar noon being zero (e.g. $\omega = 37.5°$ for 1430 hrs solar time).

Solar Noon

For any given location solar noon is the local time of day when the sun is at its highest altitude.

4.3 Radiation Measurement Concepts

Pyranometer

A pyranometer is an instrument for measuring the solar irradiance on a plane surface from a solid angle of 2π. When the solar radiation coming from the solid angle of the sun's disk is obscured from the instrument, a pyranometer can be used to determine the irradiance on a plane surface of diffuse solar radiation.

Pyrheliometer

A pyrheliometer is an instrument normally used to measure the irradiance on a plane surface of radiation of all wavelengths from a small solid angle. When orientated towards the sun, a pyrheliometer can be used to determine the direct solar irradiance.

Pyrgeometer

A pyrgeometer is an instrument for measuring the irradiance on a plane surface of longwave radiation.

Pyrradiometer

A pyrradiometer is an instrument for measuring the net irradiance on a plane surface of radiation of all wavelengths from a solid angle of 2π.

Solarimeter

A solarimeter is a specific type of pyranometer based upon the Moll-Gorczynski thermopile design.

4.4 General Concepts

Instantaneous (quasi steady state) Efficiency

The instantaneous (quasi steady state) efficiency of a solar collector is defined as the ratio of the average useful power extracted from the collector to the average solar radiation flux incident at the aperture under specified quasi steady-state conditions.

Conversion Factor (η_0)

The conversion factor of a collector is the instantaneous efficiency of the collector when the mean fluid temperature is equal to the ambient air temperature and under specified quasi steady state conditions.
The conversion factor is often termed "Effective Tau Alpha Coefficient" or "Eta Zero".

Time Constant

The time constant of a collector is the time period required for the temperature of the fluid leaving the collector to achieve 63.2% of its rise, following a step increase in solar irradiance.

Useful Power

The useful power extracted from a collector is the thermal power transferred by the heat transfer fluid.

Quasi Steady-State

Quasi steady-state describes the state of the solar collector test when the flowrate and temperature of the heat transfer fluid entering the collector are substantially constant and the variations in the outlet temperature of the heat transfer fluid are only due to small variations of the solar radiation flux incident at the collector aperture (see section 8.2.6.2 for a quantitative definition).

Surrounding Air Temperature

The surrounding air temperature is the temperature of the air in the vicinity of the collector.

Surrounding Air Speed

The surrounding air speed is the speed of the ambient air in the vicinity of the collector.

Wind Speed

The wind speed is the speed of the air measured in accordance with the recommendations of the World Meteorological Organisation, normally measured ten metres above ground level.

Mean Fluid Temperature

The mean fluid temperature of the heat transfer fluid is defined as the average of the fluid temperature at the inlet and outlet of the collector. The mean fluid temperature is normally determined by measurement of the fluid temperature at the inlet to the collector and the fluid temperature rise or drop across the collector. The mean fluid temperature is then determined from the following equation:

$$T_m = T_i + 0.5 \, (T_e - T_i)$$

Air Mass

The air mass is the ratio of the mass of the atmosphere in the actual earth-sun path to the mass which would be traversed if the sun were directly overhead at sea-level.

5. COLLECTOR MOUNTING AND LOCATION

5.1 General

5.1.1 The collector should be mounted in a manner such that no injury to persons may be caused (for example no injury should be caused from falling glass in the event of failure).

5.1.2 The size of the collector(s) to be tested should be large enough so that the performance characteristics determined will be indicative of those that would occur when the collector is part of an installed system. The total aperture area of the collector(s) should be at least 1 m².

5.1.3   The collector should be mounted such that the angle of the tilt of the aperture from the horizontal may be fixed at 45° ± 5°.

5.1.4   The collector should be mounted such that the lowest edge of the collector is greater than 0.5 metres above the local ground or floor surface.

5.1.5   The collector mounting should not obstruct the aperture of the collector.

5.1.6   The structure on which the collector is mounted should not significantly affect the effective back or side insulation of the collector. An open type structure is recommended.

5.1.7   Figure 1 shows the method of mounting the collector at a typical outdoor test facility.

Fig. 1 - Method of mounting the solar collectors at the outdoor test facility, University College, Cardiff.

5.2 Outdoors

5.2.1   The collector should be securely mounted such that it can safely withstand forces arising from wind gusts.

5.2   The collector may be mounted in a fixed position facing South, or it may be moved to track the sun in azimuth.

5.2.3   The collector should be located where there will be no significant energy reflected onto the collector from surrounding buildings, or any other surfaces in the vicinity for the duration of the tests. This requirement will be satisfied if the ground and immediately adjacent surfaces are diffuse reflectors with a reflectivity of less than 0.2.

5.2.4   The collector should be located where there will be no significant obstructions in the foreground of the collector subtending an angle greater than 15° to the horizontal.

5.2.5   The collector should be located such that there will be no significant net interchange of thermal radiation with surrounding surfaces. Surfaces which are at a higher or lower temperature than the ambient air temperature should not be capable of affecting the thermal input to the collector by more than 0.5% of the solar input.

5.2.6   The collector should be located such that a shadow will not be cast onto the collector at any time during the test period.

5.2.7   The collector should be located such as to allow the free passage of air in front of and behind the collector but not to induce air movements in the vicinity of the collector (e.g. by creating a chimney effect). The collector should therefore be mounted at least 1 metre away from vertical walls.

5.2.8   If the collector is to be located during testing on the top of a building, the collector should be located at least 2 metres away from the roof edge.

5.3 Indoors (for thermal loss measurements)

5.3.1   The collector should be located in a closed room.

5.3.2   The temperature of the air surrounding the collector should not vary by more than ± 1°C about its mean value. This requirement may be met by using a large sized test chamber or by air conditioning.

5.3.3   The thermal emittance of the enclosing walls should be greater than 0.90. This requirement will generally be met if the wall material is plaster, glass, wood brick or plastic. Most paints have an emissivity greater than 0.80.

5.3.4   The radiation temperature of the room should not deviate from the air temperature by more than 3K. This requirement will necessitate the shielding of space heaters and other warm surfaces and cold window surfaces. This requirement applies to both the front and back view of the collector. Problems may be encountered in air conditioned buildings if testing is carried out near to an outside wall.

5.3.5   The shortwave irradiance (wavelengths $< 4 \mu m$) on the collector aperture should not exceed 1 Wm^{-2}. This may be confirmed by the use of a pyranometer with a zero check as described in Section 7.1.4.4.

5.3.6   The collector should be free from draughts or other air movements when testing under still air conditions.

6. THE TEST INSTALLATION

6.1 Configurations

6.1.1   Typical test installations are shown in Figs. 2 to 6 as follows:
Fig. 2   An installation for instantaneous efficiency measurements outdoors (pictorial view).
Fig. 3   An installation for indoor thermal loss measurements (pictorial view).
Fig. 4   A closed loop arrangement for instantaneous efficiency measurements outdoors (schematic).
Fig. 5   An open loop arrangement for instantaneous efficiency measurements (schematic).
Fig. 6   An arrangement for indoor thermal loss measurements (schematic).

6.1.2   An open loop system may have the advantage that a constant head device can maintain a more constant flow rate than a simple circulating pump.

Fig. 2 - Outdoor solar collector test facility at the Institute of
Applied Physics, TNO, Delft, Holland.

6.1.3 Fluid loops requiring a continuous supply of fluid are not recommended.

6.1.4 The heat transfer fluid used for collector testing may be water or a fluid recommended by the collector manufacturer. The specific heat and density of the fluid used should be known to within ± 1% over the range of fluid temperatures used during the tests.

6.1.5 Some fluids may need to be periodically changed to ensure that their properties remain well defined.

6.1.6 When the test loop is freshly charged with water the system should be heated to its maximum operating temperature in order to expell the air dissolved in the water.

6.1.7 Filters should be placed upstream of the fluid measuring device, the pump and elsewhere in accordance with normal practices (nominal filter size 200 μm).

6.1.8 An air separator and air vent should be placed at a suitable point in the system and at the outlet of the collector.

Fig. 3 - Transportable solar collector test facility at the KFA Physics Institute, Jülich, Germany (the facility is suitable for both outdoor and indoor solar collector testing).

6.1.9   A length of transparent tube should be installed in the fluid loop such that air bubbles will be observed if present.

6.1.10 Safety pressure relief valves should be used in closed loop systems.

6.1.11 The pump should be located in the fluid loop in such a position that the heat which it dissipates in the fluid does not impair either the control of the panel inlet temperature or the measurement of the fluid temperature rise through the collector.

6.2   Pipework and Fittings

6.2.1   The piping used in the loop should be resistant to corrosion and suitable for operation at temperatures up to 95°C. If non aqueous fluids are used, then compatibility with system materials should be confirmed.

6.2.2   Pipe lengths should be generally kept short. In particular the piping between the outlet of the fluid temperature regulator and the inlet to the collector should be minimized to reduce the effects of the environment on the fluid inlet temperature.

6.2.3   Pipework between the temperature sensors and the inlet to the collector should be protected with weather resistant insulation and solar reflecting foil in order to ensure that the effects of the environment cannot change the true inlet temperature or the measured temperature rise by more than ± 0.05°C. Insulation of the pipework may influence the effective thermal capacity of the collector test loop.

6.3   Temperature Regulation of the Fluid

The fluid temperature at the inlet to the collector should be stable to ± 0.1 K. This may be achieved by various methods. For example:

6.3.1   Thermostatic Bath

A thermostatic bath may be used to control the temperature of the fluid but this should be continuously stirred to ensure uniformity of the temperature. In order to prevent overheating of the bath during outdoor tests it may be necessary to pre-cool the fluid before returning it to the thermostatic bath. The use of only a single thermostatic bath with a pre-cooler may restrict the number of test points which can be obtained in any given day because a period of time is required to change the bath temperature from one value to another.

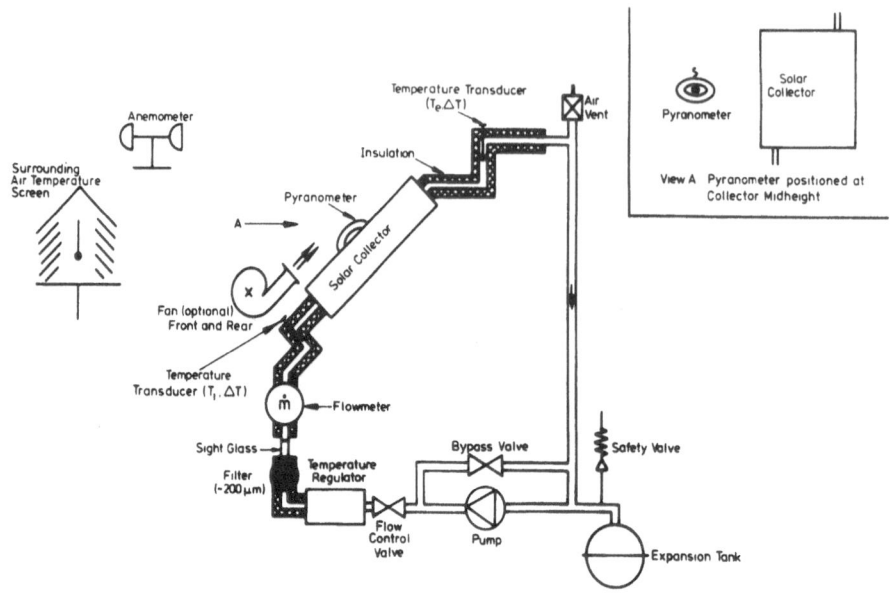

Fig. 4 - Schematic representation of a closed loop test installation for outdoor instantaneous efficiency measurements

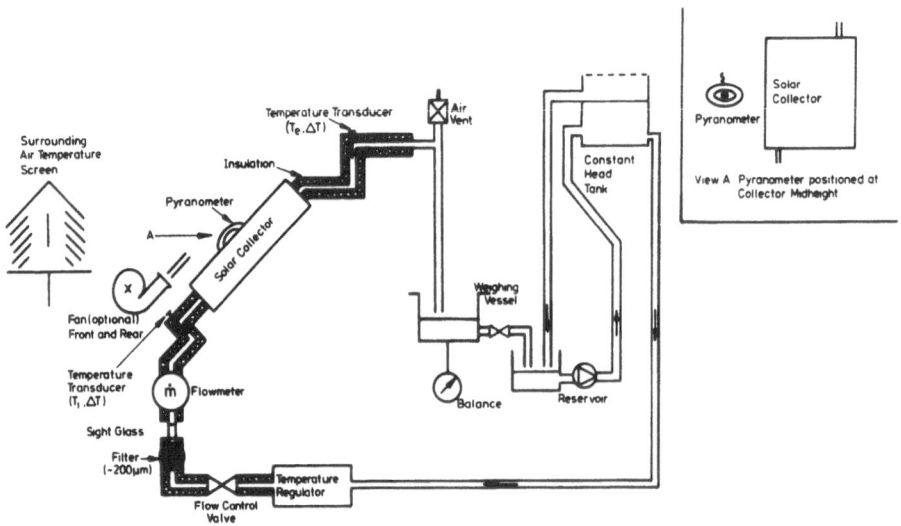

Fig. 5 - Schematic representation of an open loop test installation for outdoor instantaneous efficiency measurements

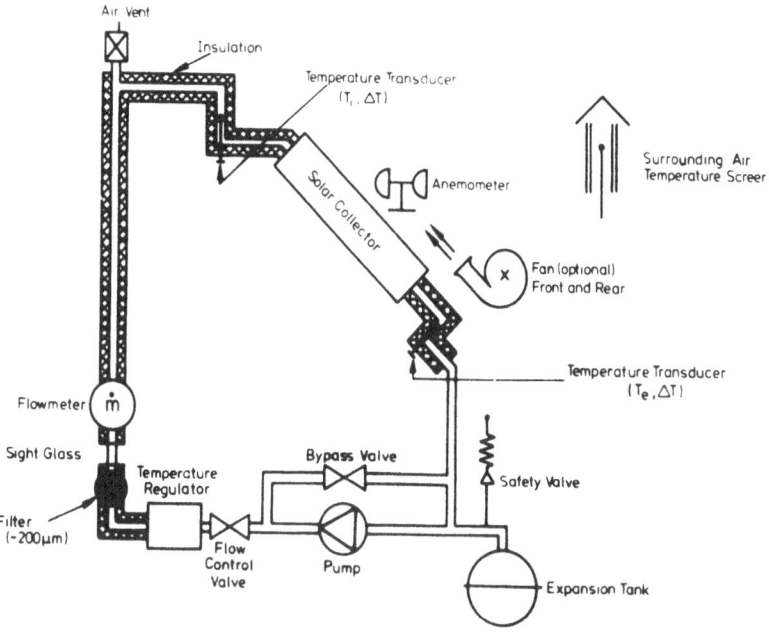

Fig. 6 - Schematic representation of a test installation for indoor
   thermal loss measurements

6.3.2 Mixing

The controlled mixing of a high temperature fluid source and a low temperature
fluid source may be used to provide temperature regulation. This method has
the advantage of enabling rapid changes from one collector inlet temperature to
another.

6.3.3 Inline Temperature Control

An inline electric heater may be used for providing temperature control of the
fluid close to the inlet to the collector. Boiling of the fluid on the surface of the
electric element may occur if the power dissipation in the element is high and
hence only small temperature corrections should be made in this way.
An electric heater with feedback control may be found to be particularly useful
on test facilities that have long pipe runs between the primary temperature regulation vessel and the collector.

7. INSTRUMENTATION FOR USE IN COLLECTOR TESTING

The test methods identified for the determination of the thermal performance of
solar collectors require the measurement of 6 primary experimental variables.
These are:
a)   solar radiation
b)   the temperature of the heat transfer fluid at the collector inlet (fluid inlet
     temperature)
c)   the temperature difference of the heat transfer fluid between the collector
     outlet and the collector inlet
d)   the ambient air temperature
e)   the flow rate of the collector heat transfer fluid
f)   the surrounding air speed.

In addition, two other variables are required to be determined; these are:
g)   the angle of incidence of the direct solar radiation at the collector aperture
h)   the pressure drop across the collector.

In the light of the experience gained during the European collaborative collector
testing programme, recommendations are made in this section with regard to the
measurement of these variables. In addition to these recommendations, consideration should be given to the World Meteorological Organisation guidelines (Ref.3).

7.1 Solar Radiation

7.1.1 Transducer

7.1.1.1   The transducer for the measurement of solar irradiance should be a
pyranometer (some types of which are termed solarimeter) classified by the

World Meteorological Organisation as 1st or 2nd class. These classifications require that the pyranometer have the characteristics given in Table 1.

TABLE 7.1 - Classification of pyranometers

	1st Class	2nd Class
Sensitivity (W.m^{-2})	$< \pm 1$	$< \pm 5$
Stability (maximum % change per year)	± 1	± 2
Temperature (maximum % error due to changes of ambient temperature)	± 1	± 2
Selectivity (maximum % error due to departure from assumed spectral response)	± 1	± 2
Linearity (maximum % error due to non-linearity not accounted for)	± 1	± 2
Time constant (maximum)	25 sec	1 minute
Cosine response (maximum % deviation from that assumed, taken at sun elevation of 10° on a clear day)	± 3	$\pm 5 - 7$
Azimuth response (maximum % deviation from that assumed taken on a clear day)	± 3	$\pm 5 - 7$

7.1.1.2   The pyranometer should be mounted such that the detector is located in the plane of the collector aperture and to one side of the collector, at the mid height. The mounting position is shown pictorially in Fig. 7.

Fig. 7 - Solar radiation measurement instruments

7.1.1.3   If the detector of the pyranometer is asymmetric the orientation should be such that the cables from the pyranometer point North when the pyranometer faces South.

7.1.1.4   The cables should be shielded from direct solar radiation and screened from electromagnetic interference.

7.1.1.5   Prior to collector testing the glass dome of the pyranometer should be cleaned and checked to ensure that it is free of condensation.

7.1.2   Total Incident Solar Radiation Measurements

The instrument installed as described in Section 7.1.1 will be suitable for determining the total solar irradiance in the plane of the collector.

7.1.3   Diffuse Solar Radiation Measurement

7.1.3.1   The diffuse solar irradiance in the plane of the collector may be determined by shading the detector of the pyranometer from the direct solar radiation in the solid angle subtended by the sun's disk. This may be accomplished by using an opaque disk of approximately 50 millimetres diameter attached to a slender rod on a direct line between the detector and the sun. The disk should be held at a distance such that just the dome of the pyranometer is shaded.

7.1.3.2   The same pyranometer may be used for the diffuse radiation measurements as for the total incident solar radiation mèasurements. The measurements should then be taken prior to a test period and after a test period.

7.1.3.3   Alternatively, a second pyranometer may be mounted in a similar manner as that described in Section 7.1.1, but fitted with a shade ring of a suitable width such that the detector is shielded from direct solar radiation during the period in which tests are performed. A typical pyranometer and shade ring is shown pictorially in Fig. 7.

7.1.3.4   Diffuse radiation may be determined from the difference between global radiation measurements made with a pyranometer on a horizontal surface and direct radiation measurements made with a pyrheliometer pointed at the sun.

7.1.4   Calibration and Accuracy

7.1.4.1   For collector testing the repeatability of the measurements may be of more importance than the absolute accuracy of the measurements. It is therefore recommended that the pyranometer be recalibrated each year by the manufacturer or by a national meteorological office. In the latter case the pyranometer should be recalibrated upon receipt from the manufacturer and any discrepancies between the two calibrations noted.

7.1.4.2   The output of a pyranometer is affected by the temperature of the pyranometer body. The calibration constant of the Kipp and Zonen pyranometer, for example, may vary by approximately 0.1 to 0.2% K^{-1}. Therefore the pyranometer should be mounted such as to ensure a free passage of ambient air over it in order that the calibration can be related to the ambient air temperature.

7.1.4.3   The variation of pyranometer calibration with tilt angle is thought to be small (< 1%). However, in order to facilitate intercomparison of test results measured in Europe, it is recommended that all measurements be made with the instrumented tilted to 45° ± 5°.

7.1.4.4   The zero offset should be checked by placing a light tight box over the pyranometer. A pyranometer will not always give a zero reading outdoors at night because of the low values of effective sky temperature which sometimes occur. Low sky temperatures depress the zero reading of pyranometers.

7.1.5   Data Recording

7.1.5.1   The output of the pyranometer should be connected to a calibrated, continuous trace, high impedance chart recorder, which can produce an accuracy of ± 1% of the reading.

7.1.5.2   A saving in manpower during testing and data analysis can be achieved by the use of digital data logging systems. A chart recorder for reference purposes is also recommended, however, as this enables a permanent record to be kept of the radiation intensity level over the test period and deviations from the steady state condition will be apparent.

7.1.5.3   The instrument data recording system and all joins in connecting leads should be checked for thermally induced voltages or interference. The check should be made in the vicinity in which the system will be operating in order to investigate interference.

7.1.5.4   If a chart recorder is used the measured value should be the mean value indicated on the trace. This mean value may be obtained by visual inspection of the trace or by manually digitizing the trace to obtain the mean.

7.1.5.5   An integrator may be used to obtain a mean instantaneous value, provided that a continuous trace is available to ensure quasi steady state conditions. The integrator should have an accuracy of ± 1% for the readings taken.

7.2   Temperature Measurement

Three temperature measurements are required for solar collector testing. These are the fluid inlet temperature, the fluid temperature difference between the outlet and inlet of the collector and the ambient air temperature. The required accuracy and the environment for these measurements are different and hence

the transducer and associated equipment required may be different.
General comments upon various transducers for temperature measurement are given below.

7.2.1   Absolute Temperature

7.1.1.1   Mercury in Glass Thermometers

7.2.1.1.1 Mercury in glass thermometers are available graduated at each 0.1°C or 0.05°C. The range of such thermometers is usually small, typically 30°C and hence more than one thermometer may be required for collector testing.

7.2.1.1.2 Most thermometers are calibrated for total immersion in the fluid and hence errors will be introduced if the thermometer is used when only partially immersed in the fluid stream. Good thermal contact with the fluid stream is essential.

7.2.1.1.3 The thermometer should be regularly calibrated over its operating range against a reference thermometer.

7.2.1.1.4 One disadvantage of mercury in glass thermometers is that a "hard copy" of the measurement is not normally produced, so frequent visual monitoring is necessary.

7.2.1.2   Thermocouples

7.2.1.2.1 Thermocouples are widely used in the European Community for solar collector testing. A thermocouple is a junction between two dissimilar metals which when joined together produce a voltage which is a nonlinear function of temperature (T). This function may be approximated by:

$$E = a_0 + a_1 T + a_2 T^2 \qquad\qquad (7.2\text{-}1)$$

Suitable types of thermocouple junction are Copper/Constantan, Iron/Constantan and Chrome/Alumel.

7.2.1.2.2 As the voltage generated by thermocouple junctions may be affected by the methods of joining the two metals, each thermocouple should be calibrated against a reference thermometer over the range of temperatures for which it is to be used and a calibration curve obtained.

7.2.1.2.3 The need for a complete recalibration at frequent intervals may be avoided by immersing the thermocouples in a suitable insulating material so that the effects of strain hardening and oxidation are reduced. Annual calibration checks at a few selected points on the calibration curve should be sufficient to verify the calibration.

7.2.1.2.4 The connections between a thermocouple and the readout device will produce unwanted voltages unless compensated for either thermally or electrically as shown schematically in Figs. 8 and 9, respectively.

7.2.1.2.5 All connecting leads should be screened from electromagnetic interference, and suitable connections should be used to minimise the risk of unwanted voltages.

7.2.1.2.6 The influence of heat leakage along connecting leads should not produce detectable errors.

7.2.1.3   Platinum Resistance Thermometers

7.2.1.3.1 Platinum resistance thermometers (PRT's) have an especially stable and reproducible resistance temperature relationship. For this reason PRT's offer an accurate method of measuring temperatures.

7.2.1.3.2 The temperature resistance relationship for a platinum resistance thermometer is given by:

$$\frac{R_T}{R_0} = 1 + a_1 \mid T - a_2 (T/100 - 1) (T/100) \mid \qquad (7.2\text{-}2)$$

where:
R_T is the thermometer resistance at temperature T,
R_0 is the thermometer resistance at $0°C$, and
a_1 and a_2 are constants for the individual thermometer.

7.2.1.3.3 Calibration of the thermometer, associated connecting leads, bridge circuit and readout device should be performed approximately once a year. Calibration should be over the range of $0°C$ to $100°C$ to obtain either a calibration curve or determine the constants a_1 and a_2 in equation 7.2-2.

7.2.1.4   Others

The use of temperature transducers other than those detailed in Sections 7.2.1.1, 7.2.1.2 and 7.2.1.3 cannot be recommended as they have not been widely used in the European Community for the testing of solar collectors and hence not fully evaluated.

7.2.2   Temperature Difference

The temperature difference of the heat transfer fluid between the outlet and inlet of the solar collector is required to be measured to an accuracy of within ± 0.1 K. Two types of transducer are recommended:

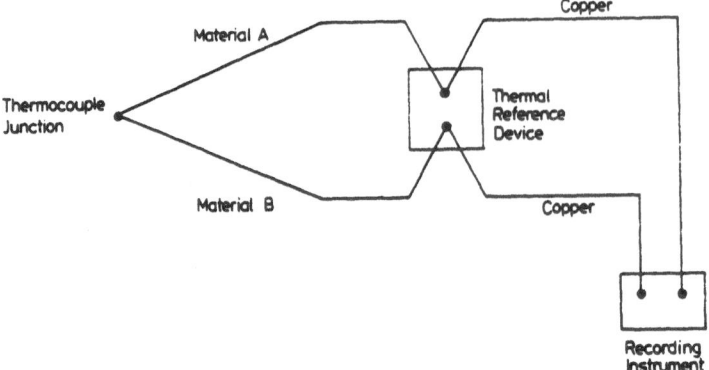

Fig. 8 - Schematic representation of the arrangement for thermal compensation
of the junctions between the recording instrument and thermocouple

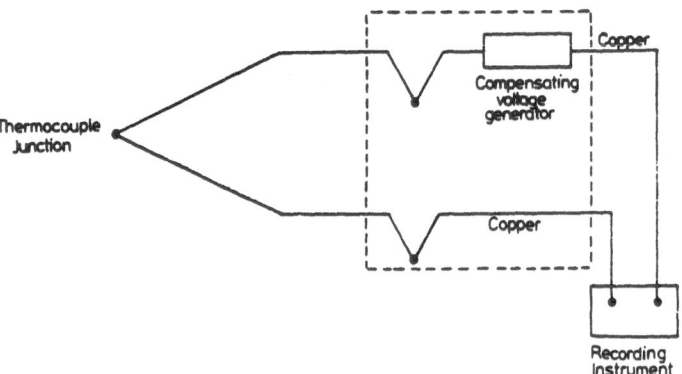

Fig. 9 - Schematic representation of the arrangement for electronic compensation
of the junctions between the recording instrument and thermocouple

a) thermopiles (differential thermocouples)
b) a matched pair of platinum resistance thermometers.

7.2.2.1   Thermopiles

7.2.2.1.1 Thermocouples detailed in Section 7.2.1.2 may be connected in series
to provide a higher output voltage and arranged differentially as shown in Fig.10,
to measure temperature difference. In order to provide sufficient resolution, the
minimum number of thermocouples recommended for such an arrangement is 6
(3 by 3).

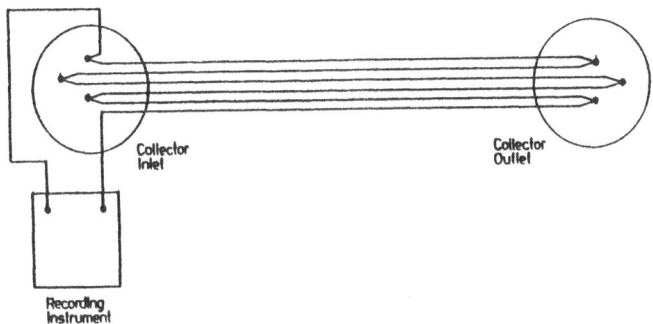

Collector
Inlet

Collector
Outlet

Recording
Instrument

Fig. 10 - Schematic representation of a thermopile arrangement for
measuring the temperature difference across the solar collector.

7.2.2.1.2 The thermocouples in a thermopile should be electrically insulated
from each other.

7.2.2.1.3 A zero check for the differential thermopile arrangement should be
made by placing both thermopiles in a fluid bath at temperatures in the normal
test range. Alternatively the check may be performed in the collector test configuration by replacing the collector with a short piece of well-insulated piping.
Depending on the quality of the thermocouple and of the recording instrumentation it may be necessary to derive a correction curve as a function of the absolute fluid temperature and/or the ambient temperature.

7.2.2.1.4 For temperature difference measurements it is important to know the
slope of dE/dT for the different absolute temperatures since from eq.7.2-1 the
following equation may be obtained:

$$\frac{dE}{dT} = a_1 + 2a_2 T \qquad\qquad (7.2\text{-}3)$$

i.e. the same voltage will represent a different temperature difference for different
values of absolute temperature. For example for Chromel/Alumel thermocouples

$$\left.\frac{dE}{dT}\right|_{0°C} = 40 \ \mu V.K.^{-1}$$

$$\left.\frac{dE}{dT}\right|_{70°C} = 41 \ \mu V.K.^{-1}$$

i.e. a 2.5% variation.

The thermopiles may be calibrated for temperature difference by placing one thermopile in a thermal reference and the other in a fluid at a higher temperature measured by a reference thermometer. The indicated temperature difference should be monitored by the data recording device to be used in collector testing. This test may be repeated for small temperature differences over a range of absolute temperatures so that the calibration curve for dE/dT may be obtained.

7.2.2.2   Platinum Resistance Thermometers

7.2.2.2.1 Platinum resistance thermometers (PRT's) as described in Section 7.2.1.3 may be arranged differentially to measure temperature differences.

7.2.2.2.2 A zero check for PRT's may be performed in a similar manner to that described in Section 7.2.2.1.3.

7.2.2.3   Simple Calibration Check

A calibration procedure is summarized in Section 7.7 which is intended to provide a simple overall check on the temperature difference and flow rate measuring instruments.

7.2.3   Heat Transfer Fluid Inlet Temperature (T_i)

7.2.3.1   The temperature of the heat transfer fluid at the collector inlet should be measured to an accuracy of within ± 0.1°C.

7.2.3.2   The temperature transducer used should be selected and calibrated in accordance with the recommendations of Section 7.2.1.

7.2.3.3   The transducer should be positioned at no more than 200 millimetres from the collector inlet. If it is necessary to position it outside this range, a test should be made to verify that the fluid temperature measurement is not affected.

7.2.3.4   To ensure mixing of the fluid at the position of temperature measurement, a bend in the pipework or an orifice should be placed upstream of the transducer and the transducer probe should point upstream (see Fig. 11).

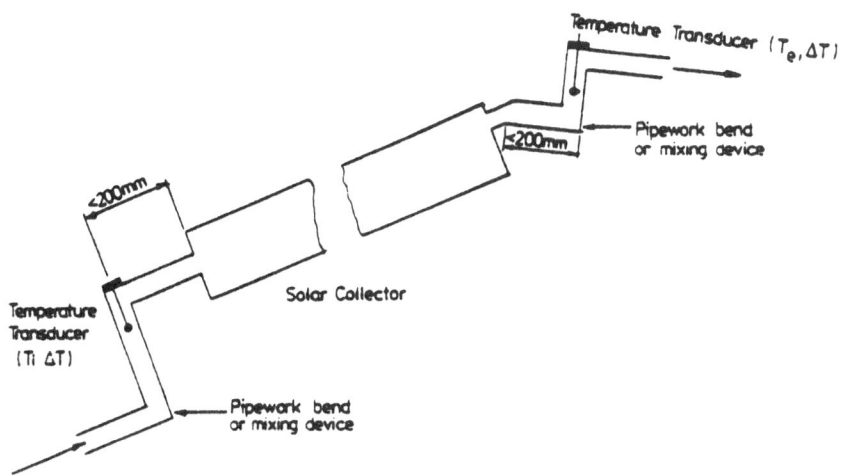

Fig. 11 - Schematic representation of the recommended transducer positions for measuring the heat transfer fluid temperature.

7.2.3.5   The transducer should be coupled to a calibrated chart recorder to provide a continuous permanent record. Alternatively a data logging system or precision voltmeter should be used and the measurements sampled/recorded at frequent intervals.

7.2.4   Heat Transfer Fluid Temperature Difference (ΔT)

7.2.4.1   The temperature difference of the heat transfer fluid between the collector outlet and the collector inlet should be measured within an accuracy of 0.1 K. As temperature differences are to be measured in the range of only 1.5 K to 15 K achieving the required accuracy is important.

7.2.4.2   The temperature transducers used should be selected and calibrated in accordance with the recommendations of Section 7.2.2.

7.2.4.3   The transducers should be positioned at no more than 200 millimetres from the collector inlet and outlets. If it is necessary to position one of the transducers outside this range tests should be made to verify that the fluid temperature difference measurement is not affected.

7.2.4.4   To ensure mixing of the fluid at the position of temperature measurement a bend in the pipework or an orifice should be placed upstream of each transducer and the transducer probes should point upstream (see Fig. 11).

7.2.4.5   The transducer should be coupled to a calibrated chart recorder to provide a continuous permanent record. Alternatively a data logging system or precision voltmeter should be used and the measurements sampled/recorded at frequent intervals.

7.2.5   Surrounding/Ambient Air Temperature (T_a)

7.2.5.1   The ambient or surrounding air temperature should be measured to an accuracy of within ± 0.5°C.

7.2.5.2   The temperature transducer used should be selected and calibrated in accordance with the recommendations of Section 7.2.1.

7.2.5.3   For outdoor measurements the transducer should be shaded from direct and reflected solar radiation by means of a white painted, well ventilated shelter, such as a meteorological screen, or by two concentric vertical metal pipes. The shelter should be placed at the mid-height of the collector but at least 1.2 m above the local ground surface to ensure that it is away from the influences of ground heating. The shelter should be positioned to one side of the collector within 10 m of it. A typical radiation screen for the transducer is shown in Fig. 12.

Fig. 12 - Radiation screen for the measurement of ambient air temperature

7.2.5.4   For indoor measurements the transducer should also be shielded in order to minimize radiation exchange. This may be achieved by placing the transducer inside two vertical concentric metal pipes. The transducer and shield should be positioned at the collector mid-height within 2 m of the collector.

7.2.5.5   If air is forced over the collector by a fan the air temperature monitored should be the mean of the forced air measured at the top and bottom of the collector.

7.2.5.6   A permanent record of the ambient air temperature during the tests should be made by connecting the transducer to a calibrated chart recorder. Alternatively a data logging system or precision voltmeter should be used and the measurements sampled/recorded at frequent intervals.

7.3   Fluid Flow Rate

7.3.1   Transducers

7.3.1.1   Variable Area Meters

7.3.1.1.1 Variable area meters are a simple, reliable and widely used flow rate meter. The flow rate is determined by the position of a float in a tapered glass tube.

7.3.1.1.2 The measurement is dependent upon the viscosity and density of the fluid which may vary considerably over the range of fluid temperatures and flow rates required for collector testing. Determination of the fluid flow rate to within ± 1% may therefore be difficult using this type of flow meter.

7.3.1.2   Magnetic Flow Meters

7.3.1.2.1 Magnetic flow meters measure the volume flow rate and hence the density of the fluid should be known over the operating temperature range of the fluid.

7.3.1.2.2 Magnetic flow meters are unaffected by the viscosity of the fluid and are capable of accuracies within ± 1%. The accuracy of this type of meter deteriorates as the flow rate decreases, therefore transducers with a low flow rate threshold are recommended.

7.3.1.3   Positive Displacement Meters

7.3.1.3.1 Positive displacement meters measure volume flow rate and hence the density of the fluid should be known over the operating temperature range of the fluid.

7.3.1.3.2 In some models the fluid displaces a piston as it flows through the meter and hence an oscillating pressure drop may be imposed.

7.3.1.3.3 Some types of positive displacement meter totalize the flow and hence are not recommended as variations in the flow rate about the mean value are not immediately apparent.

7.3.1.4 Turbine Meters

7.3.1.4.1 Turbine meters also measure the volume flowrate and hence the density of the fluid should be known over the temperature range of the fluid.

7.3.1.4.2 Turbine meters are widely used for collector testing within the European Community as accuracies of within ± 1% may be achieved providing the device is suitably calibrated over the flowrate range.

7.3.1.4.3 It is recommended that to prevent unwanted signals, special attention should be given to ensuring that the fluid is free of entrained air and contaminants.

7.3.1.5 Metering Pumps

Metering pumps provide an accurate method of producing and measuring the fluid flowrate for collector testing. The principal disadvantage of metering pumps is their high cost.

7.3.1.6 Weighing Devices

A simple and accurate method of determining the fluid flowrate is to divert the fluid downstream of the collector into a vessel and measure the mass of fluid delivered during a measured time period.

7.3.1.7 Calorimetric Devices

7.3.1.7.1 Systems are under development in the European Community for measuring the calorimetric flowrate directly, i.e. the product of mass flowrate, specific heat and the temperature rise of the fluid.
The advantage of these systems is that the specific heat and density of the fluid used during the test do not need to be known explicitly. Only a little experience with such devices has been obtained to date.

7.3.1.7.2 Such devices may consist of an inline electrical heater with variable input power situated downstream of the collector in the fluid loop. Accurate temperature transducers immediately upstream and downstream of the heater give the measurement of the temperature rise of the fluid across the heater. A wattmeter measures the power input to the heater. The whole device is usually encased in approximately 50 millimetres of insulation.

7.3.1.7.3 The device is operated by varying the power input into the heater until the temperature rise across the heater is the same as that across the collector. If the heat loss from the device is negligible, the power input to the heater as measured by the wattmeter is equal to the useful power extracted by the collector. The device should be calibrated for heat losses as a function of absolute temperature.

Alternatively a constant power input to the device may be used and the useful power extracted by the collector determined by a comparison of the fluid temperature rises across the collector and heater.

7.3.2 Accuracy and Calibration

7.3.2.1 The mass flow rate of the heat transfer fluid should be measured to an accuracy of within ± 1%.

7.3.2.2 It is recommended that the flow meter be calibrated before each series of collector tests using the heat transfer fluid specified for the testing of the collector.

7.3.2.3 The flow meter should be calibrated over the range of fluid flow rates and temperatures to be used during collector testing.

7.3.2.4 The recommended method of calibration is to collect the fluid downstream of the flow rate transducer over a period of approximately 300 seconds. The mass of the fluid collected should then be accurately determined and the average flow rate over the period obtained. It is important that during this period the flow rate is maintained constant to within ± 1%. This may be achieved by means of a constant head device.

7.3.2.5 A calibration procedure is summarized in Section 7.7 which is intended to provide a simple overall check on the flow rate and temperature difference measuring instruments.

7.3.2.6 The properties of water for use in solar collector testing when the heat transfer fluid is water are given in Table 7.2.

7.3.3 Data Recording

Data recording equipment should provide a continuous accurate hard copy or display of the fluid flowrate. The resolution should be such that deviations of ± 1% in the flowrate may be observed.
Alternatively, if it can be verified that the test facility pump will produce a reproducible flowrate constant to ± 1%, an initial measurement of the flowrate for a test may be sufficient.

TABLE 7.2 - Properties of water for analysis of collector testing results

Temperature T°C	Density $\rho(\text{kg } l^{-1})$	Specific Heat $c_f(\text{kJ kg}^{-1}\text{K}^{-1}$	Kinematic Viscosity $\nu(\text{cS})$	Dynamic Viscosity $\mu(\text{N s m}^{-2})$
5	0.9999	4.204	1.501	1501
10	0.997	4.193	1.300	1300
15	0.990	4.186	1.137	1136
20	0.9982	4.183	1.004	1002
25	0.9970	4.181	0.8927	890
30	0.9956	4.179	0.8005	797
35	0.9940	4.178	0.7223	718
40	0.9922	4.179	0.6561	651
45	0.9902	4.181	0.5999	594
50	0.9881	4.182	0.5505	544
55	0.9852	4.183	0.5085	501
60	0.9833	4.185	0.4709	463
65	0.9804	4.188	0.4386	430
70	0.9775	4.191	0.4092	400
75	0.9747	4.194	0.3837	374
80	0.9718	4.198	0.3612	351
85	0.9690	4.203	0.3406	330
90	0.9653	4.208	0.3222	311
95	0.9615	4.213	0.3058	294

7.4   Surrounding Air Speed

7.4.1   Considerations

7.4.1.1   The speed of the surrounding air in the vicinity of the collector should be measured to an accuracy of within $\pm$ 0.5 m.s^{-1} for both indoor and outdoor testing.

7.4.1.2   The effect of the surrounding air speed is to increase the heat transfer from the aperture cover to the environment.

7.4.1.3   The significance of the direction of movement of the surrounding air has not been well established. Until some better understanding becomes available it may be considered sufficient during outdoor testing to correlate results with the surrounding air speed only.

7.4.1.4   Because the direction is known to affect the collector heat losses, testing under conditions of forced air flow may be advantageous in order to permit intercomparison of results. The air should be directed parallel to the collector, and the forced air speed measured, not the meteorological wind speed.

7.4.1.5   The anemometer used for collector testing should be capable of giving reliable measurements over the range of 0.2 to 10 m.s.$^{-1}$.

7.4.2 Transducers

7.4.2.1 Cup Anemometers

A cup anemometer is a device for measuring the total horizontal component of air speed and is suitable for outdoor collector testing when no forced air movement is employed.

A cup anemometer does not indicate the direction of the air movement.

7.4.2.2 Vane Anemometer (Air Meters)

A vane anemometer consists of a number of vanes supported on short radial arms and mounted on a spindle. It is a directional transducer which measures the air speed in the direction of the spindle axis.

7.4.2.3 Hot Wire/Hot Thermistor Anemometers

These instruments contain a hot element which is cooled by the air. The resistance of the element changes with temperature and hence with air speed. Some models are suitable for outdoor and indoor testing. Hand held devices are especially suited for determining the air speed at various positions around the collector. Most models are directional transducers.

7.4.2.4 Vortex Anemometers

Vortex anemometers which sense vortex formation frequencies created by the air flowing past an obstruction may be used for measuring the air speed both indoors and outdoors. Most models are directional transducers.

7.4.2.5 Pressure Tubes (Pilot Tubes)

Pressure tube anemometers are suitable for the measurement of air speed both indoors and outdoors. The pressure tube is a directional transducer.

7.4.3 Measurement Position

7.4.3.1   All outdoor measurements should be made within a distance of 10 metres of the collector.

For outdoor testing when no forced air movement is employed the air speed may be measured using either a directional or non-directional transducer. If directional measurements are made the horizontal component of the air speed across the collector should be recorded.

7.4.3.2   For outdoor and indoor testing under conditions of forced air movement the air speed may vary over the collector area. A series of measurements should therefore be taken at 50 mm in front of the collector aperture at positions equispaced over the collector area. An average measurement should then be determined.

7.4.3.3   Non-directional measurements may be made using a directional transducer if a vane is used to keep the transducer pointing in the direction of the air movement.

7.4.4   Measurement Frequency

7.4.4.1   Under natural outdoor conditions the surrounding air speed is seldom constant and gusting frequently occurs. The measurement of an average air speed is therefore required during the testing period. This may be obtained either by an arithmetic average of sampled values or by a time integration over the test period.

7.4.4.2   Under conditions of forced air movements it may be necessary to obscure part of the collector aperture from solar radiation in order to take sample measurements of the air speed.
For this procedure the transducer and its mounting should obscure less than 5% of the collector aperture for less than 10% of the test period.

7.4.5   Calibration and Accuracy

7.4.5.1   The air speed should be measured to within ± 0.5 m.s.$^{-1}$.

7.4.5.2   The anemometer should be recalibrated at yearly intervals.

7.4.6   Data Recording

7.4.6.1   Anemometers which produce an electrical output may be coupled to a chart recorder or data logging system to provide a permanent record of the air speed and/or direction.

7.4.6.2   The use of counters coupled to anemometers will provide an integrating capability. The total air run during the test (measured in metres) may be calculated and the mean air speed determined by timing the test period. These devices however give no indication of the variable nature of the air speed.

7.5   Angle of Incidence of Direct Solar Radiation

7.5.1   The angle of incidence of direct solar radiation at the collector plane should be determined to ensure that it is less than 40° during the collector test period.
The angle of incidence may be determined either by measurement or by calculation.

7.5.2   The angle of incidence may be determined by calculation knowing the time of the collector test, the collector tilt angle, the collector azimuth angle and the geographical position of the test site. (Refer to Ref. 4).

7.5.3   A device for measuring the angle of incidence can be produced by mounting a pointer normal to a flat plate. Graduated concentric rings may then be used to measure the length of the shadow cast by the pointer and hence the incidence angle. The device should be positioned in the plane of the collector at one side. A typical measuring device is shown in Fig. 13.

Fig. 13 - Angle of Incidence Indicator.

7.6   Collector Fluid Pressure Drop

7.6.1   Considerations

Generally the pressure drop across a collector is small but some types of collector have a significant pressure drop and hence the pressure drop acros a solar collector should be determined.
The fluid pressure drop between the collector inlet and outlet should be determined by means of a static pressure tapping close to the collector inlet and outlet.
The configuration of the static pressure hole and the method by which it is produced may influence the accuracy of the measurement more than the type of transducer used.

7.6.2   Transducers

7.6.2.1   Manometers

Liquid manometers are suitable for the measurement of pressure difference. Consideration should be given to ensuring that the density of the fluid in the manometer is accurately known and that the height of the fluid columns may be determined to a sufficiently good resolution. For example inclining the manometer may enhance the resolution.

7.6.2.2 Differential Pressure Transducers

Differential pressure transducers are suitable for the measurement of pressure difference. As these devices normally provide an electrical output they are suitable for use with data logging systems and with normal data recording equipment.

7.6.3 Static Pressure Orifice

7.6.3.1 The static pressure orifice should be installed normal to the flow. (An orifice inclined towards the flow will measure a higher pressure.)

7.6.3.2 The orifice size should be small. A hole diameter of 1 mm is normal.

7.6.3.3 The orifice edge should be square and free from burrs. Failure to remove the burrs may result in pressure measurement errors.

7.6.3.4 A smooth straight piece of pipe should be incorporated before the pressure tappings at the collector inlet and outlet. A straight pipe length of 10 pipe diameters is normal.

7.7 Overall Calibration Error

7.7.1 Summary

In any collector testing installation perhaps the most difficult variables to measure accurately are the fluid flow rate and the temperature rise or drop of the fluid as it passes through the collector. Calibration of the transducers and instrumentation which measure these parameters is often both time consuming and expensive. The test method presented here is intended to provide a simple overall check on the flow rate and temperature difference measuring instruments used for testing the thermal performance of a solar collector. The procedure may be used to check outdoor test loops, indoor heat loss rate facilities and test loops in solar simulators.

7.7.2 Apparatus

The equipment required in addition to that already available in the test facility consists of an insulated electrical water heater in a small water vessel, a watt meter, and a means of varying the electrical supply to the water heater (e.g. a variable transformer).
The heater should be installed in place of a collector in the test facility and the pipe fittings insulated in the same way as for collector testing. The heater unit should be shielded from solar radiation if necessary. The flow rate and temperature measuring devices should be installed in their normal positions in the test loop.
The power supply to the heater should be stabilised for best results.
The watt meter should be calibrated to an accuracy of within ± 1%.

7.7.3 Measurements

The following steady state measurements should be taken for each calibration point.

(i)   the fluid temperature at the inlet to the heater unit;
(ii)   the fluid temperature difference across the heater unit (ΔT);
(iii)   the fluid flowrate;
(iv)   the ambient or surrounding air temperature;
(v)   the power input to the heater.

Procedure

With zero input power to the heater unit, calibration measurements should be made for the following approximate values of the temperature difference between the mean fluid temperature in the heater and ambient air temperature

$$(T_m - T_a) = 0, \ 15, \ 30, \ 45, \ 60 \ K$$

The procedure detailed above should be repeated with electrical input powers of 500, 1000 and 1500 watts.

Presentation of the Results

The errors in the facility instrumentation calibration may be determined for each calibration point using the following expression:

$$\text{Error} = \frac{\dot{m}c_f\Delta T - (P - \dot{Q}_\varrho)}{(P - \dot{Q}_\varrho)}$$

where:

P = electrical input power (watts)
$\dot{Q}_\varrho$ = rate of heat loss from the heater unit (watts)
   = $\dot{m}c_f\Delta T$ for $P = 0$

The results may be presented graphically in the form of variation of error with ΔT and variation of error with $(T_m - T_a)$.

7.8 Others

In addition to the measurements and the instrumentation described in previous paragraphs the following measurements may also provide useful information regarding the conditions during the test period.

7.8.1 Longwave Radiation

The variations in longwave radiation from the sky and ground may have an effect on the thermal performance of a solar collector.
The longwave irradiance in the collector plane may be measured using a Pyrgeometer. A Pyrradiometer may also be used and the longwave irradiance determined by subtraction of the solar irradiance measured using a pyranometer from the

total irradiance measured using the pyrradiometer.

7.8.2 Humidity

The hymidity of the air in the vicinity of the collector may be measured using a hygrometer or determined using a combination of wet and dry bulb thermometers.

8. TEST METHODS

8.1 Scope

The instantaneous efficiency in various working conditions is considered to be the most suitable parameter to characterize the thermal performance of a collector. Three test methods have been identified as being suitable for the determination of the instantaneous efficiency of a collector in the European environment. These are:

(i)   the determination of an outdoor instantaneous efficiency curve of a collector in quasi steady state conditions;
(ii) the determination of an instantaneous efficiency curve of a collector from combined indoor-outdoor testing;
(iii) the determination of the outdoor instantaneous efficiency of a collector under specified reference test conditions.

The above methods have in common the same experimental procedure in order that the results should be essentially the same. The reproducibility of the environmental conditions during outdoor testing may be poor and hence the conditions during testing have to be restricted in order that the effects on the collector performance of surrounding air speed, solar irradiance levels, angle of incidence and ambient temperature are within a known band.

It is these restrictions on the conditions during testing that are different for each method. A method may therefore be chosen to suit the purpose of the test, the local climate or the time available for testing.

Many of the specified test conditions are common for the 3 methods with the notable exception that in the second method of testing given above the conditions of the outdoor environment are in part replaced by performing some tests indoors. The overall heat loss rate and heat loss coefficient of the collector determined indoors with this method are important characteristics of the collector and should be determined for all collectors tested.

In addition, test methods are presented here for determining the following collector characteristics:

(i)   an effective thermal capacity of a collector;
(ii) the time constant of a collector;
(iii) the influence on the collector performance of the angle of incidence of direct solar radiation at the collector aperture;
(iv) the pressure drop across a collector.

These collector characteristics are considered to be complementary to the instantaneous efficiency of the collector.

8.2 Determination of an Outdoor Instantaneous Efficiency Curve of a Collector in Quasi Steady-State Conditions

Summary

The method is near to the well-known National Bureau of Standards (NBS) test method (ref. 5), ASHRAE 93.77 standard (ref. 6), and the Italian UNI standard (ref. 10).
The objective is to determine an instantaneous efficiency curve outdoors in the conditions and environment for which the collector is designed.
Test conditions have to be restricted in order that the influence on the collector performance of wind, solar radiation intensity, angle of incidence of direct solar radiation, and ambient temperature, remain within a known band as these effects are not taken into consideration in the computation of the instantaneous efficiency. As a result of these restrictions, tests in most of Central and Northern Europe often take several months to complete because suitable climatic conditions occur only rarely.

8.2.2 Test Installation

Test installations suitable for the determination of the outdoor instantaneous efficiency of a collector in quasi steady-state conditions are shown in Figs. 4 and 5.

8.2.3 Measurements

The following measurements should be obtained in accordance with the recommendations of Section 7 and recorded on the format sheet given in Section 11:
(i)   the total solar irradiance;
(ii)   the diffuse solar irradiance;
(iii) the temperature of the heat transfer fluid at the collector inlet;
(iv)   the temperature difference of the fluid between the outlet and inlet of the collector;
(v)   the surrounding air temperature;
(vi)   the flow rate of the heat transfer fluid;
(vii) the surrounding air speed;
(viii) the angle of incidence of direct solar radiation.

8.2.4 Conditioning of the Collector

8.2.4.1   The collector should be visually inspected and any damage recorded.

8.2.4.2   Before each day's testing period, the heat transfer fluid should be circulated at approximately 80°C for 30 minutes to expell moisture from the collector components.

8.2.4.3   The fluid should be inspected for entrained air or particles by means of the transparent tube built into the fluid loop pipework.

8.2.4.4   The collector should be vented of trapped air by means of an air valve if fitted or by circulating the fluid at a high flow rate as necessary.

8.2.4.5   The collector aperture cover plate should be thoroughly cleaned.

8.2.4.6   In order to obtain a quasi steady-state the heat transfer fluid should be circulated through the collector for 15 minutes at the fluid temperature to be used for the series of test points. The collector should be illuminated by the sun during this period.

8.2.5   Test Conditions

8.2.5.1   The heat transfer fluid should flow from the bottom to the top of the collector.

8.2.5.2   The fluid flow rate should be 0.02 kg s^{-1} ± 10% per square metre of collector aperture area unless otherwise recommended by the collector or fluid manufacturer.

In some types of collector design the fluid flow may be close to the transition region between laminar and turbulent flow. This may result in instability of the internal heat transfer coefficient and a subsequent influence on the collector performance.

8.2.5.3   The temperature rise of the fluid during passage through the collector should be less than 15 K but greater than 1.5 K. The use of temperature difference measurements of less than 1.5 K is not recommended because of the problems of instrument accuracy.

8.2.5.4   The total solar irradiance at the plane of the collector should be greater than 600 W m^{-2}.

8.2.5.5   The angle of incidence of direct solar radiation at the collector aperture should be less than 40°.

8.2.6   Test Period

8.2.6.1   The test period for a data point should be 30 minutes. During this period the measurements to be obtained as detailed in Section 8.2.3 should be monitored and recorded.

8.2.6.2   A test period should be considered to be representative of quasi steady-state conditions if the following conditions are satisfied during the 30 minute test period:
(i)   the total solar irradiance does not vary by more than 50 W m^{-2};
(ii)   the temperature of the heat transfer fluid at the collector inlet does not vary by more than ± 0.1 K;
(iii)   the surrounding air temperature does not vary by more than ± 1 K;
(iv)   the heat transfer fluid mass flow rate does not vary by more than ± 1%;
(v)   the heat transfer fluid temperature difference between the outlet and inlet of the collector does not vary by more than 0.1 K.

8.2.6.3   To ensure that the data used is not affected by conditions prior to the test period, the data values recorded in the format sheet should be the integrated or averaged values of the measurements taken over the last 15 minutes of the 30 minute period.

8.2.7   Test Procedure

8.2.7.1   The solar collector is to be tested over a range of operating temperatures in conditions detailed in Section 8.2.5 to give an instantaneous efficiency curve as detailed in Section 8.2.8. In order that data points over the full range of the curve may be obtained, the range of fluid inlet temperatures used should be from ambient or below, up to the maximum temperature that the collector or test facility can withstand. With water as the heat transfer fluid a maximum inlet temperature of 90°C is usually sufficient.

8.2.7.2   It is important that a wide range of fluid inlet temperatures be used as the heat losses from the collector, which are dependent upon the mean fluid temperature, are an important factor in the thermal performance of the collector.

8.2.7.3   At least 4 fluid inlet temperatures should be used to give fluid temperatures equispaced over the collector range. One of the fluid inlet temperatures chosen should be within ± 2 K of the surrounding air temperature.

8.2.7.4   More than 1 data point should be obtained for each fluid inlet temperature and if conditions permit an equal number should be taken prior to solar noon and after solar noon.

8.2.7.5   A total of at least 16 data points should therefore be obtained.

8.2.8   Computation and Presentation of the Results

The instantaneous efficiency η, may be calculated from the following equation:

$$\eta = \frac{\dot{Q}_u}{A_a G}$$ (8.2-1)

where $\dot{Q}_u$ may be calculated from

$$\dot{Q}_u = \dot{m} \, c_f \, \Delta T$$ (8.2-2)

A value of c_f appropriate to the mean fluid temperature should be used.

If m is obtained from volumetric flowrate measurement then the densitw should be determined for the temperature of the fluid in the flow meter.
The instantaneous efficiency should be presented graphically as a function of the reduced temperature difference T*, using the format sheet given in Section 11

where

$$T^* = U_0 \left(\frac{T_m - T_a}{G} \right)$$ (8.2-3)

$$U_0 = 10 \text{ W m}^{-2} \text{ K}^{-1}$$

$$T_m = (T_i + \frac{\Delta T}{2})$$

Graphical presentation should be made by statistical curve fitting using the least square method as detailed in Section 10 to obtain an instantaneous efficiency curve of the form:

$$\eta = \eta_0 - a_1 T^* - a_2 (T^*)^2 \qquad (8.2\text{-}4)$$

Typical instantaneous efficiency curves for a single glazed collector, double glazed collector and a selective surface collector are shown in Figs. 14, 15 and 16 respectively.

The associated conditions present during testing should be recorded in the format sheet of Section 11. These are an important part of the test results because they provide a record of the conditions for which the instantaneous efficiency curve applies.

8.3 Determination of an Instantaneous Efficiency Curve of a Collector from Combined Indoor-Outdoor Testing

8.3.1 Summary

The instantaneous efficiency curve of a collector may be obtained by combining the instantaneous efficiency of the collector determined outdoors when the mean fluid temperature is equal to that of the surrounding air, with the measured heat losses from the collector determined indoors. The method is similar to the Bundesverband Solar Energie (BSE) method (Ref. 7).

The advantage of this method is that few outdoor measurements are required, enabling testing to be performed more quickly than full outdoor tests. The environmental conditions indoors are less variable than outdoors and hence the reproducibility of the heat loss measurements should be better.

An efficiency curve determined indoors may not be completely representative for outdoor collector performance and correlations between outdoor and combined indoor-outdoor determined efficiencies are still being evaluated. The influence of the collector internal heat transfer coefficient is also under investigation.

8.3.2 Test Installation

Test installations suitable for determining the outdoor efficiency when the mean fluid temperature is that of the surrounding air, are shown in Figs. 4 and 5. Figure 5 is representative of the test installation recommended by the BSE. A test installation suitable for determining the heat loss rate from a collector indoors is shown in Fig. 6.

8.3.3 Measurements

The measurements, which should be obtained in accordance with the recommendations of Section 7 and recorded in the format sheet given in Section 11, are those detailed in Section 8.2.3.

8.3.4 Conditioning of the Collector

The collector should be conditioned as recommended in Section 8.2.4.
During the determination of the collector indoor heat loss rate under still air conditions, the time for circulating the heat transfer fluid prior to commencement of tests at a given temperature (see Section 8.2.4.6) should be at least 60 minutes as the response of a collector under these conditions is slower.

8.3.5 Test Conditions

8.3.5.1 Conversion Factor (η_0) Measurements (outdoor)

8.3.5.1.1 The heat transfer fluid should flow from the bottom to the top of the collector.

8.3.5.1.2 The fluid flow rate should be 0.02 kg s^{-1} ± 10% per square metre of aperture area, unless otherwise recommended by the collector or fluid manufacturer.

8.3.5.1.3 The temperature rise of the fluid passing through the collector should be less than 15 K but greater than 1.5 K. The use of temperature difference measurements of less than 1.5 K is not recommended because of the problems of instrument accuracy.

8.3.5.1.4 The surrounding air speed should be greater than 4 m s^{-1} during the test to minimize the influence of the radiative heat sink temperature. This requirement may be met by the use of a fan blowing air from the bottom to the top of the collector.

8.3.5.1.5 The total solar irradiance at the plane of the collector should be greater than 600 W m^{-2}.

8.3.5.1.6 The angle of incidence of direct solar radiation at the collector aperture should be less than 40°.

8.3.5.2 Indoor Heat Loss Rate Measurements

8.3.5.2.1 In order that the temperature profile of the absorber should be as uniform as possible and similar to that present during conversion factor (η_0) measurements the flow direction of the heat transfer fluid should be from the top to the bottom of the collector. (In the case of collectors with symmetrical absorbers this may be achieved by inverting the collector and maintaining the same collector pipework.)

8.3.5.2.2 The fluid flow rate should be 0.02 kg s^{-1} ± 10% per square metre of collector aperture area, unless otherwise recommended by the collector or fluid manufacturer. The fluid flowrate should be identical to the flow rate used for the conversion factor (η_0) measurement. In some types of collector design the fluid flow may be close to the transition region between laminar and turbulent flow. This may result in an instability of the internal heat transfer coefficient and a subsequent influence on the collector performance.

8.3.5.2.3 The temperature difference of the fluid between the inlet and outlet of the collector should be less than 15 K but greater than 1.5 K.

8.3.5.2.4 The shortwave irradiance (wavelengths less than 4 μm) at the collector aperture should not exceed 1 W m^{-2}. (This may be confirmed by the use of a solarimeter with a zero check as described in Section 7.1.4.4).

8.3.5.2.5 The radiant temperature of the room walls should not vary from the surrounding air temperature by more than 3 K.
(Note: room temperatures are often more stable at night than during the day.)

8.3.5.2.6 The surrounding air temperature should be between 15°C and 25°C.

8.3.6   Test Period

The test period for outdoor and indoor measurements should be of the same duration and under the same quasi steady state conditions as detailed previously in Section 8.2.6.

8.3.7   Test Procedure

8.3.7.1   Conversion Factor (η_0) Measurements (outdoors)

Data points should be obtained by taking the measurements detailed in Section 8.3.3 under the conditions specified in Section 8.3.5.1.
At least four data points should be taken with the temperature difference between the mean temperature of the heat transfer fluid and that of the surrounding air less than or equal to 10 K.

8.3.7.2   Indoor Heat Loss Rate Measurements

Two sets of test results should be obtained. One set with natural convection only and a second set with the surrounding air speed greater than 4 m s^{-1}. Data points should be obtained by taking the following measurements under the conditions specified in Section 8.3.5.2:
(i)   the temperature of the heat transfer fluid at the collector inlet;
(ii)   the temperature difference of the fluid between the outlet and inlet of the collector;
(iii) the surrounding air temperature;
(iv)   the flow rate of the heat transfer fluid;
(v)   the surrounding air speed.

For each set of results 2 data points should be taken at or near the following fluid inlet temperatures: 30°C, 50°C, 70°C and 90°C.

8.3.8   Computation and Presentation of the Results

8.3.8.1   Indoor Heat Loss Rate

The following computation and presentation should be performed for both the natural convection and forced convection test results:

The indoor collector overall heat loss rate may be calculated from the following equation:

$$\dot{Q}_\ell = \dot{m} \, c_f \, (T_i - T_e) \qquad (8.3\text{-}1)$$

A value of c_f appropriate to the mean fluid temperature should be used.

If $\dot{m}$ is obtained from volumetric flow rate measurement then the density should be determined for the temperature of the fluid in the flow meter.

The indoor heat loss rate should be presented graphically as a function of $(T_m - T_a)$ using the format sheet of Section 11.

A collector heat loss coefficient may be calculated from the following equation:

$$U_m = \frac{Q_\ell}{A_a(T_m - T_a)} \qquad (8.3\text{-}2)$$

The heat loss coefficient should be presented graphically as a function of the surrounding air temperature in the form

$$U_m = a_0 + a_1 (T_m - T_a) \qquad (8.3\text{-}3)$$

Numerical values of a_0 and a_1 may be determined by the use of a least squares fit of the data as detailed in Section 10.

Typical overall heat loss rate and heat loss coefficient curves are shown in Figs. 17 and 18 respectively.

8.3.8.2   Conversion Factor

The conversion factor of a collector may be calculated from the following equation:

$$\eta_0 = \frac{\dot{m} \, c_f \Delta T}{A_a G} \qquad (8.3\text{-}4)$$

Equation 8.3-4 is only valid if the net heat loss rate is zero. If the surrounding air temperature deviates from the mean fluid temperature, so that heat loss occurs, then η_0 may be determined from the following equation:

$$\eta_0 = \frac{\dot{m} \, c_f \Delta T}{A_a G} + \frac{U(T_m - T_a)}{G} \qquad (8.3\text{-}5)$$

where U is the value of the heat loss coefficient at $T_m = T_a$ obtained from equation 8.3-3.

The arithmetic mean of the η_0 data points should be calculated from:

$$\bar{\eta}_0 = \frac{\Sigma(\eta_0)}{N} \qquad (8.3\text{-}6)$$

When N is the number of data points.

8.3.8.3 Instantaneous Efficiency

Instantaneous efficiency values may be derived from the equation

$$\eta = \bar{\eta}_0 - \frac{\dot{Q}_\ell}{A_a G} \qquad (8.3\text{-}7)$$

The instantaneous efficiency may then be presented as a function of the reduced temperature difference T*, using the values of $\dot{Q}_\ell$ determined from equation 8.3-1 and $G = 800$ W m^{-2}

and where $\quad T^* = U_0 \dfrac{(T_m - T_a)}{G} \qquad (8.3\text{-}8)$

$$U_0 = 10 \text{ W m}^{-2} \text{K}^{-1}$$

$$G = 800 \text{ W m}^{-2}$$

Instantaneous efficiency curves should be presented for measurements of $\dot{Q}_\ell$ made under natural and forced convection (2 curves) using the format sheet of Section 11. All associated test conditions should be recorded in the format sheets of Section 11. These are an important part of the test results because they provide a record of the conditions for which the instantaneous efficiency curve applies.

8.3.9 Determination of the Back and Side Heat Loss of a Collector

A back and side heat loss rate and heat loss coefficient may be determined in a similar manner to that described in Section 8.3.8.1 if the aperture of the collector is covered with insulation material of overall conductance less than 0.2 W m^{-2} K^{-1}. The measurements required are those detailed in Section 8.3.7.2. Knowledge of the back and side heat loss rate may be important when estimating outdoor thermal performance by means of mathematical models.

8.4 Determination of the Outdoor Instantaneous Efficiency of a Collector under Specified Reference Test Conditions

8.4.1 Summary

In this procedure a small number of well defined conditions are specified for outdoor collector testing in order to achieve good reproducibility of the test results.

The results obtained may then be used for the rating of solar collectors. Naturally it is not possible to study the full range of the collector operation with this type of test procedure. The method is similar to that recommended for outdoor testing by AFNOR (Ref. 8).

8.4.2 Test Installation

Test installations suitable for the determination of the outdoor instantaneous efficiency of a collector under specified quasi steady state reference conditions are shown schematically in Figs. 4 and 5.

8.4.3 Measurements

The measurements, which should be obtained in accordance with the recommendations of Section 7 and recorded in the format sheet given in Section 11, are those detailed in Section 8.2.3.

8.4.4 Conditioning of the Collector

The collector should be conditioned as recommended in Section 8.2.4.

8.4.5 Test Conditions

The following test conditions should apply for all test periods:

8.4.5.1   The heat transfer fluid should flow from the bottom to the top of the collector.

8.4.5.2   The fluid flowrate should be 0.02 kg s^{-1} ± 10% per square metre of collector aperture area unless otherwise recommended by the collector or fluid manufacturer.

In some types of collector design the fluid flow may be close to the transition region between laminar and turbulent flow. This may result in an instability of the internal heat transfer coefficient and a subsequent influence on the collector performance.

8.4.5.3   The speed of the surrounding air in the vicinity of the collector should be greater than 1.5 m s^{-1} but less than 5.5 m s^{-1}.

8.4.5.4   The surrounding air temperature should be less than 30°C but greater than 5°C.

8.4.5.5   The solar irradiance during the previous hour to the test period should be within 120 W m^{-2} of the mean value during the test period.

8.4.6 Test Period

The duration and measurement conditions of the test period should be the same as detailed previously in Section 8.2.6.

8.4.7 Test Procedure

8.4.7.1   The collector should be operated in the conditions detailed in Section 8.4.5 and 8.4.6, and the measurements detailed in Section 8.4.3 taken to obtain a data point.

8.4.7.2   At least two data points should be obtained for each of the following four specified reference test conditions and with the angle of incidence of direct solar radiation $\leqslant 40°$

a)	$T_i = T_a$	$G = 800 \pm 40 \text{ W m}^{-2}$
b)	$T_i = T_a + 20°C$	$G = 800 \pm 40 \text{ W m}^{-2}$
c)	$T_i = T_a + 40°C$	$G = 800 \pm 40 \text{ W m}^{-2}$
d)	$T_i = T_a + 60°C$	$G = 800 \pm 40 \text{ W m}^{-2}$

If possible an equal number of data points prior to solar noon and after solar noon should be obtained for each fluid inlet temperature.

8.4.8   Computation and Presentation of the Results

The instantaneous efficiency η, may be calculated from the following equation:

$$\eta = \frac{\dot{Q}_u}{A_a G} \tag{8.4-1}$$

where $\dot{Q}_u$ may be calculated from:

$$\dot{Q}_u = \dot{m} \, c_f \Delta T \tag{8.4-2}$$

A value of c_f appropriate to the mean fluid temperature should be used.
If $\dot{m}$ is obtained from volumetric flowrate measurement then the density should be determined for the temperature of the fluid in the flow meter.
The efficiency values should be tabulated along with associated test conditions in the format sheet of Section 11.
The mean values of the instantaneous efficiency should be plotted graphically as a function of the reduced temperature difference T^*, in the format sheet of Section 11, for comparison with instantaneous efficiency curves if available.

Where $$T^* = U_0 \frac{(T_m - T_a)}{G} \tag{8.4-3}$$

$$U_0 = 10 \text{ W m}^{-2} \text{K}^{-1}$$

$$T_m = (T_i + \Delta T/2)$$

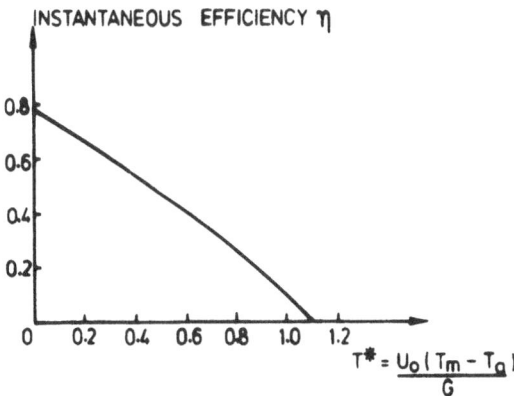

Fig. 14 - Typical thermal efficiency curve - single glazed collector

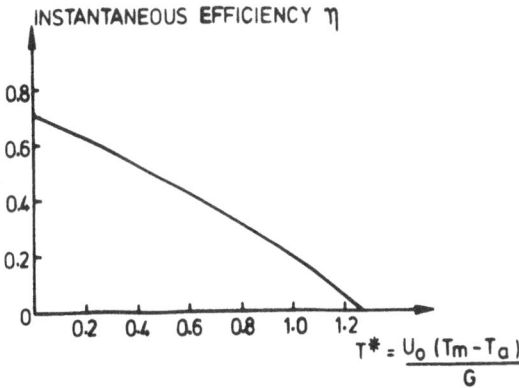

Fig. 15 - Typical thermal efficiency curve - double glazed collector

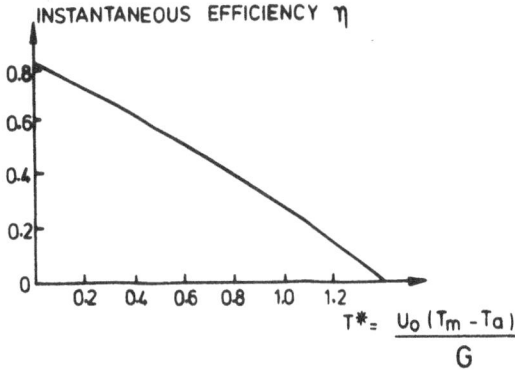

Fig. 16 - Typical thermal efficiency curve - single glazed collector
with a selective surface absorber

– 443 –

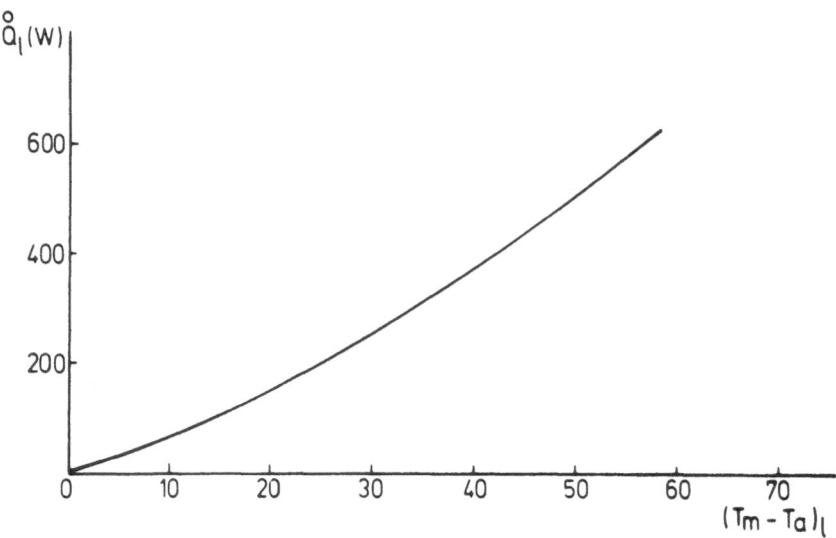

Fig. 17 - Typical overall heat loss rate curve of a collector
$(A_a = 2.0 \text{ m}^2)$

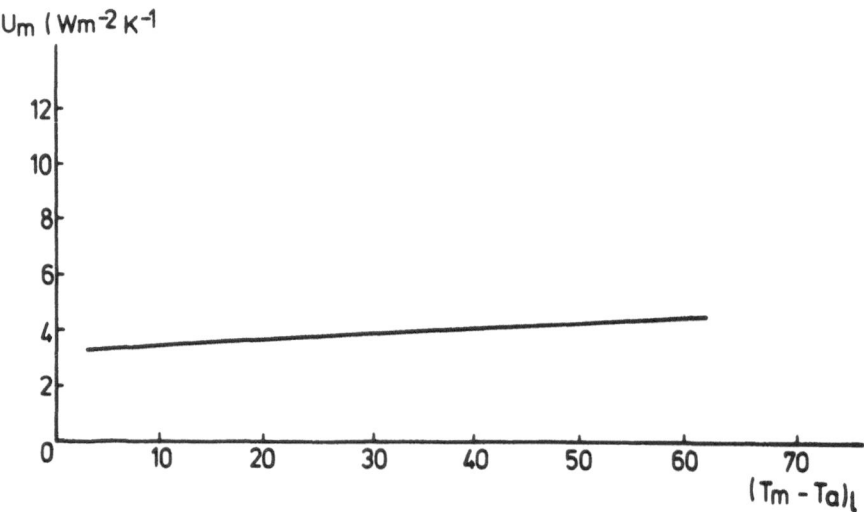

Fig. 18 - Typical heat loss coefficient curve of a collector

8.5 Complementary Test Methods

The following test methods are presented in this section to determine collector characteristics that are complementary to the instantaneous efficiency:

(i)   determination of an effective thermal capacity of a collector;
(ii)   determination of the time constant of a collector;
(iii)   determination of the influence on collector performance of the angle of incidence of direct solar radiation at the collector aperture;
(iv)   determination of the pressure drop across a collector.

8.5.1   Determination of an Effective Thermal Capacity of a Collector

8.5.1.1 Summary

The effective thermal capacity of a collector is an important parameter that can influence the overall system efficiency when a collector is incorporated into a solar heating system.

A collector can usually be considered as a combination of masses each at a different temperature. When the collector is operating and the fluid inside it changes from one temperature to another, not all the collector components will change by the same temperature difference. Hence it is useful to consider an effective thermal capacity. Unfortunately the effective thermal capacity depends on the collector operating conditions and therefore is not in general a constant.

For practical purposes however, it is useful to define a test procedure which will permit a reproducible value of effective thermal capacity to be determined, and it will be appreciated that the value obtained may be a function of the procedure adopted.

Various thermal capacity test methods are being used in the community (see Refs. 11 to 16). In addition, procedures for calculating the effective thermal capacity are given in Refs. 7 and 16, and a comparison of several of the methods is given in Ref. 17.

The method considered as suitable for recommendation is the integral method (Ref. 12). This method has the following advantages:

(i)   the method is independent of the characteristics of the test loop;
(ii)   an integrated energy balance is considered and no time derivatives are required;
(iii)   the inlet temperature of the fluid does not have to be controlled during the transient period;
(iv)   two data points for the indoor heat loss rate curve are also obtained.

8.5.1.2   Test Installation

Testing should be performed indoors where the surrounding air temperature can be maintained constant. The thermal loop should be similar to that used for the determination of the indoor heat loss rate of the collector shown in Fig. 6. In addition a facility to record directly the temperature difference between the fluid at the collector inlet and the surrounding air temperature during the tests is beneficial.

8.5.1.3 Procedure

With a fluid flow rate similar to that used for collector thermal efficiency testing the fluid should be circulated from the top to the bottom of the collector, at a constant inlet temperature until steady state conditions are reached.

The thermostat control governing the fluid temperature at the collector inlet should be set to another position approximately 10°C above the initial inlet temperature. During the subsequent temperature rise of the collector, the fluid flow rate, inlet temperature and temperature difference across the collector should be recorded.

The surrounding air temperature or alternatively the temperature difference between the fluid inlet temperature and the surrounding air should also be recorded. The measurements should be recorded continuously until a new steady state has been reached.

8.5.1.4 Theory

The transient behaviour of the collector between the two indoor steady states 1 and 2 is assumed to be represented by the following equation:

$$C \frac{dT_m}{dt} = - \dot{m}c_f \Delta T - A_a U_m (T_m - T_a) \qquad (8.5\text{-}1)$$

$$\Delta T = T_e - T_i \quad \text{(negative)}$$

Integrating equation 8.5-1 over the time period between the two steady states, we obtain:

$$C(T_{m_2} - T_{m_1}) = - \int_1^2 \dot{m} \, c_f \Delta T \, dt - A_a U_m \int_1^2 (T_m - T_a) dt \quad (8.5\text{-}2)$$

Since $T_m = T_i + \dfrac{\Delta T}{2}$ we may express

$$T_m - T_a = (T_i - T_a) + \frac{\Delta T}{2} \qquad (8.5\text{-}3)$$

Combining equations 8.5-2 and 8.5-3 and rearranging, the following equation for the collector thermal capacity is obtained:

$$C = \frac{\dot{m}c_f \int_1^2 \Delta T \, dt - A_a U_m \left| \int_1^2 (T_i - T_a) \, dt + \frac{1}{2} \int_1^2 \Delta T \, dt \right|}{(T_{m_2} - T_{m_1})} \qquad (8.5\text{-}4)$$

8.5.1.5 Determination and Presentation of the Results

From the test records $(T_i - T_a)$ and ΔT should be plotted versus time. The areas under the curves between the two steady states are

$$\int_1^2 (T_i - T_a) \, dt \quad \text{and} \quad \int_1^2 \Delta T \, dt \quad \text{respectively.}$$

The heat transfer coefficient U_m of the collector may already have been determined during indoor collector heat loss rate testing. A_aU_m however may be obtained directly from the two steady states since in steady state equation 8.5-1 gives:

$$0 = -\dot{m}c_f\Delta T - A_aU_m(T_m - T_a)$$

and hence:

$$A_aU_m = \frac{-\dot{m}c_f\Delta T}{(T_m - T_a)}$$

A_aU_m should therefore be evaluated for both steady states and a mean value taken.

It follows therefore that all the terms in equation 8.5-4 may be determined and a value of the effective thermal capacity obtained.

8.5.2   Determination of the Time Constant of a Collector

8.5.2.1   Summary

The time constant of a collector is a characteristic of the transient response of the collector. A value of the time constant may be determined by monitoring the temperature of the heat transfer fluid at the collector outlet after a step increase in the solar irradiance at the collector aperture. The method is similar to methods presented in Refs. 6 and 8.

8.5.2.2   The fluid loop should be the same as that used for thermal efficiency testing, examples of which are given in Figs. 4 and 5. Additional instrumentation is required to monitor the temperature of the fluid at the collector outlet. It is advantageous that the collector be mounted so that it may be steered to track the sun.

Alternatively in order to achieve standard conditions of testing, the test may be performed indoors where the surrounding air temperature can be maintained constant and using simulated solar radiation.

8.5.2.3   Procedure

Testing should be performed under clear sky conditions with solar irradiance at the plane of the collector aperture greater than 750 W m^{-2}.

The aperture of the collector should be shielded from the solar radiation input by means of a solar reflecting screen. The heat transfer fluid should be circulated at ambient temperature through the collector at the same flow rate as that used during collector thermal efficiency testing.

When the collector has reached steady state and with the fluid inlet temperature constant at the temperature of the surrounding air, the screen should be removed. The surrounding air temperature and the temperature of the heat transfer fluid at the collector outlet should be recorded until the collector assumes another steady state condition. A steady state condition is considered to have been achieved when the fluid outlet temperature variation is less than 0.5°C in 10 minutes.

8.5.2.4 Determination and Presentation of the Results

The temperature difference $(T_e - T_a)$ between the fluid at collector outlet and the surrounding air temperature should be plotted versus time, commencing from the initial steady state at time t_0.
From the test records of the fluid outlet temperature, the time t_2 when the second steady state is reached should be noted.
From the plot of $(T_e - T_a)$ vs time the time t_1 should be determined when $(T_e - T_a)$ has increased by 63.2% of the overall increase between time t_0 and t_2.
The time constant of the collector following the step increase in solar irradiance at time t_0 is given by the following equation:

$$\tau = t_1 - t_0 \qquad\qquad (8.5\text{-}5)$$

8.5.3 Determination of the Influence on Collector Performance of the Angle of Incidence of Direct Solar Radiation at the Collector Aperture

8.5.3.1 Summary

The angle of incidence of direct solar radiation (θ) has an effect on the efficiency of a solar collector. The effect is most pronounced on collectors with honeycomb structures and those with double glazed aperture covers.

8.5.3.2 Procedure

Measurements should be taken to determine the instantaneous efficiency of the collector using the method of testing detailed in Section 8.3. Testing should be performed on one clear sky day to obtain 4 sets of data as follows:

- . 1 measuring point prior to solar noon $\qquad$ $40° < \theta < 50°$
- . 2 measuring points at solar noon $\qquad$ $\theta < 15°$
- . 1 measuring point after solar noon $\qquad$ $40° < \theta < 50°$

The angle of incidence of direct solar radiation should be determined using the methods described in Section 7.5.

8.5.3.3 Presentation of the Results

An angle modifier may be defined as

$$K(\theta) = \frac{\eta(\theta > 0°)}{\eta(\theta = 0°)} \qquad\qquad (8.5\text{-}6)$$

which may be approximated for most collectors to

$$K(\theta) = \frac{\eta(\theta > 0°)}{\eta(\theta < 15°)}$$

A mean value of $K(\theta)$ for an angle of incidence in the region $40° < \theta < 50°$ may therefore be determined.

If a more detailed knowledge of the influence of the angle of incidence is required the procedure detailed above should be repeated for a number of values of solar incidence angle so that a graphical presentation of $K(\theta)$ may be obtained as shown in Fig. 19.

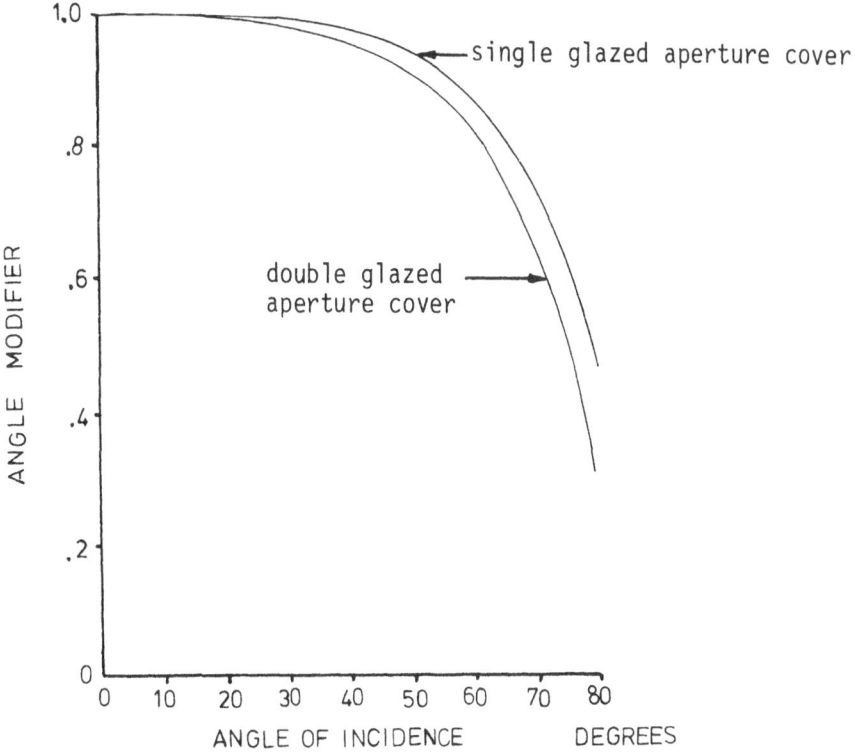

Fig. 19 - Angle of incidence modifier for single and double glazed collectors.

8.5.4 Determination of the Pressure Drop Across a Collector

8.5.4.1 Introduction

The pressure drop across a collector may be of importance to designers of solar collector systems. The fluid to be used in the collector should be that used for the thermal performance tests detailed in Sections 8.2, 8.3 and 8.4.

In order that a representative envelope of pressure drops may be determined, a number of fluid flow rates should be used.

8.5.4.2 Test Installations

Test installations suitable for the determination of the pressure drop across a collector are those previously described and shown schematically in Figs. 4, 5 and 6. Instrumentation may be reduced to that required to perform the measurements as specified in Section 8.5.4.3.

8.5.4.3 Measurements

The following measurements are to be obtained in accordance with the recommendations of Section 7:
(i)   the fluid temperature at the collector inlet;
(ii) the fluid flow rate;
(iii) the pressure drop between the collector inlet and outlet connections.

8.5.4.4 Conditioning of the Collector

The fluid to be used should be free from contaminating particles.
Filters in the fluid loop should be cleaned or renewed.
The collector should be vented of air by means of an air valve or other suitable means such as increasing the fluid flow rate.

8.5.4.5 Test Conditions

The collector should be positioned and mounted in the same manner as for thermal performance tests. (See Section 5.)
The fluid flow rate should be held constant to within ± 1% of the nominal value during test measurements.
The fluid inlet temperature should be held constant to within ± 5°C during test measurements.

8.5.4.6 Test Procedure

8.5.4.6.1 The pressure drop between the collector inlet and outlet connections should be determined with the fluid flowing from the bottom to the top of the collector, using a fluid inlet temperature within ± 5°C of that of the surrounding air and at the following mass flow rates:
(i) $0.010 \pm 10\%$ kg s^{-1} per square metre of collector aperture area
(ii) $0.05 \pm 10\%$ kg s^{-1} ”
(iii) $0.075 \pm 10\%$ kg s^{-1} ”
(iv) $0.100 \pm 10\%$ kg s^{-1} ”
(v) $0.150 \pm 10\%$ kg s^{-1} ”

Alternatively five flow rates should be taken over the flow rate range recommended by the collector manufacturer.

8.5.4.6.2 The tests detailed above should be repeated with the fluid flow direction from the top to the bottom of the collector.

8.5.4.6.3 The fittings used to measure the fluid pressure may themselves cause a drop in pressure. A zero check on the pressure drop should be determined by removing the collector from the fluid loop and repeating the tests above with the pressure measuring fittings suitably connected together.

8.5.4.7 Computation and Presentation of the Results

The pressure drop should be presented graphically as a function of the fluid flow rate for each of the tests performed in Section 8.5.4.6. The graphical format should be that given in the format sheet of Section 11.

9. TEST METHODS UNDER DEVELOPMENT

9.1 Scope

Within the community there are many methods of determining the performance of solar collectors which have not yet been fully evaluated. It is considered that by making definite recommendations with regard to these procedures, the natural course of research and development may be restricted. This section therefore only provides a resume of such methods.

The following methods of testing are included in this section:

. (9.2)   determination of the instantaneous efficiency of a collector under transient conditions;
. (9.3)   determination of the thermal performance of a collector using a solar simulator;
. (9.4)   determination of the solar absorptance of the collector absorber plate surface;
. (9.5)   determination of the thermal emittance of the collector absorber plate surface.

9.2 Determination of the Instantaneous Efficiency Curve of a Collector under Transient Conditions

9.2.1 Summary

The method is based upon the following heat balance for the collector under transient conditions.

Total power collection = useful power extraction + rate of heat gain of the collector

i.e. $\dot{Q}_T = \dot{Q}_u + C \dfrac{d(T_m)}{dt}$ (9.2-1)

where   C = effective thermal capacity of the collector.

The method requires that the effective thermal capacity of the collector should also be determined (see Section 8.5.1). The advantage of the method is that the number of sunny days required to obtain sufficient data points for determination of a performance characteristic is reduced.

9.2.2 Procedure

The test is similar to that proposed in Ref. 9. The fluid inlet temperature is allowed to rise steadily and monotonely during the test. The measurements are similar to those described in Section 8.2.3 except that in addition the value of $d(T_m/dt)$ must also be determined.

It is important that the fluid inlet temperature rise, i.e. $d(T_i/dt)$ should be constant during a test so that a constant fluid temperature difference between the collector inlet and outlet will be present.

9.2.3 Presentation of the Results

The useful power collection may be determined from:

$$\dot{Q}_u = \dot{m} \, c_f \Delta T \qquad (9.2\text{-}2)$$

The total power collection of the solar collector may then be determined from equation 9.2-1 and corresponding instantaneous efficiencies from

$$\eta = \frac{\dot{Q}_T}{A_a G} \qquad (9.2\text{-}3)$$

An efficiency curve should then be determined as detailed in Section 8.2.8.

9.3 Determination of the Thermal Performance of a Collector Using a Solar Simulator

9.3.1 Summary

The use of a solar simulator for the thermal performance testing of solar collectors presents the opportunity to test the full range of performance characteristics in a short period of time.

As each individual parameter may be varied independently the effects of various solar collector operating conditions may be studied. These conditions are also more reproducible than conditions present during outdoor testing.

However, in order to readily predict the outdoor thermal performance satisfactorily the solar simulator and indoor environment should be representative of the outdoor environment.

The main reasons for collector testing indoors using a solar simulator being unrepresentative of the outdoor environment are:

(i)   incorrect radiation spectrum;
(ii) non-uniformity of the radiation intensity over the collector;
(iii) divergence of the radiation beam;
(iv) unrepresentative diffuse component of the radiation;

(v)   incorrect field of view temperature;

(vi) unrepresentative surrounding air temperature;

(vii) unrepresentative collector inclination.

With an increasing number of solar simulators being commissioned in the European community, it is anticipated that major developments in solar simulator design and operating methods will be made in the near future and recommendations documented.

9.3.2   Procedures

The test methods described in Section 8 and others requiring solar irradiance may be performed using artificial solar irradiance.

9.4   Determination of the Solar Absorptance of the Collector Absorber Plate Surface

9.4.1   Summary

The solar absorptance α, of the solar collector absorber surface directly affects the solar energy absorbed by the collector and hence the thermal performance of the collector. At present there is no agreed standard test method to determine the solar absorptance of a surface. The solar absorptance of a surface is a function of wavelength, angle of incidence and surface temperature.

The evaluation of the solar absorptance has not been the object of concerted effort in European collector testing.

9.4.2   Procedure

The solar absorptance of a surface may be determined by measuring the reflectance ρ, of the surface spectrally over the wavelength region of approximately 0.3 μm to 2.5 μm. The reflectance may be measured by using a reflectance spectro photometer with an integrating sphere attachment. The reflectance should be determined for a range of angles of incidence and at the working temperature of the surface.

It is anticipated that other test procedures are being used to determine solar absorptance values.

9.4.3   Determination and Presentation of the Results

From the measured reflectances a weighted average spectral reflectance should be determined, using an agreed spectral distribution for solar radiation to provide the weighting function. An agreed method for performing the weighting computation should be used.

Using Kirchhoff's law for an opaque body the total hemispherical solar absorptance α, may be deduced from the equation

$$\alpha = 1 - \rho$$

The solar absorptance should be presented with details of the method of test used, the reference temperature, the weighting function and the weighting computation.

9.5 Determination of the Thermal Emittance of the Collector Absorber Plate Surface

9.5.1 Summary

The (total hemispherical) thermal emittance ϵ, of a solar collector absorber surface influences the thermal losses from the collector. At present there is no agreed standard method to determine the thermal emittance. The thermal emittance of a surface is a function of wavelength; angle of emission and surface temperature.

The evaluation of the thermal emittance has not been the object of concerted effort in European collector testing.

9.5.2 Test Methods

Many test methods are in use for measuring the thermal emittance of a surface. Three methods are outlined here:

The thermal emittance may be determined by measurement of the spectral hemispherical reflectance of the surface over the wavelength region of 2.5 μm to approximately 25 μm for various angles of incidence. A weighted average spectral reflectance should then be determined using the spectral distribution of a blackbody radiator at the measurement temperature to provide the weighting function. The thermal emittance may then be deduced by the application of Kirchhoff's law from the equation:

$$\epsilon = 1 - \rho$$

The thermal emittance may be determined by calorimetric methods. The sample is usually isolated in a vacuum chamber with the surrounding walls at a known constant temperature. The sample may then be heated to a known temperature above the wall temperature and either the rate of heat input to maintain a known steady state temperature or the rate of cooling of the sample measured. The thermal emittance may then be determined by considering the energy balance of the sample.

The thermal emittance may also be determined by direct measurement of the radiant flux emitted. The surface should be heated to a known temperature above that of its surroundings and the radiant flux emitted measured directly with a detector exhibiting a flat response over the thermal radiation wavelength region. The thermal emittance may then be obtained by comparison with that from a known sample at the same temperature.

9.5.3 Presentation of the Results

The total hemispherical thermal emittance should be presented with details of the method of test used, the reference temperature and the weighting function where necessary.

10. DATA ANALYSIS

10.1 Presentation

The recommended format for presenting the data and results obtained from solar collector testing is given in the format sheet of Section 11.

10.2 Curve Fitting

10.2.1 Curve fitting is required for graphical presentation of data in the format sheet. In particular the methods of testing detailed in Sections 8.2 and 8.3 require that a second order curve fit should be made to the experimental test data of collector efficiency (η) against reduced temperature (T^*).
The assumed relationship is:

$$\eta = \eta_0 - a_1 (T^*) - a_2 (T^*)^2 \qquad (10\text{-}1)$$

10.2.2 The recommended method of curve fitting data is the method of least squares, i.e. that which minimizes the sum of the squares of the errors:

$$S = \sum_{i=1}^{i=N} e_i^2 \qquad (10\text{-}2)$$

where N is the number of data points and e_i is the error of the curve fit, defined for collector efficiency data as:

$$e_i = \eta_i - (\eta_0 - a_1 (T^*)_i - a_2 (T^*)_i^2 \qquad (10\text{-}3)$$

and where η_1 represents the experimental value corresponding to $(T^*)_i$.
The values of η_0, a_1 and a_2 in equation 10-1 may be obtained by differentiating equation 10-2 with respect to η_0, a_1 and a_2 respectively and solving the three equations obtained.
It should be noted that by defining an error of fit as given in 10-3 an assumption in the distribution of the errors between η and T^* is made.

10.2.3 Weighting of curve fits may be made if required. For example the influence of data with large curve fit errors may be reduced by assuming a curve fit which minimizes the sum of the squares of the errors.

$$S = \sum_{i=1}^{i=N} (e_i \cdot \exp \cdot (-e_i))^2$$

10.2.4 An indication of the closeness of the data to the fitted curve may be obtained by calculating the estimated standard deviation of the errors of the curve fit (σ_e) from the following equation:

$$\sigma_e = [\frac{S}{N-1}]^{\frac{1}{2}}$$

10.2.5 At least 16 data points are required to obtain sufficient confidence in the curve fit.

10.3 Further Analysis

The thermal performance of a collector is influenced by the environment in which it is operating. During the testing of a collector the collector may be operated in various conditions within the range of the specified test conditions. From the measurements of surrounding air speed, ambient air temperature and diffuse solar irradiance, the influence of these parameters on the collector efficiency may be obtained if data is selected with reference to these parameters and corresponding collector efficiency curves determined.

11. FORMAT SHEET

The format sheet presented in this section is at present being used within the European Community for presenting the results of solar collector performance tests.
(Note: The nomenclature of the format sheet is temporarily not wholly consistent with the nomenclature of the main text of this document.)

CEC *Commission of the European Communities*

SOLAR COLLECTOR TESTING PROGRAMME

Performance Tests
Format Sheets

REVISED VERSION 1978

EUROPEAN SOLAR COLLECTOR TESTING PROGRAMME
FORMAT SHEETS
TESTS TO CEC RECOMMENDATIONS

Collector Reference No ....

TESTS PERFORMED BY: ....

Address ....

Date .... Tel .... Telex ....

1. Description of Solar Collector

    1. 1 NAME OF COLLECTOR AND MANUFACTURER ....

    ....

    1.2 COLLECTOR TYPE (please complete as applicable)

    FLAT PLATE [ ]   EVACUATED TUBE [ ]   OTHER........ [ ]

      - Number of covers ..... Number of tubes ..... Cover material ....

      - Cover thickness ...... Tube diameter .... Aperture dimensions ......

    1. 3 ABSORBER

      - Material ....

      - Surface treatment ....

      - Manufacturing process ....

      - Weight empty .... kg   Water content ....kg

      - Dimensions ....

      ....

    1. 4 THERMAL INSULATION AND CASING

      - Thermal insulation: Thickness .... mm

      - Material ....

      - Casing: Material ....

      - Total weight of collector with water .... kg

      - Gross dimensions ....

1. 5 LIMITATIONS

- Maximum temperature of operation ....

- Maximum pressure ....

- Acceptable heat transfer fluids ....

  ....

1.6 SCHEMA OF SOLAR COLLECTOR

2. Instantaneous Efficiency

2.1 METHOD ....

....

2.2 SCHEMA OF TEST INSTALLATION

2.3 INSTRUMENTATION ....

    – Incident radiation..

    – Diffuse radiation ....

    – Fluid mass flow ....

    – Ambient temperature. ....

    – Fluid absolute temperature ....

    – Differential fluid temperature ....

    – Wind velocity ....

    – Data recording. ....

    – Pressure drop transducers ....

2.4 PHOTO OF SOLAR COLLECTOR TEST RIG

3. Instantaneous Efficiency Curve

THE INSTANTANEOUS EFFICIENCY η IS DEFINED BY $\eta = \dfrac{\dot{Q}_u}{A_a G}$

$\dot{Q}_u$: useful power extracted (W)

G : solar irradiance at collector aperture (Wm^{-2})

A_a : aperture area (m^2)

Aperture area used for curve = m^2

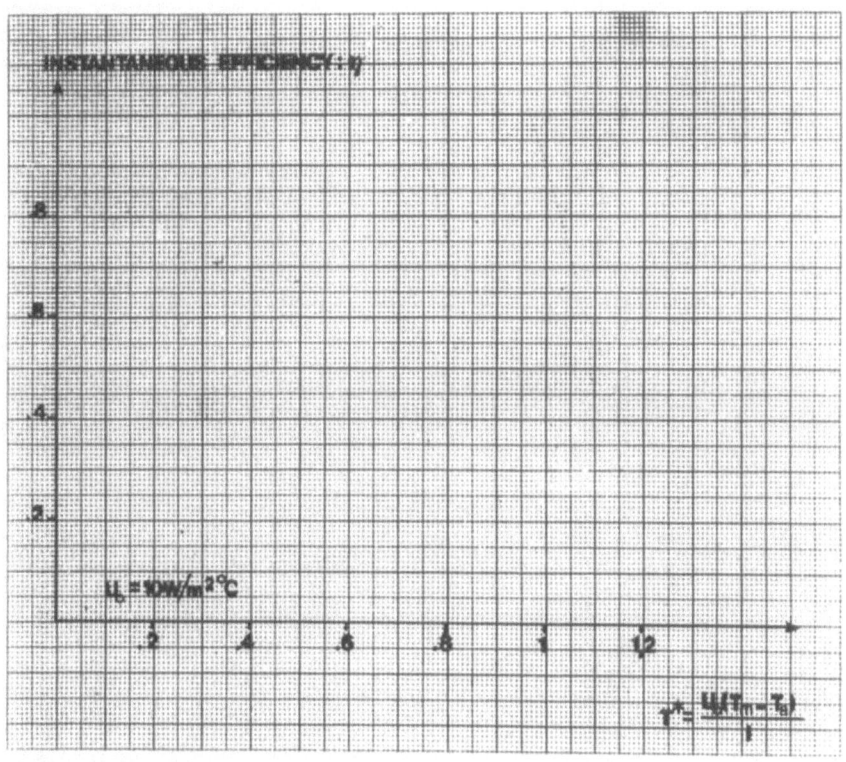

LINEAR PERFORMANCE CHARACTERISTIC $\eta = \eta_0 - a_1 T^*$

SECOND ORDER FIT TO DATA $\eta = \eta_0 - a_1 T^* - a_2 G(T^*)^2$

$\eta =$

4. Instantaneous Efficiency : Experimental Data

FLUID $- - - - - - - - - - - - - - - -$ $C_f - - - - Jkg^{-1}K^{-1}$ SLOPE $- - - - - - - -$

LATITUDE $- - - - - - - -$
LONGITUDE $- - - - - - - -$
COLLECTOR AZIMUTH $- - - - - - -$
LOCAL TIME AT SOLAR NOON $- - - - - -$

DATE	LT	G	G_d/G	T_a	u_s	T_i	$(T_e - T_i)$	T_m	$\dot{m}$	T^*	η
D-M-Yr	Hrs-Mins	Wm^{-2}	%	°C	ms^{-1}	°C	K	°C	$kg\,s^{-1}$	Km^2W^{-1}	—

APERTURE AREA $- - - - - - - -$ m^2

5. Heat Losses

5.1 GLOBAL THERMAL LOSSES

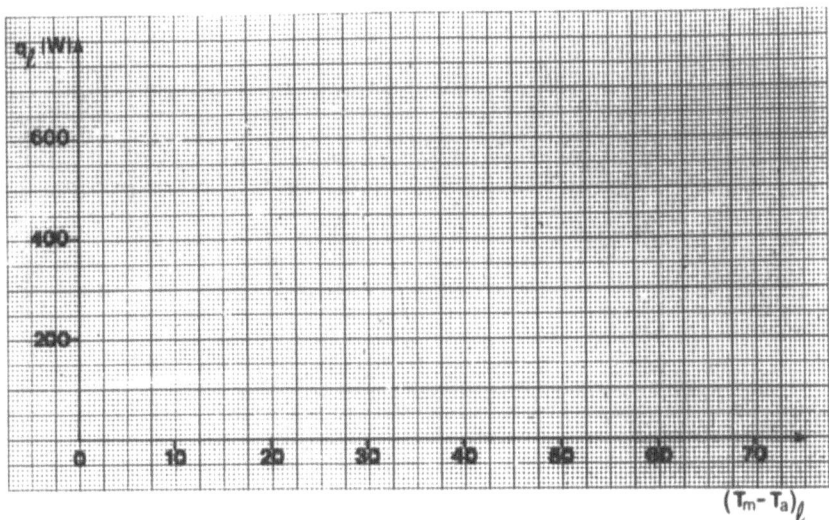

The heat losses q_ℓ are expressed as a function of the termperature difference $(T_m - T_a)_\ell$

5.2 GLOBAL HEAT TRANSFER COEFFICIENT

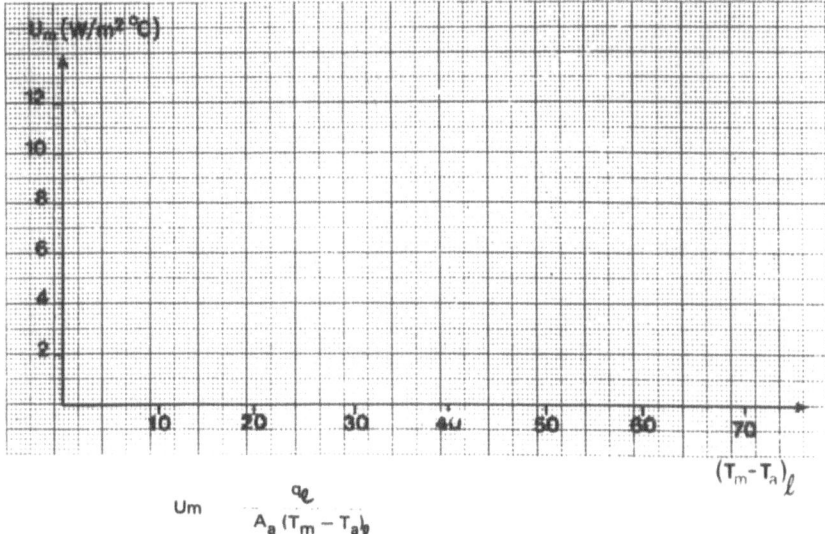

$$Um \qquad \frac{q_\ell}{A_a\,(T_m - T_a)_\ell}$$

5.3 HEAT LOSSES: EXPERIMENTAL DATA

FLUID: – – – – – – – – – – C_f – – – – – $Jkg^{-1}K^{-1}$ SLOPE: – – – – – – – –

indoor ☐
outdoor ☐

DATE	LT	T_a	u_s	T_i	(T_i-T_e)	T_m	$\overset{\bullet}{m}$	$(T_m-T_a)_\ell$	$\overset{\bullet}{Q}_\ell$	U_m		
D-M-Yr	Hrs-Mins	°C	ms^{-1}	°C	K	°C	kgs^{-1}	K	W	$Wm^{-2}K^{-1}$		

APERTURE AREA – – – – – – – – – m^2

– 464 –

6. Pressure Drop

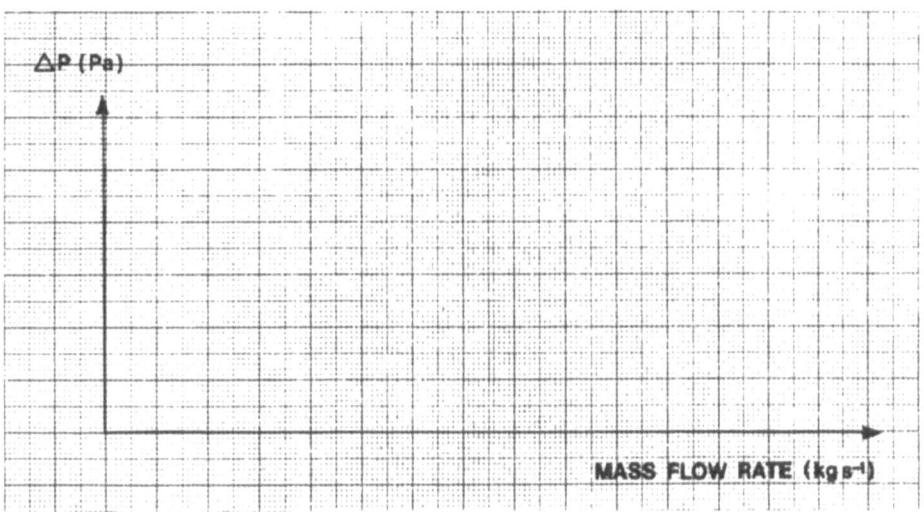

The pressure drop $\triangle P$ across the collector is measured with water at ambient temperature

7. **Other Methods or Special Remarks**

(Give a short description of methods and essential results)

8. Nomenclature

Symbol	Meaning	Units
A_a	Aperture area of collector	m^2
A_g	Gross area of collector	m^2
A_p	Absorber plate area of collector	m^2
c_f	Specific heat capacity of fluid	$Jkg^{-1}K^{-1}$
G	Solar irradiance at collector aperture	Wm^{-2}
G_d	Diffuse solar irradiance	Wm^{-2}
LT	Local time	Hours – Mins
$\dot{m}$	Mass flow rate of heat transfer fluid	$kg\ s^{-1}$
$\dot{Q}_\ell$	Overall heat loss rate	W
$\dot{Q}_u$	Useful power extracted from collector	W
T_a	Ambient air temperature	$^\circ C$
T_e	Collector outlet (exit) temperature	$^\circ C$
T_i	Collector inlet temperature	$^\circ C$
T_m	Mean temperature of fluid in collector	$^\circ C$
T^*	Reduced temperature	Km^2W^{-1}
u_s	Surrounding air speed	ms^{-1}
U_m	Heat loss coefficient	$Wm^{-2}K^{-1}$
ΔP	Pressure drop across collector	Pa
η	Collector thermal efficiency	dimensionless
η_o	Efficiency when $T^* = 0$	dimensionless

REFERENCES

1. SI-Le Système International d'Unités. Offilib, F75005, BIPM, Paris.
2. Units and Symbols in Solar Energy. Solar Energy, Vol.21, pp.65-68.
3. Guide to Meteorological Instrument and Observing Practices - World Meteorological Organisation (1971).
4. Solar Energy Thermal Processes - Duffie, J.A. and Beckman, W.A. - Wiley (1974).
5. Method of Testing for Rating Solar Collectors Based on Thermal Performance (NBSIR 74-635).
6. Method of Testing to Determine the Thermal Performance of Solar Collectors (ASHRAE 93-77).
7. Usability of Solar Collectors - Bundesverband Solarenergie (BSE) (1978).
8. Capteurs Solaires, Mesure des Performances Thermique - AFNOR 50-501 (1977).
9. Determination of the Instantaneous Efficiency of a Solar Collector by Transient Method - E. Aranovitch et al., JRC-Ispra (1976).
10. Collettori Solari, Criteri e Raccomandazioni - Ente Nazionale Italiano di Unificazione, UNI - Fascicolo No. 4184.
11. Methods for the Determination of the Effective Heat Capacity of a Solar Collector - E. Aranovitch and M. Ledet, JRC-Ispra (1976).
12. Integral Method for the Determination of the Apparent Heat Capacity of a Solar Collector - J. Bougard, Faculté Polytechnique de Mons (1978).
13. Method for the Measurement of the Heat Capacity of Solar Collectors - L. Cuypers and W.L. Dutre, Katholieke Universiteit Leuven (1977).
14. Simplified Measurement of the Heat Capacity of a Solar Collector - J. Removille, Luxatom (1978).
15. Measurement of the Solar Collector Heat Capacity by an Oscillatory Method - W.L. Dutre and D. Gilliaert, Katholieke Universiteit Leuven (1978).
16. Methods for the Determination of the Apparent Heat Capacity of a Solar Collector - J. Bougard, Faculté Polytechnique de Mons (1978).
17. Determination of the Thermal Capacity of CEC 4 Collector by Several Methods - J. Bougard et al., Faculté Polytechnique de Mons (1979).

STANDARD PROCEDURES FOR TERRESTRIAL PHOTOVOLTAIC PER-
FORMANCE MEASUREMENTS

Specification No. 101 (Issue 2)

Report EUR 7078EN - 1981

edited by K. Krebs
Commission of the European Communities
Joint Research Centre - Ispra Establishment
21020 Ispra (Va) - Italy

ABSTRACT

This specification lays down standard procedures for determining the electrical
performance of photovoltaic solar cells, modules and arrays for use in terrestrial
applications. Measuring methods and instrumentation are described for testing
in natural sunlight and with a solar simulator. In order to minimize discrepancies
caused by variations in the spectral distribution of the incident radiation, this
specification requires performance ratings to be related to a ".standard sunlight"
distribution and irradiances to be measured by specially calibrated reference solar
cells.
Included in this document are procedures for the measurement of temperature
coefficients and internal resistance and also requirements for simulators and instrumentation.
In Appendix I methods are described for the calibration of reference cells.
Appendix II contains information about short circuit error computations.
Appendix III describes methods for the measurement of relative spectral response.
In Appendix IV we present a list of definitions of relevant meteorological and
photovoltaic terms with the aim of encouraging the use of a common terminology
for photovoltaic testing.

CONTENTS

1. Purpose
2. Scope
3. Approach

G. Beghi (ed.), Performance of Solar Energy Converters: Thermal Collectors and Photovoltaic Cells, 469–500.

4. Standard Sunlight
5. Measurement in Natural Sunlight
6. Measurement in Simulated Sunlight (Steady State)
7. Measurement in Simulated Sunlight (Pulsed)
8. Measurement of Temperature Coefficients
9. Determination of Internal Series Resistance
10. Reference Solar-Cells
11. Simulators
12. Test Mounts
13. Instrumentation

TABLES

FIGURES

1. PURPOSE

The purpose of this Specification is to lay down standard procedures for deter-mining the electrical performance of photovoltaic solar cells, modules and arrays for use in terrestrial applications.

2. SCOPE

These procedures apply to:
a) single solar cells with or without a protective cover
b) sub-assemblies of solar cells
c) flat modules
d) complete arrays consisting of two or more modules.

They are not applicable to solar cells designed for operation in concentrated sun-light, to modules or arrays embodying concentrators, or to hybrid collectors, which, in addition to generating electricity, transfer heat to fluids for use in thermal systems.

These procedures can be applied to all devices with known light intensity dependence.

Methods are presented for testing in natural sunlight and with a solar simulator. Definitions of relevant meteorological and photovoltaic terms, some of which are used in this Specification, are given in Appendix IV. It is recommended that this terminology be used in reports and correspondence.

3. APPROACH

In current practice the photovoltaic performance of a solar cell, module or array is determined by exposing it at a known temperature to stable sunlight, natural or simulated, and tracing its current-voltage characteristic while at the same time measuring the magnitude of the incident irradiance. The measured performance is then corrected to Standard Test Conditions (for definition see Appendix IV). The corrected power output at a particular working voltage is commonly referred to as the rated or "peak" power for the purpose of performance rating.

As the response* of a solar cell is wavelength-dependent, its performance is significantly affected by the spectral distribution of the radiation, which in the case of natural sunlight varies with location, weather, time of wear and time of day. If the irradiance is measured with a thermopile-type radiometer, which is not spectrally selective, the measured conversion efficiencies can vary by as much as 15% from day to day at the same place, due to changes in the spectral distribution.

The procedures laid down in this Specification are designed to minimize such discrepancies by relating the performance rating to a standard terrestrial solar spectral irradiance distribution, hereafter referred to as "standard sunlight". This is done by measuring irradiance with a reference solar cell or module, that is to say a cell or module having essentially the same configuration and relative spectral response as the test cells or modules, which has been calibrated at $25 \pm 2°C$ in terms of short-circuit current per unit of "standard sunlight" irradiance (AW^{-1} m^2) by an approved Solar Cell Calibration Agency. The reference cell automatically takes account of variations in spectral distribution. Because of this, location and weather conditions are not critical when the reference cell method is used for outdoor performance measurements, so long as direct sunlight is available. Moreover, since the time constants of radiation monitor and test cell are the same, fluctuations in solar intensity can be accepted, provided they do not occur in the course of the measurement.

If the performance of a cell, module or array is related to a known standard sunlight distribution, it is possible for a user or array designer, using the spectral response of the cells, to compute within a reasonable tolerance its performance in

---

* The term "response" is commonly used but "responsivity" is strictly correct.

light of any other known spectral irradiance distribution.

The procedures laid down in this Specification are also intended to minimize errors arising from faulty temperature measurement and voltage and current instrumentation.

4. STANDARD SUNLIGHT

The standard solar spectral irradiance distribution ("standard sunlight"), for the purpose of this Specification, is obtained by multiplying the spectral irradiance values in Table I (see also Fig. 1) by the factor 1000/834.6 to normalize the total irradiance (area under the curve) to 1000 Wm^{-2}. The distribution given in Table I, for direct sunlight at Air Mass 1.5, is the same as that specified in ERDA/NASA 1022/77/16, Revised Terrestrial Photovoltaic Measurement Procedures (June 1977). It is derived from the Labs and Neckel Air Mass Zero (AM0) data. The parameters used in converting the AM0 to AM1.5 data are as follows:

. precipitable water, cm   2
. ozone, cm   0.34
. aerosol scattering parameters:
      Angstrom exponent, ϵ 1.3
      Turbidity coefficient, t   0.12

5. MEASUREMENT IN NATURAL SUNLIGHT

Measurements in natural sunlight shall be carried out only when an irradiance of at least 800 Wm^{-2} is available and the total radiation (sun + sky) will not fluctuate in the course of a measurement by more than ± 1%. The test procedure shall be as follows:

5.1 Mount the test sample* near to the reference solar cell so that their active surfaces are co-planar and normal to the direct solar beam, within ± 10°. Connect to the instrumentation as shown in Fig. 2 (detailed mounting and instrumentation requirements are given in Sections 12 and 13). The entire test sample and reference cell shall be fully irradiated by the direct sunlight and no collimators shall be used. Reflected sunlight falling on the active surface of the sample shall be kept to a minimum.

5.2 Ensure that the temperature of the reference cell is 25 ± 2°C. When testing a single cell, maintain its temperature at the same level.

5.3 Record the current-voltage characteristic and temperature of the test sample

---

* For simplicity, the term "test sample" is here and hereafter used to denote a single solar cell, module, panel or array.

at the same time as the short-circuit current and temperature of the reference cell. In the case of a module, panel or array, where it is not practicable to control the temperature, shade it from the sun and wind before the measurement until it is at the ambient air temperature and then carry out the measurements immediately after removing the shade. In most cases, the thermal inertia of the assembly will limit the temperature rise during the first few seconds to less than 2°C and its temperature will remain reasonably uniform.

5.4 Correct the measured current-voltage characteristic to Standard Test Conditions by applying the following equations:

$$I_2 = I_1 + I_{sc} \left[\frac{I_{sr}}{I_{mr}} - 1 \right] + \alpha(T_2 - T_1) \tag{5.1}$$

$$V_2 = V_1 - R_s(I_2 - I_1) - KI_2(T_2 - T_1) + \beta(T_2 - T_1) \tag{5.2}$$

where:

I_1, V_1	are co-ordinates of the measured characteristic
I_2, V_2	are the corresponding co-ordinates of the corrected characteristic
I_{sc}	is the measured short-circuit current of the test sample
I_{mr}	is the measured short-circuit current of the reference solar cell
I_{sr}	is the short-circuit current of the reference solar cell in standard sunlight at 1000 Wm^{-2}
T_1	is the measured temperature of the test sample
T_2	is the standard temperature (25 ± 2°C)
α and β	are the current and voltage temperature coefficients of the test sample for standard sunlight at 1000 Wm^{-2} (β is negative)
R_s	is the internal series resistance of the test sample
K	is a curve correction factor.

Procedures for the measurement of the temperature coefficients α and β are given in Section 8 and a method of determining R_s is described in Section 9.
The curve correction factor K may be obtained from current-voltage characteristics measured on the test sample at different temperatures. A typical value for silicon cells is 1.25 x 10^{-3} Ohms/°C. With a cell giving, for example, a current of 950 mA at maximum power, this gives an additional voltage transformation of about 1 mV/°C per cell at the maximum power point, i.e. a change of about 2% for $(T_2 - T_1) = 10°C$. When $(T_2 - T_1)$ is under 2°C, as it should be if the specified procedure is followed, the curve correction term in the above voltage equation can generally be ignored.

6. MEASUREMENT IN SIMULATED SUNLICHT (STEADY STATE)

For measurements in simulated sunlight, the simulator shall meet the requirements of Section 11. When using a steady state simulator, the test procedure shall be as follows:

6.1 Mount the reference solar cell with its active surface in the test plane. The normal of the solar cell should be parallel within ± 5° to the centre-line of the beam. Connect to the instrumentation as shown in Fig. 2.

6.2 Ensure that the temperature of the reference cell is 25 ± 2°C.

6.3 Set the irradiance to produce from the reference solar cell its calibrated short-circuit current under Standard Test Conditions.

6.4 Remove the reference cell and mount the test sample in the same position normal (within ± 5°) to the beam centre-line. (Alternatively, if the beam is sufficiently wide and uniform, mount the reference cell and test module side-by-side.) Connect to the instrumentation as shown in Fig. 2.

6.5 Without changing the simulator setting, record the current-voltage characteristic and temperature of the test sample. When testing a module, panel or array, shade it from the simulator so that it is uniformly at the ambient air temperature and then carry out the measurement immediately after removing the shade. As in natural sunlight, the thermal inertia of the assembly will in most cases limit the temperature rise during the first few seconds to less than 2°C and its temperature will remain reasonably uniform.

6.6 If the measured temperature of the test sample is other than 25 ± 2°C, correct the curve to 25° by applying the appropriate temperature coefficients of current and voltage.

7. MEASUREMENT IN SIMULATED SUNLIGHT (PULSED)

The procedure, when using a pulsed solar simulator, is as follows:

7.1 Mount the test sample near to the reference cell so that their active surfaces are co-planar and normal (within ± 5°) to the centre-line of the beam at a distance corresponding to an irradiance of 1000 Wm^{-2} ± 5%. Connect to the instrumentation as shown in Fig. 2. (The electrical instrumentation is normally provided as part of the simulator.)

7.2 Record the current-voltage characteristic and correct it to 25°C and the nominal 1000 Wm^{-2}. The time interval between data points should be sufficiently long to ensure that the response time of the device under test and the rate of data collection do not introduce errors. The corrections are usually carried out automatically in a data processing unit - the temperature correction from manual inputs of the appropriate voltage and current temperature coefficients and the ambient air temperature (which is taken to be the temperature of the test sample) and the irradiance correction from the measured short-circuit current of the reference cell. The accuracy of the correction procedure should be verified periodically by measuring directly the performance of a single cell at selected levels of irradiance and temperature and comparing the results with corresponding extrapolated data.

8. MEASUREMENT OF TEMPERATURE COEFFICIENTS

The temperature coefficients of current (α) and voltage (β) vary with irradiance and, to a lesser extent, with temperature. However, for the purpose of correcting current-voltage characteristics to Standard Test Conditions (Section 5.4), the values corresponding to an irradiance of 1000 Wm^{-2} and temperatures of up to 20°C above ambient are appropriate. The coefficients are best measured in simulated sunlight, using a single representative solar cell of the same area and configuration as those in the relevant module, panel or array. The procedure with a steady state simulator is as follows:

8.1 Attach a suitable lightweight temperature sensor to the front contact of the test cell.

8.2 Mount the test cell on a temperature-controlled block with vacuum hold-down device, using the attached sensor to provide the control signal.

8.3 Position the test cell and a suitable reference solar cell side-by-side with their active surfaces in the test plane and at normal incidence (within ± 5°) to the centre-line of the simulator beam. Connect to the instrumentation as shown in Fig. 2.

8.4 Set the irradiance to produce from the reference solar cell its calibrated short-circuit current under Standard Test Conditions.

8.5 With the test cell stabilized at or near ambient temperature, measure its short-circuit current (I_{sc}) and open-circuit voltage (V_{oc}).

8.6 Stabilize the test cell at a temperature T_2 about 10°C above the ambient temperature T_1 and then at a temperature T_3 about 20°C above ambient. Repeat I_{sc} and V_{oc} measurements at each temperature.

8.7 Calculate α_c and β_c, the temperature coefficients for single cells, by a least square fit to the measured temperature dependence of V_{oc} and I_{sc}, preferably on more than one sample.

8.8 For a module, panel, array or other assembly of cells, calculate the temperature coefficients as follows:

$$\alpha = n_p \alpha_c$$
$$\beta = n_s \beta_c$$

where:

n_p is the number of cells in parallel and
n_s the number in series.

The procedure with a pulsed simulator is similar, except that, in this case, the
short-circuit current signal from the reference cell is used to correct the data for
small variations from the nominal irradiance of 1000 Wm^{-2} (see Section 7.2).
The use of a pulsed simulator is preferred because there is no additional heat
input to the cell during the measurement.

A similar procedure may be used to determine α and β at other irradiances and
temperatures. At sub-ambient temperature, precautions must be taken to prevent
moisture condensing on the active surface of the test cell, for instance, by directing a dry nitrogen gas jet onto the surface or by enclosing the cell in a vacuum
chamber.

9. DETERMINATION OF INTERNAL SERIES RESISTANCE

R_s may be simply determined in simulated sunlight by the method described by
Wolf and Rauschenbach*. The procedure is as follows (see Fig. 3):

9.1 Trace the current-voltage characteristic of the test sample at two different
irradiances (the magnitudes of which do not have to be known) and at the same
temperature (within ± 5°C) in the range 23° to 27°C.

9.2 Choose a point on each characteristic, preferably near the "knee" of the
curve, where the current is the same increment ΔI below the short-circuit current.

9.3 Measure the voltage displacement ΔV between these two points.

9.4 Calculate R_s as follows:

$$R_s = \frac{\Delta V}{I_{sc1} - I_{sc2}} \tag{9.1}$$

where: I_{sc1} and I_{sc2} are the two short-circuit currents.

---

* M.Wolf and H.Rauschenbach, "Series Resistance Effects on Solar Cell Measurements", Advanced Energy Conversion, Vol.3, pp.455-479, April-June 1963.

9.5 Repeat steps 9.2 to 9.4, using a characteristic taken at a third irradiance and the same temperature in combination with each of the first two curves. Take the mean of the three values of R_s thus calculated.

10. REFERENCE SOLAR CELLS

10.1 Selection

Cells selected for calibration as references shall be stable devices having essentially the same relative spectral response as the test samples. The matching shall take account of the spectral response spread in production cells and the effects, if any, of the module window and encapsulant. At least two master reference cells shall be selected to minimize mismatch errors (see Section 11.2). For production testing, the relative spectral response of a typical batch of production cells should be measured and the "reddest" and "bluest", representing the response spread, selected for calibration as master reference cells. Master reference cells need not be specially packaged.

10.2 Calibration of Master Reference Cells

The calibration of master reference cells, i.e. the determination of the short-circuit current per unit of standard sunlight irradiance ($AW^{-1} m^2$), shall be carried out by an approved Solar Cell Calibration Agency with the necessary facilities and expertise. Information on calibration techniques is given in Appendix I.

10.3 Working Reference Cells and Modules

To save wear and tear on master reference cells, the day-to-day monitoring and setting of irradiance levels in performance testing shall be carried out by means of working reference cells or modules. Cells shall be protectively mounted as laid down in Section 12.

A working reference cell or module shall be calibrated against the appropriate master reference cells by simultaneously measuring their short-circuit current in natural or simulated sunlight and calculating the calibration value I_{sw} as follows:

$$I_{sw} = \frac{1}{2} \left[\frac{I_{mw1} \cdot I_{sr1}}{I_{mr1}} + \frac{I_{mw2} \cdot I_{sr2}}{I_{mr2}} \right] \qquad (10.1)$$

where:

I_{mw1} and I_{mr1} are the simultaneously measured short-circuit currents of the working reference cell or module and master reference cell 1,

I_{mw2} and I_{mr2} are the simultaneously measured short-circuit currents of the same working reference cell or module and master reference cell 2,

I_{sr1} and I_{sr2} are the short-circuit currents of master reference cells 1 and 2 per unit of standard sunlight irradiance.

If all three devices are compared in a single test, $I_{mw1} = I_{mw2} = I_{mw}$, and the equation simplifies to:

$$I_{sw} = \frac{I_{mw}}{2} \left[\frac{I_{sr1}}{I_{mr1}} + \frac{I_{sr2}}{I_{mr2}} \right] \qquad (10.2)$$

10.4   Care of Reference Cells and Modules

It is recommended that master reference cells be kept in protective packing under lock and key, to preserve them against damage and degradation. They should be brought out periodically (at least weekly when in continuous use) and used to check the calibration of the working reference cell. The cause of any change in the current ratios should be investigated. In natural sunlight, such a change would indicate that one or both of the reference cells have degraded but, with a simulator, an alternative explanation could be a change in the spectral content of the light. The windows of working reference cells and modules must be kept clean and free of scratches.

11. SIMULATORS

Two types of solar simulators are commercially available for photovoltaic performance testing. The "steady state" type of, for example, filtered xenon, dichroic filtered tungsten (ELH) or modified mercury vapour with tungsten electrodes, is suitable for single cells and small modules. The pulsed type, consisting of one or two long-arc xenon flash lamps with no filters or optics, is better for large modules, panels and arrays, as it can irradiate large areas uniformly. Another advantage of this type is that there is negligible heat input to the test cells, so that they remain uniformly at the ambient temperature, which can be easily and accurately measured. The pulse-forming network and the data acquisition and processing equipment are generally supplied as part of the simulator. All simulators shall meet the following requirements:

11.1   Total Irradiance in the Test Plane

1000 Wm^{-2} (nominal), as measured with a suitable reference cell.

11.2   Spectral Match

The spectral irradiance distribution of the simulator shall match that of standard sunlight (Fig. 1) to the extent necessary to limit to less than ± 1% short-circuit current errors due to maximum possible spectral response* mismatch between the reference and test cells. The smaller the spectral response mismatch, the

---

* The term "response" is commonly used but "responsivity" is strictly correct.

greater the allowable tolerance on the simulator. The method of computing the error in a particular case is explained in Appendix II. If the error exceeds ± 1%, the simulator must be improved (e.g. by filtering) until the requirement is met.

11.3   Uniformity

The irradiance in the test plane, as measured by a solar cell detector, shall be uniform over the area of the test sample to within ± 2%* of the average value. The largest dimension of the detector shall be less than one-half the smallest dimension of the cell being measured. In the case of modules and arrays, the detector shall be no bigger than a single component cell.

When reflected light can affect the uniformity of irradiance, a representative module shall be placed in the appropriate position in the test plane and the uniformity checked with the detector placed over each cell position.

11.4   Temporal Stability

In the case of steady state simulators, the irradiance in the test plane, as measured by a solar cell detector, shall not fluctuate by more than ± 1% during the time it takes to make a complete performance measurement.

11.5   Beam Subtense Angle

The angle subtended by the apparent source of the simulator on a point on the test cell shall be less than 30°.

11.6   Simulators should be checked at regular intervals and also whenever a lamp, filter or optic is changed, to ensure that the above requirements continue to be met. With new eauipçent, it is recommended that monthly checks be carried out initially, the interval being extended as experience shows this to be permissible.

12. TEST MOUNTS

Single test cells and unpackaged master reference cells shall be mounted on temperature-controlled blocks embodying a vacuum hold-down device, a suitable thermocouple and four terminal contacts (current + and −, voltage + and −). The mount for working reference cells shall consist of a robust temperature-controlled block, equipped with a suitable thermocouple (preferably attached to the upper contact bar) and protected by an easily cleaned distorsion-free window, allowing the cell an unobstructed view over a solid angle of 2π steradians. Test mounts of working reference cells should preferably be installationally interchangeable.

----

* ± 3% permissible for production line testing.

13 INSTRUMENTATION

The temperature of test cells, panels and standard cells shall be measured to an accuracy of ± 2°C. Large panels shall be fitted with not less than three temperature sensors per m² of panel area. One sensor is sufficient for measurements with pulsed simulated sunlight.

In natural sunlight, sensors applied to the front or back of the panel may not register the cell temperature to the required accuracy and corrections (using an adequately instrumented dummy of the same design) may therefore be necessary. Voltages and currents shall be measured to an accuracy of ± 0.5% using separate voltage and current leads (see Fig. 2). Short-circuit currents shall be measured at zero voltage, using a variable bias, preferably electronic, to offset the voltage drop across the series resistor. In the case of reference cells, the short-circuit current may alternatively be determined by measuring the voltage drop across a standard 4-terminal fixed resistor, provided that the measurement is made at a voltage less than 0.03 V_{oc} within the range where there is a linear relationship between current and voltage and if necessary the reading is corrected to zero voltage. Voltmeters shall have an internal resistance of at least 20 kΩ/V. The calibration of all instruments shall be checked at frequent intervals.

Wave-length [μm]	Spectral Irradiance[a] [Wm^{-2}μm^{-1}]	(Number of photons)/(sec cm^2)[b]	Wave-length [μm]	Spectral Irradiance[a] [Wm^{-2}μm^{-1}]	(Number of photons)/(sec cm^2)[b]
0.295	0	----------	0.900	807.83	4.5747×10^{15}
.305	1.32	9.9792×10^{11}	.9075	793.87	2.7356×10^{15}
.315	20.96	1.7405×10^{13}	.915	778.97	2.7088×10^{15}
.325	113.48	1.0841×10^{14}	.925	217.12	2.3093×10^{15}
.335	182.23	2.4591×10^{14}	.930	163.72	4.4507×10^{14}
.345	234.43	3.5699×10^{14}	.940	249.12	9.7273×10^{14}
.355	286.01	4.5903×10^{14}	.950	231.30	1.1441×10^{15}
.365	355.88	5.8232×10^{14}	.955	255.61	5.8437×10^{14}
.375	386.80	6.9247×10^{14}	.965	279.69	1.2950×10^{15}
.385	381.78	7.3599×10^{14}	.975	529.64	1.9783×10^{15}
.395	492.18	8.5893×10^{14}	.985	496.64	2.5345×10^{15}
.405	751.72	1.2539×10^{15}	1.018	585.03	9.0087×10^{15}
.415	822.45	1.6254×10^{15}	1.082	486.20	$1.8141 \cdot 10^{16}$
.425	842.26	1.7619×10^{15}	1.094	448.74	$3.0761 \cdot 10^{15}$
.435	890.55	1.8777×10^{15}	1.098	486.72	$1.0335 \cdot 10^{15}$
.445	1077.07	2.1817×10^{15}	1.101	500.57	$8.2066 \cdot 10^{14}$
.455	1162.43	2.5396×10^{15}	1.128	100.86	$4.5607 \cdot 10^{15}$
.465	1180.61	2.7161×10^{15}	1.131	116.87	$1.8592 \cdot 10^{14}$
.475	1212.72	2.8347×10^{15}	1.137	108.68	$3.8673 \cdot 10^{14}$
.485	1180.43	2.8948×10^{15}	1.144	155.44	$5.3137 \cdot 10^{14}$
.495	1253.83	3.0058×10^{15}	1.147	139.19	$2.3515 \cdot 10^{14}$
.505	1242.28	3.1451×10^{15}	1.178	374.29	$4.6631 \cdot 10^{15}$
.515	1211.01	3.1530×10^{15}	1.189	383.37	$2.4856 \cdot 10^{15}$
.525	1244.87	3.2182×10^{15}	1.193	424.85	$9.7029 \cdot 10^{14}$
.535	1299.51	3.3983×10^{15}	1.222	382.57	$7.1250 \cdot 10^{15}$
.545	1273.47	3.5013×10^{15}	1.236	383.81	$3.3230 \cdot 10^{15}$
.555	1276.14	3.5338×10^{15}	1.264	323.88	$6.2418 \cdot 10^{15}$
.565	1277.74	3.6040×10^{15}	1.276	344.11	$2.5654 \cdot 10^{15}$
.575	1292.51	3.6919×10^{15}	1.288	345.69	$2.8742 \cdot 10^{15}$
.585	1284.55	3.7666×10^{15}	1.314	284.24	$5.3696 \cdot 10^{15}$
.595	1262.61	3.7871×10^{15}	1.335	175.28	$3.2209 \cdot 10^{15}$
.605	1261.79	3.8169×10^{15}	1.384	2.42	$2.9831 \cdot 10^{15}$
.615	1255.43	3.8695×10^{15}	1.432	30.06	$5.5317 \cdot 10^{14}$
.625	1240.19	3.8992×10^{15}	1.457	67.14	$8.8455 \cdot 10^{14}$
.635	1243.79	3.9456×10^{15}	1.472	69.89	$7.0321 \cdot 10^{14}$
.645	1233.96	3.9961×10^{15}	1.542	240.85	$7.9947 \cdot 10^{15}$
.655	1188.32	3.9677×10^{15}	1.572	226.14	$5.4969 \cdot 10^{15}$
.665	1228.40	4.0195×10^{15}	1.599	220.46	$4.8178 \cdot 10^{15}$
.675	1210.08	4.1171×10^{15}	1.608	211.76	$1.5719 \cdot 10^{15}$
.685	1200.72	4.1311×10^{15}	1.626	211.26	$3.1027 \cdot 10^{15}$
.695	1181.24	$4.1418 \cdot 10^{15}$	1.644	201.85	$3.0638 \cdot 10^{15}$
.6980	973.53	$1.2483 \cdot 10^{15}$	1.650	199.68	$9.9992 \cdot 10^{14}$
.700	1173.31	6.4301×10^{14}	1.676	180.50	$4.1124 \cdot 10^{15}$
.710	1152.70	$4.1324 \cdot 10^{15}$	1.732	161.59	$8.2262 \cdot 10^{15}$
.720	1133.83	$4.1199 \cdot 10^{15}$	1.782	136.65	$6.8025 \cdot 10^{15}$
.7277	974.30	2.9610×10^{15}	1.852	2.01	$5.0932 \cdot 10^{15}$
.730	1110.93	$8.8089 \cdot 10^{11}$	1.955	39.43	$1.8535 \cdot 10^{15}$
.740	1086.44	$4.0700 \cdot 10^{15}$	2.008	72.58	$2.9643 \cdot 10^{15}$
.750	1070.44	$4.0493 \cdot 10^{15}$	2.014	80.01	$4.6397 \cdot 10^{14}$
.7621	733.08	$4.1577 \cdot 10^{15}$	2.057	72.57	$3.3654 \cdot 10^{15}$
.770	1036.01	$2.6980 \cdot 10^{15}$	2.124	70.29	$5.0424 \cdot 10^{15}$
.780	1018.42	$4.0123 \cdot 10^{15}$	2.156	64.76	$2.3306 \cdot 10^{15}$
.790	1003.58	$3.9999 \cdot 10^{15}$	2.201	68.29	$3.2869 \cdot 10^{15}$
.800	988.11	$3.9902 \cdot 10^{15}$	2.266	62.52	$4.7856 \cdot 10^{15}$
.805U	860.28	$2.2067 \cdot 10^{15}$	2.320	57.03	$3.7393 \cdot 10^{15}$
.825	932.74	$7.0375 \cdot 10^{15}$	2.338	53.57	$1.1884 \cdot 10^{15}$
.830	923.87	$1.9356 \cdot 10^{15}$	2.356	50.01	$1.1027 \cdot 10^{15}$
.835	914.95	$1.9288 \cdot 10^{15}$	2.388	31.93	$1.5673 \cdot 10^{15}$
.8465	407.11	$3.2212 \cdot 10^{15}$	2.415	28.10	$9.8088 \cdot 10^{14}$
.860	857.46	$3.0707 \cdot 10^{15}$	2.453	24.96	$1.2367 \cdot 10^{15}$
.870	843.02	$3.7067 \cdot 10^{15}$	2.494	15.82	$1.0422 \cdot 10^{15}$
.875	835.10	$1.8448 \cdot 10^{15}$	2.537	2.59	$5.0182 \cdot 10^{14}$
.8875	817.12	$4.5865 \cdot 10^{15}$			

TABLE I - Air Mass - 1.5 Spectral Distribution

[a] The total irradiance under this curve amounts to 834.6 W m^{-2}

[b] Number of photons/sec-cm^2 in this wavelength interval between the corresponding wavelength and the one preceding it. Calculated using the average wavelength and irradiance for each wavelength interval.

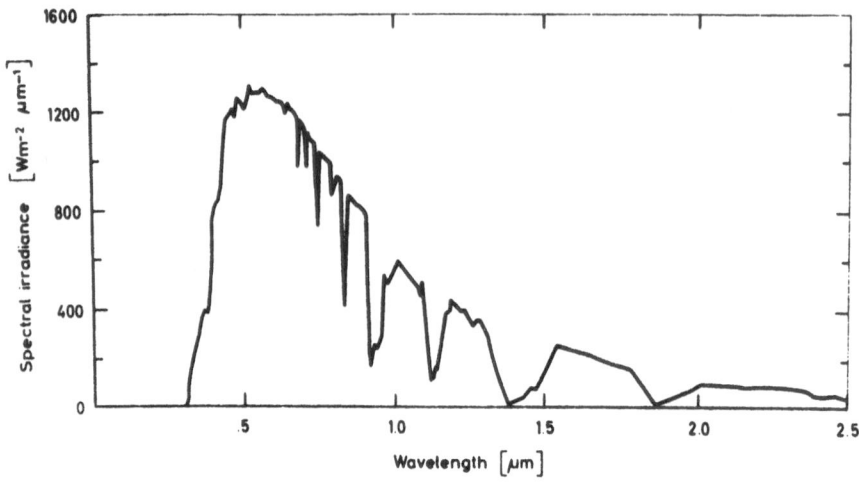

Fig. 1 : Air mass 1·5 spectral distribution

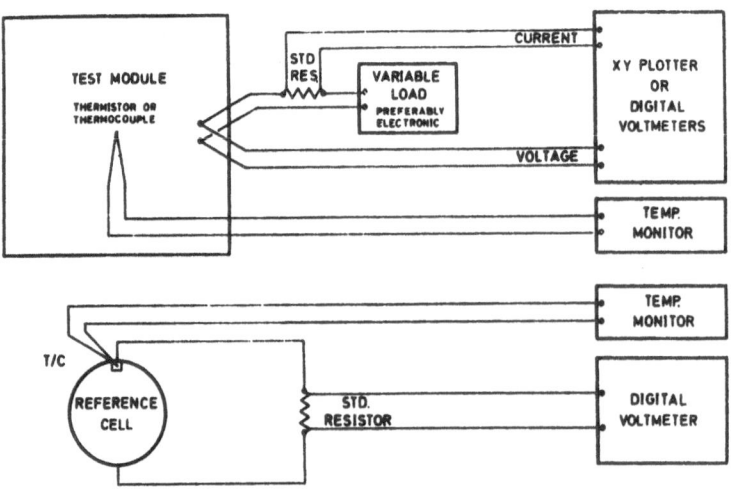

Fig. 2 : Performance test connections

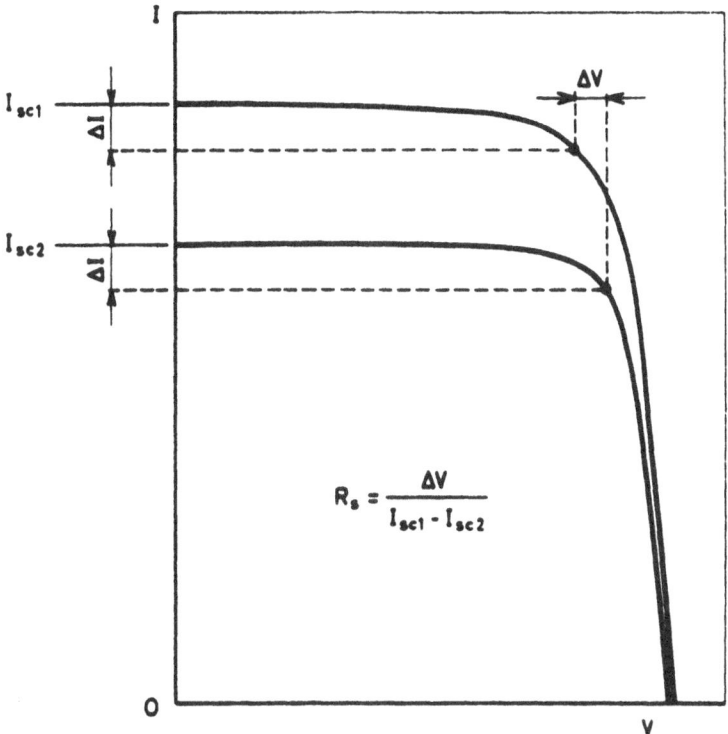

Fig. 3 : Determination of R_s

APPENDIX I - REFERENCE CELL CALIBRATION PROCEDURES

There are several possible techniques for calibrating master reference cells for terrestrial applications. Three of the better known are the following:

A.I.1   The "Normal Incidence" Method

A normal incidence pyrheliometer (NIP) and the reference cell, suitably temperature-controlled and masked by a collimator from the diffuse sky radiation, are kept pointed at the sun while simultaneous measurements are made of irradiance and short-circuit current. The "calibration value", in terms of short-circuit current per unit of "standard sunlight" irradiance ($AW^{-1} m^2$) is computed from the ratio of short-circuit current to irradiance, adjusted to the following atmospheric conditions:

. precipitable water vapour	2 cm
. turbidity coefficient	0.12
. air mass	1.5
. ozone	0.34 cm

The average of at least three calibration values on at least two different days is taken as the definitive calibration.

With this method, care must be taken to ensure that the NIP and reference cell collimators are accurately aligned to the sun (within ± 0.5°) and have the same field of view.

The following conditions for this method are specified in ERDA/NASA 1022/77/ 16, Revised Terrestrial Photovoltaic Measurement Procedures (June 1977):

a)   clear blue sky, with no observable cloud formations with a 15° half-angle cone surrounding the sun;

b)   irradiance between 750 and 900 Wm^{-2}, as measured by the NIP;

c)   atmospheric conditions sufficiently stable so that the variation in cell current is less than 0.5% during any 30 sec measurement period;

d)   the product of the optical air mass and atmospheric turbidity during measurement must be less than 0.25, the turbidity being determined from measurements at 500 nm. Alternatively, the ratio of uncollimated to collimated short-circuit current (using the NIP collimation angle of 5°42') must be less than 1.2;

e)   the optical air mass between the reference cell and the sun must be between 1 and 2.

A.I.2   The "Global Method"

The reference cell, uncollimated, is mounted horizontally on a temperature-controlled block, co-planar with a horizontal pyranometer and simultaneous readings of irradiance and short-circuit current are taken in global sunlight.

The spectral irradiance distribution of the global sunlight is measured within a short time of the calibration. The calibration value ($AW^{-1} m^2$), referred to standard sunlight, is then computed from the measured short-circuit current, the measured irradiance, the spectral response of the cell, the measured spectral irradiance distribution and the standard sunlight distribution. To minimize errors, the value is averaged over at least three different days.

The method of computation is as follows:

Let:

I_{sr} = short-circuit current of the reference cell in standard sunlight

I_{sg} = short-circuit current of the reference cell in global sunlight, as measured

s_λ = absolute spectral response of the reference cell at wavelength λ

$k_1 s_\lambda$ = relative spectral response of the reference cell at wavelength λ, as measured

$E_{s\lambda}$ = absolute spectral irradiance at wavelength λ of standard sunlight

$E_{g\lambda}$ = absolute spectral irradiance at wavelength λ of the global sunlight in which the short-circuit current was measured

$k_2 E_{g\lambda}$ = relative spectral irradiance at wavelength λ of the global sunlight, as measured

E_{glob} = measured irradiance of the global sunlight.

Then:

$$I_{sr} = \int s_\lambda E_{s\lambda} d\lambda \tag{A.1}$$

$$I_{sg} = \int s_\lambda E_{g\lambda} d\lambda \tag{A.2}$$

$$E_{glob} = \int E_{g\lambda} d\lambda = \frac{1}{k_2} \int (k_2 E_{g\lambda}) d\lambda$$

or:

$$I_{sg} = \frac{1}{k_2} \int s_\lambda (k_2 E_{g\lambda}) d\lambda = \frac{E_{glob}}{k_2 E_{g\lambda} d\lambda} \int s_\lambda (k_2 E_{g\lambda}) d\lambda \tag{A.3}$$

Dividing Eq.(A.1) by Eq.(A.3)

$$I_{sr} = I_{sg} \frac{\int (k_2 E_{g\lambda}) d\lambda}{E_{glob}} \cdot \frac{\int s_\lambda \cdot E_{s\lambda} \cdot d\lambda}{\int s_\lambda (k_2 E_{g\lambda}) d\lambda}$$

Since s_λ appears in both numerator and denominator, relative values of spectral response embodying the same constant may validly be used. Thus:

$$I_{sr} = I_{sg} \frac{\int (k_2 E_{g\lambda}) d\lambda}{E_{glob}} \cdot \frac{\int (k_1 s_\lambda) E_{s\lambda} \cdot d\lambda}{\int (k_1 s_\lambda)(k_2 E_{g\lambda}) d\lambda} \tag{A.4}$$

Thus, the calibration value I_{sr} (in A)/1000 Wm^{-2} may be computed from the tabulated values of $E_{s\lambda}$ and measured values of I_{sg}, $k_1 s_\lambda$, $k_2 E_{g\lambda}$ and E_{glob}. Similarly, the calibration value may be computed for any known spectral irradiance distribution.

The global method is simple and free from collimator alignment and field of view errors, but it requires a good test site and the following conditions for accurate and consistent calibration:

a)   clear, sunny weather, with the diffuse irradiance not greater than 0.25 of the global;

b)   global irradiance not less than 800 Wm^{-2};

c)   solar elevation not less than 54°;

d)   radiation sufficiently stable to allow the spectral irradiance distribution to be measured;

e)   no atmospheric pollution;

f)   prevailing good weather, so that measurements can be taken on three suitable days without undue delay.

A.I.3   Combined Space and Terrestrial Reference Cell Calibration

In contract to the "normal incidence" and the "global" methods which aim at approaching standard sunlight conditions by using actual terrestrial sunlight, this method requires a precisely matched sun simulator. Therefore, this method can be called the sun simulator method.

Moreover, the sun simulator method uses a set of spectrally different AMO reference solar cells out of an accepted reference standard base, instead of a pyrheliometer or pyranometer.

The requirements on a sun simulator for this application may be summarized as follows:

.   the spectral match of the sun simulator to extra-terrestrial sunlight (AMO) shall be so close that spectral mismatch errors resulting from the different spectral response of the full variety of cells to be measured at constant intensity setting of the simulator are less than 1%;

.   the stability of the simulator during the measuring period shall be such that a reproducibility of cell calibration of better than ± 0.3% is guaranteed.

For the calibration of reference cells at standard terrestrial sunlight conditions, the following steps are required:

a)   Set the simulator to AMO sunlight intensity and verify the spectral match to AMO by:

.   measuring the relative spectral energy distribution,

.   comparing the short-circuit current values of spectrally different accepted AMO reference standards measured at constant intensity setting of the simulator with the original calibration data;

b)   Measure the AMO short-circuit of the cells to be calibrated;

c)   Measure the relative spectral response of the cells to be calibrated. Make sure that errors due to non-linear intensity dependence and enhancement/quenching effects are eliminated;

d)   Place standard sunlight filter which simulates the atmospheric absorption/ scattering in the simulator light beam and measure the standard sunlight short-circuit current I_{sr} of the cell to be calibrated;

e) For verification, compute the difference between the AM0 and standard sunlight short-circuit currents of the reference cells thus:

$$I_{or} - I_{sr} = \int s_\lambda (E_{0\lambda} - E_{s\lambda}) d\lambda \doteq \int s_\lambda A_\lambda d\lambda \qquad (A.5)$$

where:

I_{or} = short-circuit current of the reference cell in AM0 sunlight
I_{sr} = short-circuit current of the reference cell in standard sunlight
s_λ = absolute spectral response of the reference cell at wavelength λ
$E_{0\lambda}$ = absolute spectral irradiance at wavelength λ of AM0 sunlight
$E_{s\lambda}$ = absolute spectral irradiance at wavelength λ of standard sunlight
A_λ = the difference between these spectral irradiances at wavelength λ, which corresponds to the atmospheric attenuation.

Then:

$$I_{or} = \int s_\lambda E_{0\lambda} d\lambda = \frac{1}{k} \int (ks_\lambda) E_{0\lambda} d\lambda$$

or

$$k = \frac{1}{I_{s0}} \int (ks_\lambda) E_{0\lambda} d\lambda$$

where:
ks_λ = relative spectral response of the reference cell at wavelength λ.

Thus:

$$I_{sr} = [1 - \frac{\int (ks_\lambda) A_\lambda d\lambda}{\int (ks_\lambda) E_{0\lambda} d\lambda}] I_{or} \qquad (A.6)$$

This value has to be compared with the value measured in d). A calibration is accepted only if the two values differ by less than 1%.

The advantages of the sun simulator method are:
. it is independent of weather conditions, seasons, location;
. it is continuously available under controlled laboratory conditions;
. spectral measurements of the light source and of the solar cells are necessary for control purposes only whereas the actual calibration is based on a simple short-circuit current measurement, thus reducing the complexity of the measurement and the influence of spectral errors.

Reference

K. Bogus, J.C. Larue and R.L. Crabb, Proc. 1st E.C. Photovoltaic Solar Energy Conf., Luxembourg, p.754 (1977).

APPENDIX II - COMPUTATION OF SPECTRAL MISMATCH ERROR IN SHORT-CIRCUIT CURRENT MEASUREMENTS

Let:

I_1 = short-circuit current of the reference cell in standard AM1.5 sunlight at 1000 Wm^{-2}

I_2 = short-circuit current of the reference cell, as measured under the simulator

$s_{1\lambda}$ = absolute spectral response of the reference cell at wavelength λ

$k_1 s_{1\lambda}$ = relative spectral response of the reference cell at wavelength λ

I_3 = short-circuit current of the test sample in standard sunlight at 1000 Wm^{-2}

I_4 = short-circuit current of the test sample, as measured under the simulator

$s_{2\lambda}$ = absolute spectral response of the test sample at wavelength λ

$k_2 s_{2\lambda}$ = relative spectral response of the test sample at wavelength λ

$E_{s\lambda}$ = absolute spectral irradiance at wavelength λ of standard AM1.5 sunlight, adjusted to a total irradiance of 1000 Wm^{-2}

$k_3 E_{s\lambda}$ = relative spectral irradiance of standard AM1.5 sunlight at wavelength λ

$E_{t\lambda}$ = absolute spectral irradiance of the simulated sunlight at wavelength λ

$k_4 E_{t\lambda}$ = relative spectral irradiance of the simulated sunlight at wavelength λ.

Then:

$$I_1 = \int s_{1\lambda} \cdot E_{s\lambda} \cdot d\lambda$$

$$I_2 = \int s_{1\lambda} \cdot E_{t\lambda} \cdot d\lambda$$

$$I_3 = \int s_{2\lambda} \cdot E_{s\lambda} \cdot d\lambda$$

$$I_4 = \int s_{2\lambda} \cdot E_{t\lambda} \cdot d\lambda$$

Integration of the products of the measured relative spectral responses and the relative spectral irradiances yields the following parameters:

$$A_1 = \int k_1 s_{1\lambda} \cdot k_3 E_{s\lambda} \cdot d\lambda = k_1 k_3 I_1$$

$$A_2 = \int k_1 s_{1\lambda} \cdot k_4 E_{t\lambda} \cdot d\lambda = k_1 k_4 I_2$$

$$A_3 = \int k_2 s_{2\lambda} \cdot k_3 E_{s\lambda} \cdot d\lambda = k_2 k_3 I_3$$

$$A_4 = \int k_2 s_{2\lambda} \cdot k_4 E_{t\lambda} \cdot d\lambda = k_2 k_4 I_4$$

The simulator irradiance is set to produce the calibrated short-circuit current from the reference cell, thus

$$I_1 = I_2 \qquad \text{and} \qquad \frac{k_3}{k_4} = \frac{A_1}{A_2}$$

The error in measuring the short-circuit current of the test sample is then given by:

$$\frac{\Delta I}{I_3} = \frac{I_4 - I_3}{I_3} \cdot 100\% = [\frac{A_1 A_4}{A_2 A_3} - 1] \cdot 100\%$$

Thus, in general, the error may be computed from the integrated products of the relative spectral responses of the reference cell and the test sample and the relative spectral irradiances of the simulator and standard sunlight.

If the procedure for selecting two master reference cells and calibrating a working reference cell against them is followed, as laid down in section 10, the maximum error from spectral mismatch may be computed by assuming that the spectral response of the reference cell is at one extreme of the spectral response spread and the spectral response of the test sample is at the other. The calculated error should then be halved, since in practice the irradiance will be set by the working reference cell, representing the middle of the spectral response spread.

APPENDIX III - MEASUREMENT OF RELATIVE SPECTRAL RESPONSE

The relative spectral response of a solar cell is measured by placing it on a temperature-controlled mount, irradiating it uniformly from a monochromatic source and measuring the short-circuit current and the irradiance at fixed wavelength intervals over the response range. The currents are then divided by the irradiances or a proportional parameter and plotted as a function of wavelength. Alternatively, the irradiance may be kept constant (for instance by varying the width of a monochromator exit slit), in which case the relative spectral response is obtained directly from the current readings.

The irradiance monitor may be a vacuum thermocouple, a pyroelectric radiometer or other suitable detector. Another alternative is a reference solar cell whose spectral response, covering the required range, has been pre-calibrated by a recognized standards laboratory. In this case, the spectral response of the test cell is computed as follows:

$$P_{t\lambda} = P_{r\lambda} \cdot \frac{I_{mt\lambda}}{I_{mr\lambda}}$$

where:

$P_{t\lambda}$ = the spectral response of the test cell at wavelength λ

$P_{r\lambda}$ = the spectral response of the reference cell at the same wavelength

$I_{mt\lambda}$ = the measured short-circuit current of the test cell at wavelength λ

$I_{mr\lambda}$ = the measured short-circuit current of the reference cell at the same wavelength.

Figures 4 and 5 show two possible test arrangements, the first embodying a quartz prism monochromator and the second a filter wheel as the monochromatic source. In both cases, the light source is a 1000 W tungsten halogen lamp operated from a stable supply at a colour temperature of 3200 K. The test cell and the irradiance monitor are mounted on opposite sides of a rotatable temperature-controlled block, so that either may be presented to the monochromatic beam in precisely the same place. Alternatively, they may be mounted on a slide with suitable positioning stops.

The filter wheel carries seventeen narrow-band filters centered at wavelengths ranging from 350 nm to 1150 nm in 50 nm steps and arranged so that each can be indexed in turn between the light source and the test cell or irradiance monitor. It is important that the filters should have negligible (under 0.2%) side-bands. The monochromator is normally used with fixed slits and manually set to the same wavelength steps.

With crystalline silicon and other cells where the response may be assumed to change linearly with irradiance, the short-circuit current of the cells (voltage drop across a standard 4-terminal fixed resistor) and the open-circuit voltage of the vacuum thermocouple or radiometer may be measured directly with a DC digital voltmeter or potentiometer. The requirements for instrumentation accuracy and the measurement of short-circuit currents laid down in section 13 apply here. If the DC method is used, the exit beam, test cell and irradiance monitor must be completely enclosed in an anti-reflective light-tight box and meticulous precautions taken to avoid thermal and other random e.m.f.s which would cause errors. Alternatively, the exit beam may be chopped at a low frequency and the output voltages amplified and rectified. In this case, it is important to ensure that the amplifiers are linear and drift-free.

With non-linear devices such as cuprous sulphide-cadmium sulphide solar cells, it is necessary to use a chopped monochromatic beam and increase the irradiance to the desired operational level (e.g. 1000 Wm^{-2}) by unmodulated bias light from a suitable steady state simulator.

Figure 6 shows a method for pulsed solar spectral response measurements. The test set-up comprises:
. a powerful photography flash lamp which provides high intensity light pulses;
. a set of 16 narrow band-pass interference filters with transmission curves centered at about 50 nm intervals from 350 to 1100 nm;
. a light-box designed to avoid light leakage, that is to ensure solar cell illumination exclusively through the filter in use;
. a sample holder ensuring reproducible sample and standard positioning;
. spectrally calibrated standard solar cell(s). This calibration is better performed in specialized photometry laboratories such as those of PTB (Germany), NPL (UK), etc.;
. a decade load resistor;
. an electronic "peak-detector";
. a digital voltmeter.

Apart from the change of light source, the measurement method remains based on the comparison of the short-circuit currents generated respectively by the cell to be measured and by the spectrally calibrated standard cell. However, in assembling the test set-up and performing the measurements, attention should be paid to the following problems:
. illumination uniformity of the test plane. Very important for absolute measurements when sample and standard are of different dimensions;
. light pulse intensity monitoring and subsequent reading correction which may either be manual or, better, automatic;
. filters transmission curves. Should be checked periodically, in particular to detect any "harmonic" transmission;
. load resistor calibration and contacts resistance;
. linearity of response I_{sc} versus light intensity at all illumination levels and all wavelengths;
. keep load resistor to minimum practical value in order to remain as close as possible to true short-circuit conditions.

Reference: J.C. Larue, Proc. 2nd E.C.Photovoltaic Solar Energy Conf.p.477 (1979).

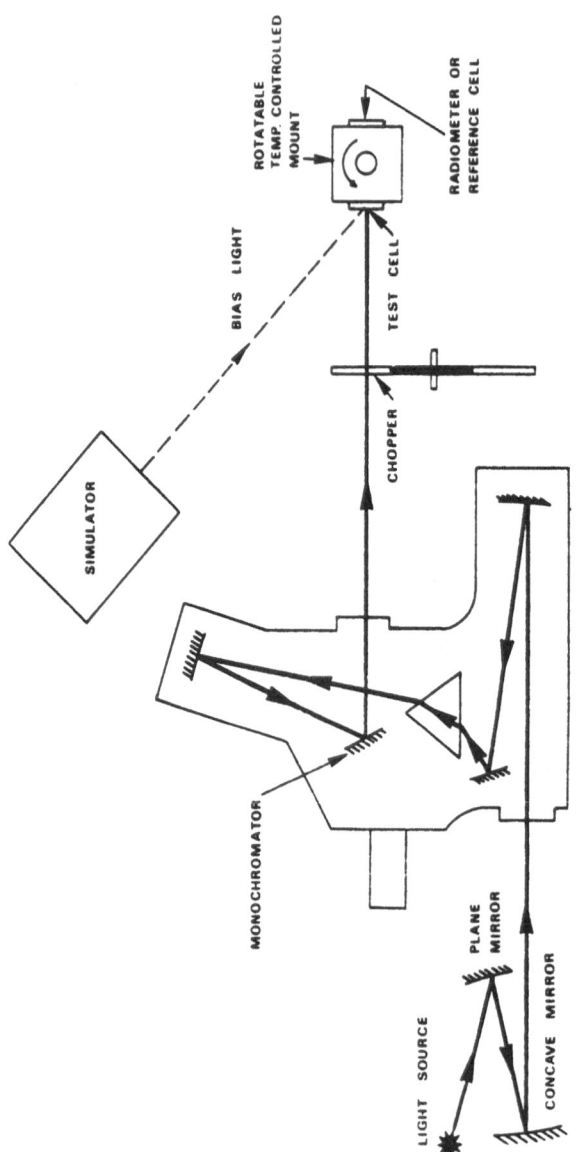

Fig. 4 : Spectral response measurement using a monochromator

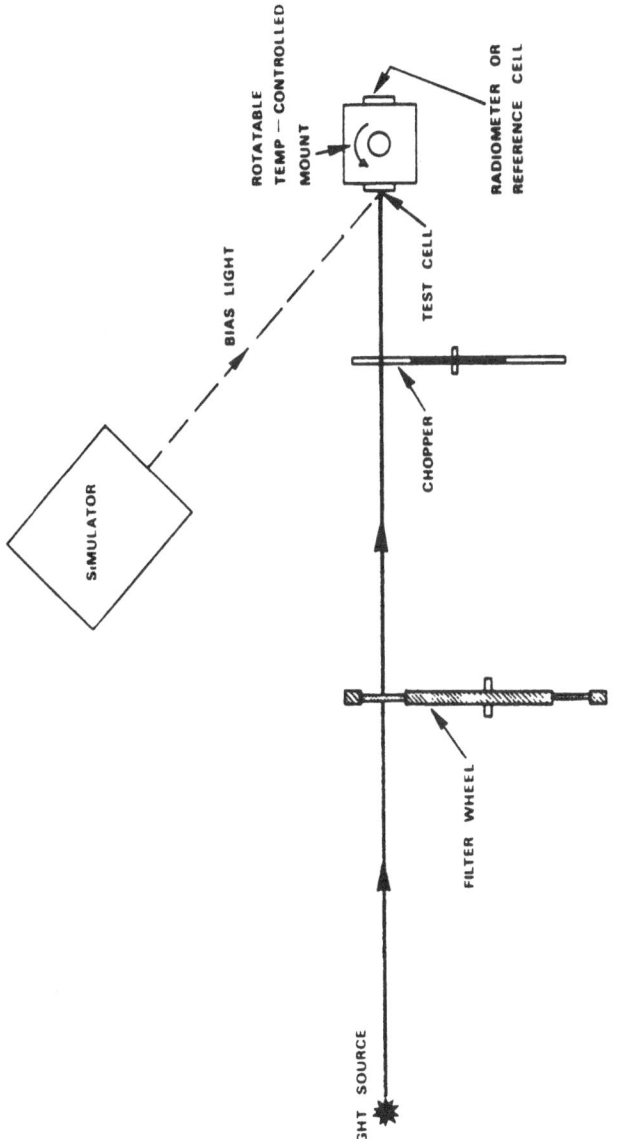

Fig. 5 : Spectral response measurement using a filter wheel

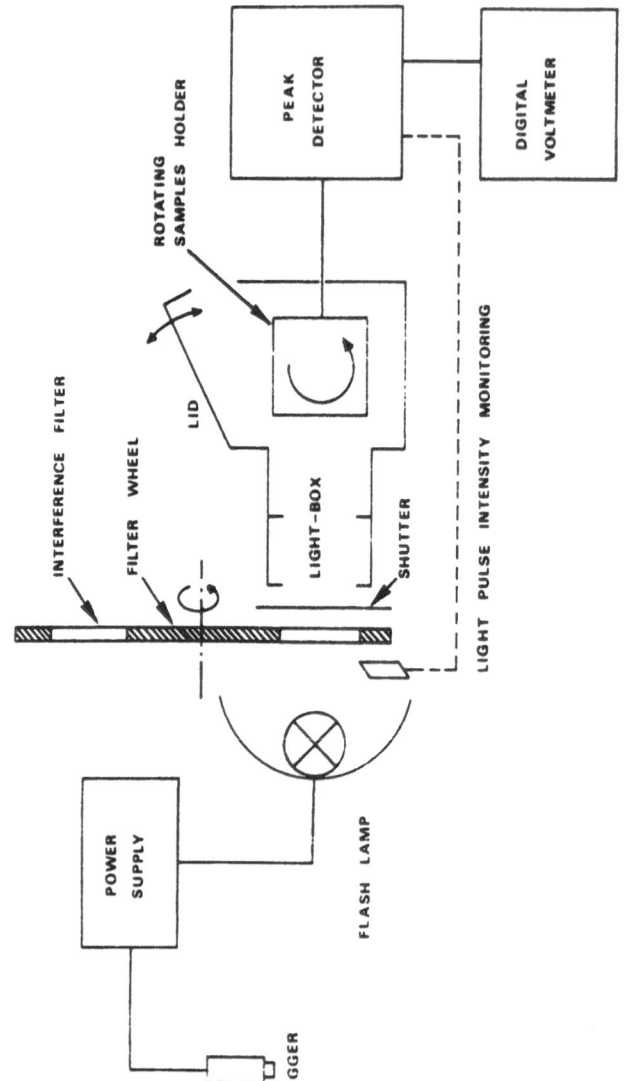

Fig. 6 : Experimental set-up for pulsed spectral response measurement

APPENDIX IV - DEFINITION OF TERMS

A.IV.1 Meteorological terminology

Air Mass

The length of path through the Earth's atmosphere traversed by the direct solar beam, expressed as a multiple of the path traversed to a point at sea level with the sun at zenith.

Solar Elevation

The angle between the direct solar beam and the horizontal plane (degrees).
Air Mass = Cosecant of the Solar Elevation.

Direct Irradiance [E_{dir}]

The radiant power from the sun (and a small area of sky surrounding it, defined by the acceptance angle of the pyrheliometer) incident upon unit surface area (W m^{-2}).

Diffuse Irradiance [E_{diff}]

The radiant power from the sky incident upon unit surface area (W m^{-2}).

Global Irradiance [E_{glob}]

The total solar radiant power incident upon unit area of a horizontal surface (W m^{-2}).
Global Irradiance = Direct Irradiance (horizontal) + Diffuse Irradiance (horizontal).

Total Irradiance [E_{tot}]

The total solar radiant power incident upon unit area of an inclined surface (w m^{-2}).

Spectral Irradiance [E_{λ}]

The irradiance (global, direct or diffuse) per unit bandwidth at a particular wavelength (W m^{-2} μm^{-1}).

Spectral Photon Irradiance

The photon flux density N_{λ} at a particular wavelength (cm^{-2} sec^{-1} μm^{-1}).
N_{λ} may be calculated from E_{λ} by $N_{\lambda} = 5.035 \times 10^{14} \lambda E_{\lambda}$ (λ in [μm]).

Spectral Irradiance Distribution

Spectral irradiance plotted as a function of wavelength.

Direct Insolation

The radiant energy from the sun (and a small area of sky surrounding it, defined by the acceptance angle of the pyrheliometer) incident upon unit surface area during a specified time period (MJ m^{-2} per hour, day, week, month or year, as the case may be).

Diffuse Insolation

The radiant energy from the sky incident upon unit surface area during a specified time period (units as for direct insolation).

Global Insolation

The total solar radiant energy incident upon unit area of a horizontal surface during a specified time period (units as for direct insolation).
Global Insolation = Direct Insolation (horiyontal) + Diffuse Insolation (horizontal) for the same time period.

Pyranometer

A radiometer normally used to measure global irradiance (or, with a shade ring or disc, diffuse irradiance) on a horizontal plane. Can also be used at an angle to measure the total irradiance on an inclined plane which in this case includes an element due to radiation reflected from the foreground.

Pyrheliometer   (sometimes called a "Normal Incidence Pyrheliometer" or NIP)

A radiometer, complete with a collimator, used to measure direct irradiance.

Turbidity

The reduced transparency of the atmosphere, caused by absorption and scattering of radiation by solid or liquid particles, other than clouds, held in suspension. As defined by Angström, the turbidity of the atmosphere is related to t, the extinction coefficient at a wavelength of 1000 nm (normally called the "turbidity coefficient") and ϵ, the wavelength exponent in the expression for the aerosol extinction function:

$$a_{D,\lambda} = t\lambda^{-\epsilon}$$

t values less than 0.10 denote a very clear condition, whereas values greater than 0.20 are a distinctly hazy condition. The average value of ϵ, which is dependent on the particle size distribution, was assumed by Angström to be about 1.3.

Precipitable Water Vapour Content

The volume of precipitable water vapour (cm^3) in a vertical column of the atmosphere 1 cm^2 in cross section.

Ozone Content

The volume of ozone (cm^3) at Standard Temperature and Pressure in a vertical column of the atmosphere 1 cm^2 in cross section.

A.IV.2 Photovoltaic Terminology

Solar Cell

The basic photovoltaic device which generates electricity when exposed to sunlight.

Module

The smallest complete environmentally protected assembly of interconnected solar cells.

Panel

A group of modules fastened together, pre-assembled and interconnected, designed to serve as an installable unit in an array.

Array

A mechanically integrated assembly of modules or panels together with support structure but exclusive of foundation, tracking, thermal control and other components, as required, to form a dc power producing unit.

Array Sub-Field

A group of solar photovoltaic arrays associated by a distinguishing feature such as physical arrangement, electrical interconnection or power conditioning.

Array Field

The aggregate of all solar photovoltaic arrays generating power within a given system.

Photovoltaic System

An installed aggregate of solar arrays generating power for a given application. A system may include the following subsystems:
a)   photovoltaic array field
b)   support foundation
c)   power conditioning and control equipment
d)   storage
e)   active thermal control
f)   land, security system and buildings
g)   conduit/wiring
h)   instrumentation.

Cell Area

The entire frontal area of the solar cell, including the contact grid (cm^2).

Module Area

The entire frontal area of the module, including borders, frame and any protruding mounting lugs (m^2).

Array Area

The entire frontal area of the array, including intermodule spacing and framework (m^2).

Cell Packing Efficiency

The ratio of the total illuminated cell area to the module area.

Module Packing Efficiency

The ratio of the total illuminated module area to the array area.

Short-circuit Current [I_{sc}]

The output current of a photovoltaic generator in the short-circuit condition at a particular temperature and irradiance, measured in accordance with Section 13 of this Specification (mA or A).

Open-circuit Voltage [V_{oc}]

The voltage generated across an unloaded (open) photovoltaic generator at a particular temperature and irradiance, as measured with a voltmeter having an internal resistance of at least 20 $k\Omega/V$ (mV or V).

Current-Voltage Characteristic

The output current of a photovoltaic generator at a particular temperature and irradiance, plotted as a function of output voltage.

Maximum Power

The power at the point on the current-voltage characteristic where the product of current and voltage is a maximum (W).

Conversion Efficiency [η]

The ratio of the maximum power to the product of area and irradiance, expressed as a percentage:

$$\eta = \frac{\text{Maximum Power}}{\text{Area x Irradiance}} \times 100\%$$

Fill Factor [FF]

The ratio of maximum power to the product of open-circuit voltage and short-circuit current:

$$FF = \frac{\text{Maximum Power}}{V_{oc} \times I_{sc}}$$

Spectral Response* [s_λ]

The short-circuit current density generated by unit irradiance at a particular wavelength (AW^{-1}), plotted as a function of wavelength.

Relative Spectral Response [$s_{\lambda rel}$]

The spectral response normalized to unity at wavelength of maximum response.
$s_{\lambda rel} = s_\lambda / s_{\lambda max}$.

Time Constant

The time required for a radiometer or photovoltaic generator to attain 63.2% of its steady state value after a step change of irradiance (μs).

Angle of Incidence

The angle between the direct solar beam and the normal to the active surface (degrees).

Reference Solar Cell or Module

A solar cell or module used to measure irradiance and to set simulator irradiance levels in terms of a standard solar spectral energy distribution ("standard sunlight"). It is a cell or module, having essentially the same configuration and relative spectral response as the test cells or modules, which has been calibrated at $25 \pm 2°C$ in terms of short-circuit current per unit of "standard sunlight" irradiance ($AW^{-1} m^2$) by an approved Solar Cell Calibration Agency.

Standard Test Conditions [STC]

Cell junction temperature, as measured at the front contact: $25 \pm 2°C$.
Irradiance, as measured with a reference solar cell: $1000 \ Wm^{-2}$.
Standard solar spectral energy distribution, as obtained by multiplying the values in Table I by the normalization factor 1000/834.6.

---

* The term "response" is commonly used but "responsivity" is strictly correct.

Rated Power * [P_L]

The power output of a photovoltaic generator at the nominal working voltage V_L under Standard Test Conditions (W).

This document was drafted by Mr. F.C. Treble.

The work was co-ordinated and edited by the CEC Joint Research Centre, Ispra Establishment, Italy.

---

* Sometimes referred to as peak power (W_p).

PHOTOVOLTAIC MODULE CONTROL TEST SPECIFICATIONS
SPECIFICATION 501
Report EUR7545EN - August 1981

edited by K. Krebs
Commission of the European Communities
Joint Research Centre - Ispra Establishment
21020 Ispra (Va) - Italy

ABSTRACT

This document lays down test specifications for photovoltaic module control
tests of the Commission of the European Communities which shall be applied
bw the Joint Research Centre for the acceptance of prototype and production
modules for pilot and demonstration projects of the Commission. It contains the
test schedule and a detailed description of 20 control tests. The purpose of
these tests is to provide data on the performance rating of photovoltaic modules
and to identify environmental factors and design features which could affect ;
their durability.

CONTENTS

G. Beghi (ed.), Performance of Solar Energy Converters: Thermal Collectors and Photovoltaic Cells, 501–522.

CT.11 Humidity-Freezing Test
CT.12 Hail Resistance Test
CT.13 Mechanical Loading Test
CT.14 Damp Heat Long Exposure Test
CT.15 High Temperature Long Exposure Test
CT.16 UV Test
CT.17 O_3 Test
CT.18 SO_2 Test
CT.19 Salt Mist Test
CT.20 Ice Formation Test

TABLES

FIGURES

APPENDIX A

Tests and Test levels

1. INTRODUCTION

This Specification lays down Control Tests of the Commission of the European Communities, which shall be applied by the Joint Research Centre for the acceptance of prototype and production modules of pilot and demonstration projects of the Commission.

The general objective of these tests is to identify key environmental factors and design features which could affect the attainment of a sufficiently long life time. These tests are also considered useful to provide information on the relative strengths and acceptabilities of design alternatives.

Concerning performance testing reference is made in this Specification to the publication "Standard Procedures for Terrestrial Photovoltaic Performance Measurements", CEC-Specification No.101 (Issue 2), EUR report 7078EN (1981).

2. DATA SHEET AND MARKING

Each module which is submitted for testing must be accompanied by a **data sheet** containing the following information:
. manufacturer's name and address
. place of manufacture
. type or model number
. type, shape and size of solar cell
. number of cells in series and in parallel
. description of the encapsulation system
. nominal weight with tolerances
. nominal operating voltage, V_L
. nominal operating cell temperature, NOCT
. maximum ratings
. minimum and maximum temperatures
. wind loading
. maximum hail diameter
. maximum system voltage
. temperature coefficients of current and voltage, α and β, measured in accordance with CEC Specification No.101 (Issue 2) (1981)
. minimum and maximum values of the following module parameters under standard test conditions (cell junction temperature of $25 \pm 2°C$ and irradiance of 1000 Wm^{-2}), measured in accordance with CEC Specification No.101 (Issue 2) (1981):
  - short-circuit current, I_{sc}
  - open-circuit voltage, V_{oc}
  - current at nominal operating voltage, I_L
. typical voltage-current characteristics measured or calculated in accordance with CEC Specification No.101 (Issue 2) (1981), at at least three irradiance levels (including 800 and 1000 Wm^{-2}) and at least three module temperatures including $25°C$ and the nominal operating cell temperature
. a scale drawing of the module, showing its main constructional features, leading dimensions (with tolerances), mounting hole or clip positions (with tolerances) and electrical terminals or leads, polarity and any built-in protective devices
. handling, mounting and electrical installation instructions (including diode protection and precautions to avoid corrosion where necessary).

Each production module shall at least carry the following clear and indelible **markings**:
. name or monogram or symbol of manufacturer
. type or model number
. polarity of terminals or leads (colour coding is permissible)
. serial number
(The date of manufacture shall be stated or traceable from the serial number).

3. TEST PROCEDURES

3.1 Summary and Test Schedule

The module shall be able to withstand exposure to the following environmental conditions:

1) temperature cycling between $-40°C$ and NOCT $+40°C$ in accordance with Test CT.10
2) humidity-freezing in accordance with Test CT.11
3) hail impact and mechanical loading in accordance with Tests CT.12 and CT.13
4) damp heat and high temperature long exposure in accordance with Tests CT.14 and CT.15
5) ultraviolet, ozone and SO_2-effects in accordance with Tests CT.16, CT.17 and CT.18
6) salt mist effects in accordance with Test CT.19
7) ice formation in accordance with Test CT.20.

Besides the electrical performance requirements as laid down in Tests CT.6 and CT.8, terminations, module twist and insulation shall meet the conditions in Tests CT.3, CT.4 and CT.5. The module shall withstand hot-spot heating due to short-circuiting at a solar irradiance of 1000 Wm^{-2} even in the case of shadowing, of cell cracks or of interconnect failures. The resistance to hot-spot effects shall be demonstrated by Test CT.9.

The complete test schedule is laid down in Tables I and II. Tests in Table I are mandatory. The output power degradation after completion of all type A tests shall not exceed 5% of the baseline electrical performance data. Module No.4 shall be kept as reference.

TABLE I - Test Schedule A

Test	Denomination of Test	Module No.			
		1	2	3	4
CT.1	Visual Inspection	o	o	o	o
5	Insulation Test	o	o	o	o
CT.6	Performance at STC	o	o	o	o
7	Determination of NOCT			o	
CT.8	Performance at NOCT			o	
9	Hot-Spot Heating Test			o	
CT.10	Temperature Cycling Test	o	o		
6	Performance at STC	o	o		
CT.11	Humidity-Freezing	o	o		
5	Insulation Test	o	o	o	
CT.6	Performance at STC	o	o	o	
1	Visual Inspection	o	o	o	

TABLE II · Test Schedule B

Test	Denomination of Test	Module No. 5	6	7	8
CT.1	Visual Inspection	o	o	o	o
2	Dimensions and Weight	o			
CT.3	Robustness and Terminations	o			
4	Mounting Twist	o			
CT.5	Insulation Test	o	o	o	o
6	Performance at STC	o	o	o	o
CT.12	Hail Resistance	o			
13	Mechanical Loading	o			
CT.14	Damp Heat Long Exposure		o		
15	High Temperature Long Exposure			o	
CT.16	UV-Test				o
6	Performance at STC				o
CT.17	O_3-Test				o
6	Performance at STC				o
CT.18	SO_2-Test				o
6	Performance at STC				o
CT.19	Salt Mist Test				o
20	Ice Formation Test	o			
CT.5	Insulation Test	o	o	o	o
6	Performance at STC	o	o	o	o
CT.1	Visual Inspection	o	o	o	o

The tests in Table II shall be used mainly to compare various module designs with respect to their behaviour under specified environmental conditions and in order to assess possible failure modes.

No module shall be accepted and/or installed at a test site that produces less than 90% of the average performance values established for the 8 modules under Standard Test Conditions.

A brief summary of tests and test levels is given in Appendix A.

3.2 Test Specifications

CT.1   Visual Inspection

Each module shall be carefully inspected under not less than 1000 lux to ensure that all the markings listed in Section 2 have been clearly made and that there are no unacceptable faults. Any suspect features (e.g. faulty soldering) shall be examined at a magnification between 8 and 30 X. Faults which shall warrant rejection are:

. broken or cracked window
. broken cell
. faulty interconnection or joint
. cells touching one another or the frame
. failure of adhesive bonds
. bubbles forming a continuous path between a cell and the edge of the module
. tacky surfaces of plastic materials
. faulty terminations
. any other fault which will lead to damage.

A note shall be made of the nature and position of any minor faults, e.g. cracks or bubbles, which may worsen in the course of subsequent tests*.

CT.2   Dimensions and Weight

The main dimensions and weight of the modules shall be checked to ensure that they are within the limits quoted on the Data Sheet.

CT.3   Robustness of Terminations Test

The objective of the test is to determine that the terminations and the attachment of the terminations to the body of the module will withstand such stresses as are likely to be applied during normal assembly or handling operations. Modules are divided into two types:
. type A : those which have wire or flying lead terminations
. type B : those which have other terminations such as tags, threaded studs, screws, etc.

1)   Procedure A

Tensile Test
With the termination in its normal position and the module held by its body, a force giving a value as stated in Table III shall be applied to the termination in the direction away from the body of the module.
The force shall be applied progressively (without any shock) and then maintained for a period of 10 s.

_____

* It is useful to complete the inspection by laser and infrared scanning of each module.

TABLE III - Tensile and Bending Forces

Total cross sectional area of conductor (mm²)	Tensile Test Force (N)	Bending Test Force (N)
S ≤ 0.05	1	0.5
0.05 < S ≤ 0.07	2.5	1.25
0.07 < S ≤ 0.20	5	2.5
0.20 < S ≤ 0.50	10	5
0.50 < S ≤ 1.20	20	10
1.20 < S	40	20

(1N = 0.102 kp)

Bending Test

With the termination in its normal position the module is held by its body in such a manner that the axis of the termination is vertical, a mass applying a force of the value given in Table III is then suspended from the end of the termination.

The body of the module is inclined (over a period of 2 s to 3 s) through an angle of approximately 90° in the vertical plane and then returned over the same period of time. The body is then inclined (over a period of 2 s to 3 s) through an angle of approximately 90° in the opposite direction and again returned over the same period of time. This operation constitutes one bend cycle.

Each termination shall be subjected to 10 cycles, five in the plane of the cells and five at 90° to the plane of the cells. The modules should remain free of the faults listed in Test CT.1.

Verify that the module is still functioning by checking short circuit current under suitable illumination before and after the test.

2) Procedure B

Tensile Test

a)   for modules with exposed terminals each termination shall be tested to the procedures as given in para.1);

b)   if the terminations are enclosed in a protective box, then the following test shall be applied.

A cable of the size and type recommended by the module manufacturer, cut to a suitable length shall be connected to the terminations inside the box using the module manufacturer's recommended procedures. The cable shall be taken through the hole of the cable gland, taking care to utilize any cable clamp arrangement provided. The lid of the box shall be securely replaced.

The module shall then be subjected to the tests given in para.1) for tensile and bending stress of wire terminations.

Torque Test

This test shall be applied to all terminations with threaded studs or screws whether contained in a box or exposed.

With the module held by its normal fixing device the torques given in Table IV shall be applied to the screws or nuts normally fitted to each terminal for a period of 10 s. During this test, a washer or metal plate with a nominal clearance hole for the screw thread shall be placed between the screw head or nut and the surface to which it is tightened.

The nuts or screws should be capable of being loosened afterwards.

TABLE IV · Torques

Nominal Thread Diameter (mm)	2.5	3	3.5	4	5	6
Torque (Nm)	0.4	0.5	0.8	1.2	2.0	2.6

(1 Nm = 0.102 kpm)

The modules should remain free of the faults listed in Test CT.1. Verify that the module is still functioning by checking short circuit current under suitable illumination before and after the test.

CT.4 Mounting Twist Test

$$\alpha = 1.2° \quad (20 \text{ mm/m})$$

One corner of the module shall be displaced from the plane of the three others by an amount x,

$$x = \alpha(L^2 + W^2)^{\frac{1}{2}} \quad (m)$$

where:

α = 0.021
L = length of module (m)
W = width of module (m).

This corresponds to a deformation angle of 1.2 degrees.
The modules should remain free of the faults listed in Test CT.1.
Verify that the module is still functioning by checking short circuit current under suitable illumination before and after the test.

CT.5 Insulation Test

1) Procedure A

A voltage of 1000 V shall be applied between the shorted output terminals and the metal frame.

If the module frame is a poor electrical conductor, the module shall be placed in contact with a metallic frame simulating an array supporting structure and the voltage shall be applied between this frame and the shorted output terminals.

The insulation resistance shall in neither case be less than 100 MΩ after application of the voltage for one minute. This test is not required for modules designed for systems with operating voltage levels less than 50 V.

2) Procedure B

The shorted output terminals shall be connected to the positive terminal of an insulation tester with a current limitation of 50 μA. The negative terminal of the tester must be connected to the electrically conducting frame of the module or a simulated metallic supporting structure. With no illumination, the voltage shall be increased at a rate not exceeding 500 $VDCs^{-1}$ to a maximum value equal to 1000 VDC plus twice the maximum system voltage declared on the Data Sheet, and then held at that level for 1 minute.

There should be no evidence of dielectric breakdown or surface tracking.

CT.6 Electrical Performance at Standard Test Conditions (STC)

$T = 25°C$	$E = 1000$ Wm^{-2}
$T = 25°C$	$E = 800$ Wm^{-2}

The voltage-current characteristics of the module shall be measured in accordance with CEC Specification No.101 (Issue 2), EUR 7078EN at 25°C and 1000 Wm^{-2} and at 25°C and 800 Wm^{-2}.

The values of I_{sc}, V_{oc} and I_L should in no case be less than the relevant minimum values quoted on the Data Sheet.

CT.7 Determination of Nominal Operating Cell Temperature (NOCT)

	$E = 800$ Wm^{-2}
$T_{air} = 20°C$	$V_w = 1$ ms^{-1}

NOCT is defined as the temperature of a cell in a module under the following nominal conditions (NC):

. irradiance   800 Wm^{-2}
. air temperature   20°C

. wind average velocity $\qquad$ 1 ms^{-1}
. mounting $\qquad$ within an array or a similar structure
oriented normal to solar noon
. electrical load $\qquad$ open circuit condition

The approach for determining NOCT is based on the observation that the temperature difference (T_{cell} - T_{air}) is largely independent of air temperature and linearly proportional to the insolation level. The plot of (T_{cell} - T_{air}) versus insolation shall be determined by conducting a minimum of two field tests for which the value of NOCT should be obtained from

$$NOCT = (T_{cell} - T_{air})_{NC} + 20°C$$

1) Preparations

Solar irradiation:
   Pyranometer calibrated against the PACRAD scale mounted on plane of modules within 0.3 m from one of its edges.
Wind:
   Both wind direction and wind speed shall be measured near to one of the sides of the module. The array shall be protected from gusty winds by a suitable fence; wind parallel to the plane of the array is not acceptable.
Air temperature:
   Measurements shall be made at the height and in the shade of the module and shall be accurate to ± 1°C. If a thermocouple is used, the thermal mass of the thermocouple shall be increased by embedding the thermocouple in a solder sphere of approximately 6 mm diameter.
Cell temperature:
   The temperature of at least two interior solar cells shall be measured to ± 1°C. Thermocouples shall be soft-soldered directly to the back of the cells. Alternatively measurements may be made by a contact-free IR-temperature measuring device, or by V_{oc}-measurements.

The module shall be located within a planar surface that extends 0.5 m beyond the module in all directions. Black aluminium panels or other modules of the same design shall be used to fill in any remaining open area of the subarray structure. The plane of the module shall be positioned so that it is normal to the sun (± 5°) at solar noon. For open field applications, the back of the subarray shall be exposed. The bottom edge of the module shall be located 0.5 m or more above the horizontal plane or ground level. For residential roof applications, the module shall be located 10 cm from the simulated roof. For hard mounting roof applications, the modules shall be located directly on the simulated roof.

There shall be no obstructions to prevent full irradiance of the module beginning a minimum of 4 hours before solar noon and up to 4 hours after solar noon. The module, front and rear, shall receive only a minimum of reflected solar energy from the surrounding area.

2) Procedure

Acquire a semi-continuous record of $(T_{cell} - T_{air})$ over a one- or a two-day period, together with insolation and other measurements as required to characterize the terrestrial environment during the testing period.

All data shall be printed out approximately every 2 minutes. In addition, solar intensity, wind speed, wind direction and air temperature shall be continuously recorded. Acceptable data shall consist of measurements made when the average wind velocity is 1 m/s ± 0.75 m/s and with gusts less than 4 m/s for a period of 10 minutes prior to and up to the time of measurement. Local air temperature during the test period shall not vary by more than 5°C and shall be in the range of 20°C ± 15°C.

Two sets of measurements can be combined into a single set provided the average air temperature of the two sets does not differ by more than approximately 5°C.

Using only acceptable data as so defined, a plot shall be constructed from a set of measurements made either prior to solar noon or after solar noon which defines the relationship between $(T_{cell} - T_{air})$ and the solar irradiance level I.

Using the plot of $(T_{cell} - T_{air})$ versus I, the value of $(T_{cell} - T_{air})$ is determined by interpolating the average value of $(T_{cell} - T_{air})$ for I = 800 Wm^{-2}.

A correction term to the preliminary NOCT for average air temperature $\overline{T}_{air}$ and wind velocity $\overline{V}_w$ is determined from Fig. 1. This value is added to the preliminary NOCT and corrects the data to 20°C and 1 m/s.

CT.8   Electrical Performance at Nominal Operating Cell Temperature (NOCT)

T = NOCT   E = 800 Wm^{-2}

The voltage-current characteristics of the module shall be measured at the nominal operating cell temperature (NOCT) and at an irradiance of 800 Wm^{-2}. Alternatively, the characteristic measured at 25°C and 800 Wm^{-2} may be transposed to the nominal operating cell temperature using the formulae (which for silicon are a good approximation):

$$I_2 = I_1 + \alpha(T_2 - T_1)$$
$$V_2 = V_1 + \beta(T_2 - T_1)$$

where:

I_1, V_1 are coordinates of the measured characteristic
I_2, V_2 are the corresponding coordinates of the transposed characteristic
T_1 is the standard temperature, 25°C
T_2 is the nominal operating cell temperature, NOCT
α, β are the current and voltage temperature coefficients of the module as quoted on the Data Sheet (β is negative).

CT.9   Hot-Spot Heating Test

Note: This test is not required if the manufacturer has used bypass diodes for a maximum number of 12 cells in series.

$$T = NOCT$$
$$I = I_{cell} \text{ at } V_R = N_s \times V_{cell}$$
$$\tau = 60 \text{ min} \qquad n = 50$$

The objective of this test is to determine the ability of the module to withstand hot-spot heating effects, as e.g. solder melting or deterioration of the incapsulation. This defect could be provoked by cracked or mismatched cells, interconnect failures or by partial shadowing or soiling.

1)   Preparations

Determine the dark current characteristics and the range of shunt resistance of the various cells within the test module.
Determine the average short circuit current I_{av} of a cell at NOCT.
Determine the maximum cell reverse voltage $V_R = N_s \times V_{av}$, where N_s is the number of cells per bypass diode, or the number of series cells per module, whichever is less. V_{av} is the average nominal operating voltage of an individual cell within the module.

2)   Procedure

1. Adjust the module temperature to NOCT ± 2°C, e.g. by IR heating, keeping the ambient temperature at 20 ± 5°C.
2. Connect within the module a cell with a high shunt resistance to a current and voltage limited power supply.
3. Adjust the illumination level and the power supply to achieve $I = I_{av}$ at $V = V_R$, keeping $T_{module} = NOCT$.
4. Keep the cell at these conditions for 60 minutes, then return to ambient temperature.
5. Repeat steps 3 and 4 50 times.
6. Repeat steps 1 to 5 for a cell with a low shunt resistance, but in this case without additional illumination (light level on the cell < 50 Wm^{-2})

7. Short circuit the module at 1000 Wm^{-2} and shadow one of its high shunt resistance cells by 10% for 60 minutes.
8. Repeat step 7 by shadowing another high shunt resistance cell.

Tests CT.1, CT.5 and CT.6 shall be repeated. The module should remain free of the faults listed in Tests CT.1 and CT.5 and the values of the electrical parameters measured in Test CT.6 should remain within 5% of the initial values.

CT.10   Temperature Cycling Test

$$T = -40°C/NOCT + 40°C$$
$$\tau = 10 \text{ min} \qquad n = 50$$

1) Preparations

The module shall be installed in a climatic chamber with automatic temperature control. During steady state temperature periods the temperature shall remain constant within ± 2°C. The maximum rate of change of the temperature shall be 2 ± 0.5°C per minute.
The air around the module shall be circulated with a velocity of not less than 2 ms^{-1}.

2) Procedure

The module shall be subjected to 50 temperature cycles between $-40° \pm 2°C$ and NOCT +40° ± 2°C. The temperature of the module shall be monitored at least during the first cycle. The module temperature shall remain stable at each extreme for a period of at least 10 minutes.

Tests CT.1, CT.5 and CT.6 shall be repeated. The module should remain free of the faults listed in Tests CT.1 and CT.5 and the values of the electrical parameters measured in Test CT.6 should remain within 5% of the initial values.

CT.11   Humidity-Freezing Test

$$T = 40°C \text{ at } \geqslant 93\% \text{ RH to } -40°C$$
$$\tau = 60 \text{ min} \qquad n = 2$$

Procedure

1. Store the module in a high humidity chamber ($\geqslant$ 93% RH) at 40°C for 48 hours.
2. Transfer the module at ambient temperature to a climatic chamber.

3. Decrease the temperature to −40°C at a maximum rate of 3°C min⁻¹.
4. Keep the module at −40°C for 60 minutes, then return to room temperature at a maximum rate of 1°C min⁻¹.
5. Repeat this cycle once.

Tests CT.1, CT.5 and CT.6 shall be repeated. The module should remain free of the faults listed in Tests CT.1 and CT.5 and the values of the electrical parameters measured in Test CT.6 should remain within 5% of the initial values.

CT.12   Hail Resistance Test

1) Preparations

Using a suitable mould, prepare ice balls of the maximum hailstone diameter specified by the module manufacturer in the Data Sheet. They should be generally spherical in shape. After forming, place the ice balls in a freezer for a minimum of 8 hours prior to use. (Note: steel balls are not a satisfactory substitute.) Adjust a suitable pneumatic or spring-actuated gun until the velocity is within ± 5% of the appropriate hailstone terminal velocity listed in Table V.

TABLE V - Size and Natural Terminal Velocity of Hail Stones

Diameter (mm)	Term. velocity (ms^{-1})	Diameter (mm)	Term. velocity (ms^{-1})
12.5	16.0	45	30.7
15	17.8	55	33.9
25	23.0	65	36.7
35	27.2	75	39.5

Figure 2 shows in schematic form a suitable pneumatic hail test equipment. The velocity is best monitored by measuring electronically the time the ice ball takes to traverse the distance between two light beams incident on photocells.

2) Procedure

Mount the module, so as to face the gun normal to the path of the ice ball, using the method of fixing prescribed by the module manufacturer.
Ensure that the temperature of all gun surfaces likely to be in contact with the ice ball are near room temperature. Take an ice ball from the freezer and shoot it so as to hit the surface of the module at the specified velocity at the first location listed in Table VI. If not otherwise specified by the manufacturer, ice balls of 25 mm diameter and a terminal velocity of 23 ms⁻¹ shall be used.
The time between removal from the freezer and shooting should be less than 1 minute. Errors of up to 10 mm from the intended target location are acceptable. Inspect the module in the impact area for signs of cell cracking or surface damage.

Repeat this procedure for the other 9 locations listed in Table VI.

The module should remain free of the faults listed in Test CT.1. Verify that the module is still functioning by checking the I-V characteristics under suitable illumination before and after the test.

TABLE VI - Locations for Ice Ball Impact

Shot No.	Location
1	One corner of the module window
2	An edge of the module, just inside the frame
3, 4	Over edges of cells, near an electrical joint
5, 6	Over points of minimum spacing between cells
7, 8	On the module window, near the points at which the module is fixed to the supporting structure
9, 10	On the module window, at points furthest from the locations selected for shots Nos. 7 and 8.

CT.13   Mechanical Loading Test

$$P \rightarrow \pm 2400 \text{ Pa}$$

1)   Preparations

Uniformly distributed loads are applied by mounting the module in a test chamber, in which the air pressure can be changed (see Fig. 3). The test chamber shall be equipped with a pressure gauge to measure the chamber pressure with a precision of $\pm 15\%$.

The air system may consist of a controllable blower, a compressed air supply, an exhaust system or a reversible blower. It is convenient to use a reversible blower, so that the positive and negative pressure differences can be applied without removing, reversing and reinstalling the module. During the test period the air pressure difference shall be kept constant. If adequate air supply is available a completely air-tight seal need not be provided at the borders of the module.

The module shall be mounted in the test chamber using the method of fixing prescribed by the module manufacturer.

If excessive air leakage occurs around the module, adequate sealing has to be provided (masking tape, sealing profiles or similar). It is also permissible to cover the entire module with a polyethylene sheet (0.05 mm thickness), mounted so that the sheet does not prevent full load being applied to the specimen.

2) Procedures

Procedure A : Determination of Load Deflection Curve (Fig. 4A)

$$P = \pm (300, 600, 900, 1200) \text{ Pa}$$
$$\tau \geqslant 10 \text{ s}$$

1. Apply a load of 600 Pa and maintain it for at least 10 s. Release the pressure difference and, after a recovery period of not more than 5 minutes, at zero load, record initial readings.
2. Apply the load in 4 steps to 300, 600, 900 and 1200 Pa, maintain at each level for at least 10 s and record the deflection readings.
3. Release the pressure difference and take readings of permanent deformation after a recovery period of 5 minutes.
4. Repeat steps 1 to 3 for negative loads.

Procedure B : Resistance to Maximum Load (Fig. 4B)

$$P = 0 \rightarrow \pm 2400 \text{ Pa} \qquad \text{rate} \leqslant 200 \text{ Pas}^{-1}$$
$$\tau \geqslant 10 \text{ s}$$

1. Increase the load continuously at a rate of less than 200 Pas^{-1} to 2400 Pa* and maintain it for at least 10 s.
2. Release the pressure difference and take readings of permanent deformation within a recovery period of 5 minutes.
3. Repeat steps 1 and 2 for a negative load of 2400 Pa.

Procedure C : Resistance to Dynamic Load (Fig. 4C)

$$P = \pm 600 \text{ Pa} \qquad n = 10$$

Apply alternating positive and negative loads of $\pm$ 600 Pa for 10 times each. The module should remain free of the faults listed in Test CT.1. Verify that the module is still functioning by checking the I-V characteristics under suitable illumination before and after the test.

---

* 2400 Pa corresponds to a wind pressure of 130 km/h (= 800 Pa) with a safety factor of 3 for gusty wind. If snow and ice loads have to be tested as well, the value of 2400 Pa is replaced by 5400 Pa (except for extreme climatic conditions where higher snow and ice loads can occur.)

CT.14   *Damp Heat Long Exposure Test*

T = 40°C

H = 93% RH   D = 10 (30) days

---

1)   Preparations

The test chamber shall be any enclosure which can maintain the temperature in the test volume constant within ± 5°C. The humidity may be generated by an automatic regulation system or by evaporation from a bath within the test volume.

2)   Procedure

The module shall be installed in the test chamber and the following test conditions shall be applied:

. temperature   40°C
. relative humidity   93 $^{+3\%}_{-2\%}$ RH
. test duration   1. phase: 10 days
                                       2. phase: 20 days

Tests CT.5 and CT.6 shall be repeated. The module should remain free of the faults listed in Tests CT.1 and CT.5 and the values of the electrical parameters measured in Test CT.6 should remain within 5% of the initial values.

CT.15   *High Temperature Long Exposure Test*

T = NOCT + 50°C   D = 30 (120) days

---

1)   Preparations

The test chamber shall be any enclosure which can maintain the temperature in the test volume constant within ± 5°C.

2)   Procedure

The module shall be installed in the test chamber and the following test conditions shall be applied:

. temperature   NOCT + 50°C
. test duration   1. phase: 30 days
                                       2. phase: 90 days

Tests CT.1, CT.5 and CT.6 shall be repeated. The module should remain free of the faults listed in Tests CT.1 and CT.5 and the values of the electrical parameters in Test CT.6 should remain within 5% of the initial values.

CT.16   Ultra Violet Test

$$T = NOCT + 30°C$$
$$F = 40 \text{ MJm}^{-2}$$

The objective of this test is to determine the ability of a module to withstand to a limited exposure to UV light. Possible defects are UV-induced deteriorations of physical, optical and electrical characteristics of the module.

Results of this test shall not be represented as equivalent to those of natural irradiation unit the degree of quantitative correlation has been established. Purpose of the method is to establish, on a relative basis, the behaviour of the module with respect to specified UV irradiation and to indicate major defects which could occur during the initial deployment of a photovoltaic system.

1)   Preparations

A light source shall be used which is able to produce a spectrally defined UV-irradiation with a uniformity of ± 15% over the module. The relative spectral irradiance shall be matched to that of standard sunlight in the λ-range between 280 and 400 nm with a maximum deviation of ± 35% for $\Delta\lambda = 80$ nm intervals. Access of ozone produced by the UV exposure shall be prevented. The module temperature shall be kept at (NOCT + 30) ± 10°C.

A total ultraviolet monitor, filtered to exclude all radiation below 280 nm and above 400 nm wavelength, shall be used to assess the ultraviolet dose. It shall be mounted co-planar with the module.

2)   Procedure

The module shall be subjected to a total UV fluence* of 40 MJm⁻². During the test the intensity acceleration factor with respect to natural UV-irradiation shall not exceed 5.

Tests CT.1 and CT.6 shall be repeated. The modules should remain free of the faults listed in Tests CT.1 and CT.5 and the values of the electrical parameters, measured in Test CT.6 should remain within 5% of the initial values.

CT.17. O_3 Test

$T = 40°C$	$H = 55\%$ RH
$O_3 = 0.5$ vpm	$D = 5$ days

---

* The indicated fluence corresponds to about 30 Mediterranian clear summer days.

Procedure

The module shall be installed in an environmental test chamber in such a way
that the air can circulate freely around the module.
The following test conditions shall be applied:
. O_3 concentration   0.5 ± 0.1 vpm
. air temperature   40 ± 2°C
. relative humidity   55 ± 5% RH
. test duration   5 days

Tests CT.1, CT.5 and CT.6 shall be repeated. The modules should remain free
of the faults listed in Tests CT.1 and CT.5 and the values of the electrical parameters measured in Test CT.6 should remain within 5% of the initial values.

CT.18   SO₂ Test

T = 25°C	H = 75% RH
SO_2 = 50 ppm	D = 5 days

Procedure

The module shall be installed in an environmental test chamber in such a way
that the air can circulate freely around the module.
The following test conditions shall be applied:
. SO_2 concentration   50 ± 20 ppm
. air temperature   25 ± 2°C
. relative humidity   75 ± 5% RH
. test duration   5 days

Tests CT.1, CT.5 and CT.6 shall be repeated. The modules should remain free of
the faults listed in Tests CT.1 and CT.5 and the values of the electrical parameters measured in Test CT.6 should remain within 5% of the initial values.

CT.19   Salt Mist Test

T = 35°C	
C = 50 g NaCl/1	D = 4 days

The objective of this test is to determine the resistance of the module to deterioration from salt mist and to give indications concerning the relative service life
in a marine atmosphere.

1) Preparations

The test chamber shall be made of materials which do not influence the corrosive effects of the salt mist. No direct spray shall impinge on the module and no drops of liquid shall fall on the test object.

The salt solution shall be prepared by dissolving 50 ± 1 g of sodium chloride in distilled or demineralized water to make up 1 ± 0.02 l of solution at $20°C$.

The pH-value shall lie between 6.5 and 7.2

The solution shall have a temperature equal to that in the chamber.

The module to be tested shall be cleaned of oil, dirt and grease as necessary until the surface is free from water break.

2) Procedure

Modules shall be supported in such a position that the active surface is approximately $15°$ to $30°$ to the vertical and parallel to the principal direction of the flow of the fog through the chamber.

The temperature in the test chamber shall be maintained at $35 \pm 2°C$. Fog concentration and flow shall be adjusted in such a way that the rate of salt deposition remains between 10 to 50 g/m^2 day.

The duration of the test shall be 4 days. The test shall be run continuously for the whole test period.

After the test the module shall be washed in running tap water for at least 5 minutes and then rinsed in distilled or demineralized water.

After drying, Tests CT.1 and CT.5 shall be repeated. The modules should remain free of the faults listed in Tests CT.1 and CT.5 and the values of the electrical parameters measured in Test CT.6 should remain within 5% of the initial values.

CT.20.   *Ice Formation Test*

$$T = +20/-10°C$$
$$\tau = 60 \text{ minutes}$$
$$n = 2$$

1) Preparations

The module shall be installed in a test chamber which allows to vary the temperature between room temperature and $-10°C$. The chamber shall be fitted with a system to spray water over the module surface at a rate of 21 $m^{-2} min^{-1}$.

2) Procedure

1.   Install the module in the test chamber in vertical position.
2.   Decrease the chamber temperature from room temperature to $-10°C$ at a

rate of 5°C/hour.

3. During the cooling-down phase spray water continuously on the front surface of the module until a temperature of +2°C is reached.

4. At temperatures below 0°C spray water intermittently to the front surface of the module until an ice layer of at least 2 mm is formed.

5. Keep the module at −10°C for 60 minutes.

6. Return to room temperature at a rate of less than 5°C/hour and repeat once again steps 2 to 5.

7. After completion of the second cycle return to room temperature at a rate of less than 5°C/hour.

The module should remain free of the faults listed in Test CT.1. Verify that the module is still functioning by checking the I-V characteristics under suitable illumination before and after the test.

APPENDIX A - TESTS AND TEST LEVELS

TESTS	TEST LEVELS
Visual inspection	. see detailed inspection list
Dimensions and weight	. check data sheet values
Robustness of terminations	. tensile force 1 to 40 N, bending force 0.5 to 20 N according to cross section of conductor, torque 0.4 to 2.6 Nm, according to thread diameter
Mounting twist	. deformation angle 1.2°
Insulation	. insulation resistance 100 MΩ, 50 μA max current at specific voltage
Performance at STC	. I-V characteristics at 25°C, 800 and 1000 Wm^{-2}
Determination of NOCT	. NOCT at 20°C, 800 Wm^{-2}, 1 msec^{-1}
Performance at NOCT	. I-V characteristics at NOCT, 800 Wm^{-2}
Hot-spot heating	. T = NOCT, I = I$_{cell}$ at V$_R$ = N$_s \cdot$ V$_{cell}$, 50 cycles as specified
Temperature cycling	. −40°C to NOCT + 40°C, 50 cycles as specified
Humidity-freezing	. 40°C at 93% RH to −40°C
Hail resistance	. 25 mm diameter hail at 23 ms^{-1}, or as specified on Data Sheet
Mechanical loading	. up to ± 2400 Pa static, ± 600 Pa dynamic, 10 cycles
Damp heat long exposure	. 40°C, 93% RH, 10 (30) days
High temperature long exposure	. NOCT + 50°C, 30 (120) days
UV-resistance	. total fluence 40 MJm^{-2} at NOCT + 30°C, as specified
O$_3$-resistance	. 0.5 vpm, 40°C, 55% RH, 5 days
SO$_2$-resistance	. 50 ppm, 25°C, 75% RH, 5 days
Salt mist	. 50 g NaCl/l, 35°C, pH 7, 4 days
Ice formation	. 20°C to −10°C, 2 cycles with water spraying, as specified

INDEX